Plumbing:
Mechanical Services

# Plumbing: Mechanical Services

# Book 1

*Fourth Edition*

## G.J. Blower
Eng Tech (CEI), MIP, LCGI, Technical Teachers Cert.

Formerly Senior Lecturer, Plumbing Mechanical
Services Section, College of North East London.

Currently an NVQ Assessor and
Training Advisor for J.T.L.

PEARSON
Prentice
Hall

Harlow, England • London • New York • Boston • San Francisco • Toronto • Sydney • Singapore • Hong Kong
Tokyo • Seoul • Taipei • New Delhi • Cape Town • Madrid • Mexico City • Amsterdam • Munich • Paris • Milan

**Pearson Education Limited**
Edinburgh Gate
Harlow
Essex CM20 2JE

and Associated Companies throughout the world

*Visit us on the World Wide Web at:*
www.pearsoneduc.com

First published 1982
Second edition published 1989
Third edition published 1995
**Fourth edition published 2002**

ISBN 0582-43228-6

**British Library Cataloguing-in-Publication Data**
A catalogue record for this book is available from the British Library

10  9  8  7  6  5  4  3
07  05  04  03

Typeset in 10/12pt Times by 35
Printed and bound in Malaysia

# Contents

# Foreword

It is said that every man is a builder by instinct and, while this may be true to a degree, the best way to become a craftsman is first to be an apprentice.

To be an apprentice or trainee means to learn a trade, to learn skills and have knowledge which will both be a means of earning a living and provide an invaluable expertise for life. But whereas this used to be the main purpose and advantage of serving an apprenticeship, nowadays many people in high positions in industry have made their way to the top after beginning with an apprenticeship.

The once leisurely pace of learning a trade has now been replaced by a much more concentrated period of learning because of the shortened length of training. In addition, there is an ever-increasing number of new materials and techniques being introduced which have to be understood and assimilated into the craftsman's daily workload. There is therefore a large and growing body of knowledge which will always be essential to the craftsman, and the aim of this craft series of books is to provide this fundamental knowledge in a manner which is simple, direct and easy to understand.

As most apprentices and trainees nowadays have the advantage of attending a college of further education to help them learn their craft, the publishers of these books have chosen their authors from experienced craftsmen who are also experienced teachers and who understand the requirements of craft training and education. The learning objectives and self-testing questions associated with each chapter will be most useful to students and also to college lecturers who may well wish to integrate the books into their teaching programme.

The needs have also to be kept in view of the increasing numbers of late entrants to the crafts who are entering a trade as adults, probably under a government-sponsored or other similar scheme. Such students will find the practical, down-to-earth style of these books to be an enormous help to them in reaching craftsman status.

L. Jaques
*Formerly Head of Department*
*Leeds College of Technology*
1982

# Preface to the fourth edition

Throughout the period this series of books has been available, the aim has been to supplement the technical knowledge of both experienced operators and craft apprentices and trainees attending colleges of further education and other approved centres. For trainee plumbers indentured under the Modern Apprenticeship Scheme, a good understanding of the basic principles of plumbing, heating and their associated disciplines has always been a prerequisite to success in the completion of a course in training for the National Vocational Qualifications. The reader should be aware, however, that technology in mechanical services is always changing. Currently this is especially so due to the influence of the New Water Regulations and the Scottish and Northern Ireland Bylaws on the work of the building services industry. It should also be noted that at the time of writing many of the familiar British Standards are being harmonised with those of Europe, and because this is an ongoing process their reference numbers may change at any time. The Standards to which reference has been made are as up to date as possible.

Readers of previous editions will note that all reference to copper, aluminium and zinc work sheet weatherings has been deleted, mainly because few plumbers undertake this work which is now almost universally the province of specialist contractors. Studies in the use of these materials are no longer required for the successful completion of the NVQ plumbing courses. It is suggested that anyone requiring information on metal sheet weatherings other than lead should make enquiries to the associations listed on page 290. Lead has always been a plumbers' material and some practical knowledge and competence in its use for weatherings is assessed at NVQ level 2. For those requiring a greater degree of skill and competence in this subject, an NVQ level 3 is available at many colleges and training centres. Also included in the text is some information on soldering using copper bits and wiping. Although limited in the field of modern plumbing, these skills still have their uses, but unfortunately, owing to lack of practice, they are becoming lost to the plumbing industry.

# 1 Safety on site and in the workshop

After completing this chapter the reader should be able to understand the basic principles of general site safety and legislation relating to health and safety in construction.

1. Employers' and employees' responsibilities relating to safe working conditions and maintenance of safety equipment. Cause of accidents and their prevention.
2. Accident reporting procedures.
3. First aid in the event of an accident.
4. Improvement and prohibition notices.
5. Recognition of potential hazards with all types of working platforms.
6. Understanding the need for using personnel protective clothing and safety equipment.
7. Safety with electrical tools and appliances. Treatment for electric shock.
8. Causes of fire and methods of fire control. Correct selection of fire extinguishers.
9. Safe use of hand tools.
10. Safe manual-lifting techniques.
11. Safety signs.
12. Safety when handling liquid petroleum gases, dangerous and toxic material.
13. The principles of mechanical handling. Safe working loads.
14. Understanding the basic principles underlying the use of simple levers and pulleys.

## Health and safety at work (HSW Act)

The Factories Act of 1833 was introduced to safeguard the health, working conditions and general well-being of children employed in the textile industry, and it is from this Act that all industrial legislation relating to health, safety and welfare can be traced.

Since this Act there have been a number of other Factories Acts, each of which has sought to improve the conditions of employees in respect of health and safety, the last and probably the most far reaching of these being the Health and Safety at Work etc. Act 1974. Under the terms of this Act severe penalties can be imposed by the courts if the various requirements are not observed. One of the main features is to make employees, as well as employers, jointly responsible for safety at places of work and for the protection of the public.

It is mandatory for companies employing five or more employees to produce a written statement setting out their safety policy, the arrangements they make to put it into effect and make all employees aware of it. Obviously there will be a wide variation in these policies and the risks and dangers vary with the type of industry and the nature of materials used. The important thing is that every employee becomes conversant with the content of such documents.

General regulations relating to health and safety are also set out in the Factories Acts, but owing to the nature of building and its hazards, special regulations are made for those employed on construction sites, these being entitled. The

Construction (Health and Welfare) Regulations 1996, which replace the original 1961 Regulations (general provisions), the Requirements of Working Places 1966, and Health & Welfare 1966. Head Protection Requirements 1989 are still in place.

There are two main bodies involved in health and safety at work:

(a)  The Health and Safety Commission (HSC)
(b)  The Health and Safety Executive (HSE)

The HSC is responsible for making recommendations and arrangements necessary or appropriate to ensure the main purpose of the Act is carried out. This body is directly responsible to the Secretary of State for Employment. The HSE is the operating arm of the HSC, its primary function being to implement the enforcement of the law relating to health and safety. It is outside the range of this book to cover in depth all the aspects concerned with safety and health legislation, and the reader is advised to contact the organisations mentioned at the end of the chapter in order to become familiar with the detailed implications of this important subject.

Accidents are nearly always caused by carelessness or lack of thought, failing to take proper precautions, ignoring instructions and failure to use safety equipment provided. In many cases accidents are caused by taking risks to get the job done quickly; such thinking is fatal. It should also be borne in mind that in many cases one's own actions may not only result in injury to oneself but also serious or even fatal injury to one's own workmates. One of the surest ways of reducing accidents is to plan and think ahead of any dangers there might be in carrying out specific tasks. A responsible person at work should always be on the look-out for possible hazards and develop a safety conscience.

It must be remembered that safety is the responsibility of employers and employees. It is the responsibility of employers to provide safety equipment, and to ensure that it is in good order and is well maintained. It is the responsibility of the employee to make sure he or she uses the equipment provided and to see that it is properly cared for whilst in use. To ignore the foregoing is in fact breaking the law.

*Control of Substances Hazardous to Health (COSHH)*

These Regulations came into effect in October 1989 and apply to all potentially dangerous substances where they are produced or used. They also apply to the production of fumes or dust in a work environment that may not in themselves be dangerous, but when workers are exposed to these substances over long periods, they may constitute a health hazard. There are some exceptions to these Regulations where very stringent legislation is already in existence, typical examples being The Control of Lead at Work Regulations 1980 and The Control of Asbestos at Work Regulations 1987.

Under the COSHH Regulations employers are required to carry out on a regular basis an assessment of the workplace, environment and the materials used or produced in order to identify possible hazards to employees. Where hazards are identified measures must be taken either to prevent or minimise the risk of exposure to employees. This may take the form of providing regular health checks, special safety equipment and clothing. It also means, for example, that all such equipment and machinery must be in good working order and regularly maintained. Records of all such maintenance must also be kept. Where health checks are required the employer must ensure that they are carried out, and must also make sure sufficient information, instruction and, where necessary, training relating to health hazards are available to employees so that they are fully aware of any risks to their health.

## Accident reporting

*Minor accidents*

All minor accidents should be recorded in an accident report book or on a special form as shown in Fig. 1.1. Should a claim be made to the Department of Health and Security details of the accident will be required.

*Serious accidents. Injuries, Diseases and Dangerous Occurrences Regulations (RIDDOR)*

An accident at work resulting in death or major injury to anyone working on site must be notified

---

**W F WARD, PLUMBING & HEATING ENGINEER**
**Accident Report Form**

**Report of an accident or injury to a person at work or on duty**
This form must be completed in all cases of accident, injury or dangerous occurrence and submitted to the Safety Officer.

Name of injured person...............................................................

Date of birth..............................................................................

Position held in organisation.......................................................

Date and time of accident ..........................................................

Particulars of injury/accident


Activity at time of injury/accident


Place of injury/accident


Details of injury/accident


First-aid treatment (if any) given


Was the injured person taken to hospital? If so, where?


Name(s) and position(s) of person(s) present when the accident occurred


Signature of person reporting incident.........................................

Date.........................................................................................

---

**Fig. 1.1**   Typical company Accident Report Form.

without delay to the HSE. This must be followed up with form **F2508** (shown in Figs 1.2(a) and (b)) within seven days.

Typical examples of injuries requiring notification are as follows:

(a) fractures other than those to fingers, thumbs or toes
(b) amputation
(c) dislocation of the shoulder, hip, knee or spine
(d) temporary or permanent loss of sight
(e) burns or penetration injury to the eye
(f) injury from electric shock resulting in loss of consciousness and the patient requiring resuscitation or admittance to hospital
(g) any other injury requiring resuscitation or admittance to hospital for more than 24 hours

**HSE**
Health & Safety
Executive

*Health and Safety at Work etc Act 1974*
*The Reporting of Injuries, Diseases and Dangerous Occurrences Regulations 1995*

# Report of an injury or dangerous occurrence

**Filling in this form**
This form must be filled in by an employer or other responsible person.

## Part A

### About you

1  What is your full name?

2  What is your job title?

3  What is your telephone number?

### About your organisation

4  What is the name of your organisation?

5  What is its address and postcode?

6  What type of work does the organisation do?

## Part B

### About the incident

1  On what date did the incident happen?

$$/\ \ \ \ /$$

2  At what time did the incident happen?
(Please use the 24-hour clock eg 0600)

3  Did the incident happen at the above address?

Yes ☐  Go to question 4

No ☐  Where did the incident happen?

☐ elsewhere in your organisation – give the
name, address and postcode

☐ at someone else's premises – give the name,
address and postcode

☐ in a public place – give details of where it
happened

If you do not know the postcode, what is
the name of the local authority?

4  In which department, or where on the premises,
did the incident happen?

## Part C

### About the injured person

If you are reporting a dangerous occurrence, go
to Part F.

If more than one person was injured in the same incident,
please attach the details asked for in Part C and Part D for
each injured person.

1  What is their full name?

2  What is their home address and postcode?

3  What is their home phone number?

4  How old are they?

5  Are they

☐ male?

☐ female?

6  What is their job title?

7  Was the injured person (tick only one box)

☐ one of your employees?

☐ on a training scheme? Give details:

☐ on work experience?

☐ employed by someone else? Give details of the
employer:

☐ self-employed and at work?

☐ a member of the public?

## Part D

### About the injury

1  What was the injury?  (eg fracture, laceration)

2  What part of the body was injured?

SPECIMEN

**Fig. 1.2**   (a) and (b) Approved HSE RIDDOR form (courtesy of the Health & Safety Executive).

3 Was the injury (tick the one box that applies)

☐ a fatality?

☐ a major injury or condition? (see accompanying notes)

☐ an injury to an employee or self-employed person which prevented them doing their normal work for more than 3 days?

☐ an injury to a member of the public which meant they had to be taken from the scene of the accident to a hospital for treatment?

4 Did the injured person (tick all the boxes that apply)

☐ become unconscious?

☐ need resuscitation?

☐ remain in hospital for more than 24 hours?

☐ none of the above.

## Part E

### About the kind of accident

Please tick the one box that best describes what happened, then go to Part G.

☐ Contact with moving machinery or material being machined

☐ Hit by a moving, flying or falling object

☐ Hit by a moving vehicle

☐ Hit something fixed or stationary

☐ Injured while handling, lifting or carrying

☐ Slipped, tripped or fell on the same level

☐ Fell from a height

How high was the fall?

☐ _____ metres

☐ Trapped by something collapsing

☐ Drowned or asphyxiated

☐ Exposed to, or in contact with, a harmful substance

☐ Exposed to fire

☐ Exposed to an explosion

☐ Contact with electricity or an electrical discharge

☐ Injured by an animal

☐ Physically assaulted by a person

☐ Another kind of accident (describe it in Part G)

## Part F

### Dangerous occurrences

Enter the number of the dangerous occurrence you are reporting. (The numbers are given in the Regulations and in the notes which accompany this form)

☐ _____

## Part G

### Describing what happened

Give as much detail as you can. For instance

• the name of any substance involved
• the name and type of any machine involved
• the events that led to the incident
• the part played by any people.

If it was a personal injury, give details of what the person was doing. Describe any action that has since been taken to prevent a similar incident. Use a separate piece of paper if you need to.

*SPECIMEN*

## Part H

### Your signature

Signature

Date
☐ / /

**Where to send the form**
Please send it to the Enforcing Authority for the place where it happened. If you do not know the Enforcing Authority, send it to the nearest HSE office.

| For official use | | | |
|---|---|---|---|
| Client number | Location number | Event number | |
| | | | ☐ INV REP ☐ Y ☐ N |

**Fig. 1.2** *continued*

*Dangerous occurrences*

These must be reported to the HSE within the same time scale as notifiable injuries. Examples of 'dangerous occurrences' are as follows:

(a)  collapse of scaffolding over 5 m high
(b)  overturning or failure of lifting equipment, e.g. cranes
(c)  explosion, collapse or bursting of any closed vessel or associated pipework

*Factory inspection and environmental health officers*

The HSW Act allows inspectors very exceptional powers in the interest of safety. They may enter premises at any reasonable time to examine and ensure compliance to the relevent statutory regulations is adhered to by employers and employees. When investigating an accident or dangerous occurrence, the inspector may order any part of the premises or tools and equipment to be left undisturbed until further inspection is completed. To enable the inspector to use any statutory powers that may be required, he or she may call upon any witnesses to assist with the necessary information needed to take further action.

*Improvement notices*

An inspector may serve an improvement notice when he or she considers any regulations are not being complied with but constitute no immediate danger. Any potential risks must be eliminated within a period specified by the inspector.

*Prohibition notices*

Such a notice may be served upon a person controlling activities and equipment which, in the inspector's opinion, constitute a serious risk to health and safety. Under these circumstances a prohibition notice may be served until all danger has been eliminated to the satisfaction of the inspector.

Appeals can be made to an Industrial Tribunal against both improvement and prohibition notices within 21 days, but a prohibition order will remain while an appeal is pending. It is advisable to seek legal advice prior to making an appeal as the result of losing can be very costly.

**General safety requirements on site**

To ensure that the correct precautions are taken and procedures adopted, the HSW Act requires that where 20 people or more are employed a safety officer must be appointed. For such small numbers, this is not necessarily a full-time position, but sufficient time must be made available to the officer to enable him or her to perform his or her duties. Among the officer's many responsibilities is the periodic inspection of the scaffolding, plant and stores to ensure there are no obvious hazards. The safety officer is also responsible for recording any accident, and, where necessary, reporting the matter to HM Factory Inspectorate. (This latter procedure is required where the accident involves a death or an injury which requires such treatment as will prevent the person returning to work for three days or more.)

On all sites when five or more people are employed a first-aid box should be provided by the employer, and this should contain sufficient and suitable equipment for the treatment of minor injuries. It should be available to all personnel and be used for first-aid requirements only, not, as sometimes is the case, a container or depository for all sorts of odds and ends from timesheets to sandwiches. On larger sites where more than 50 people are employed a qualified first-aider is required, and in many instances the safety officer undertakes this role.

**First-aid procedures**

The Construction (Health and Welfare) Regulations also specify minimum first-aid facilities, but it is only where more than 50 workers are employed on a contract that there is any requirement to appoint a qualified first-aider.

If serious injury occurs and medical aid is required urgently on site, the **recommended course of action is to summon a doctor, ambulance or other medical assistance**. Unfortunately, mechanical services tradespeople frequently work alone on construction sites when no other personnel are present or where means of telephone communication is not available. This will create difficulty in summoning assistance and could place the person requiring assistance in danger. It is therefore essential for all site workers to be familiar with basic first-aid procedures, as by prompt attention the danger of death or permanent injury can be greatly reduced. For simple and easy to read details of first-aid treatment the publications listed at the end of the chapter are recommended.

Remember that first-aid treatment is only preliminary help given to the patient prior to assistance being available from trained medical personnel.

### Basic first-aid procedures

The following text gives the basic recommendations for first-aid treatment in the event of an accident. In all cases, if there is any doubt at all about the seriousness of the patient's condition, get help and telephone for a doctor or ambulance immediately.

### Small cuts and abrasions

If possible clean the skin around the wound with a suitable anti-bacterial solution and cover with a sterile dressing or bandage. If the wound becomes painful and inflamed the casualty should consult a doctor.

### Loss of blood

If the casualty is bleeding badly owing to a large wound or cut, control the bleeding by applying a pad of sterile dressing or sterile cotton wool bandaged firmly into place. Blood tends to congeal and prompt action as described may stem the flow of blood until professional help arrives.

Arterial bleeding is usually identified by bright red blood 'spurting' from the wound. If the foregoing treatment fails to stop the bleeding, apply pressure at one of the points shown in Figs 1.3(a) and (b). As an alternative a pressure bandage may be applied but in all cases of applying pressure to an artery the maximum time limit is 15 minutes.

### Burns and scalds

The treatment of burns and scalds is very similar – both are very painful. The pain can be relieved by immersing the injured part in cool water or under a running tap (see Fig. 1.4) for a period of about 10 minutes, or less if the pain ceases. Do not attempt to remove clothing if it is sticking to the

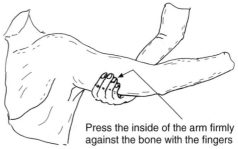

Press the inside of the arm firmly against the bone with the fingers

(a) Arterial bleeding from the arm

Apply pressure in a similar way to that on the arm. The pressure points for the legs is the inside of the thigh at the junction of the groin.

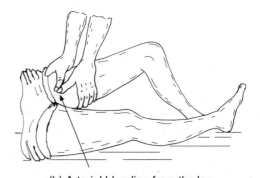

(b) Arterial bleeding from the leg

Fig. 1.3  Arterial bleeding – pressure points.

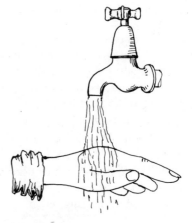

Fig. 1.4  Cooling burns or scalds prior to dressing.

wound. Cover the injured part with a dressing or bandage to insulate it from the air. The patient should be taken to hospital as quickly as possible if badly burnt.

Cover eye with sterile dressing, fixed into place with adhesive tape prior to covering with blindfold

The blindfold will reduce movement of the eyes and also prevent the casualty rubbing the injured eye

**Fig. 1.5**  Eye injury.

### Eye injuries

Eye injuries are potentially serious, especially if the eye itself is damaged. In the event of dust or a similar product getting into the eye, an eye glass may be used to wash out any particles. Prevent the casualty from rubbing the eye as this can make matters worse. If an eye wash is unsuccessful cover the eye with a soft pad, preferably cotton wool, and bandage into place, see Fig. 1.5. If the injury is caused by a blow or something is embedded in the eye, such as metallic swarf or material that could cut it, get the patient to hospital as soon as possible.

Eye injury due to contact with chemicals can be treated in a similar way. Wash the eye for 10 to 15 minutes under a running tap or with the head immersed in a bowl of water and blinking the injured eye. If this fails to give relief seek expert medical attention as quickly as possible.

### Persons suffering electric shock

An electric shock can be dangerous and in some cases can cause death if immediate action is not taken. The casualty must be removed from the power source and if the heart has stopped resuscitation procedures must be applied. The severity of shock depends on the voltage and length of time the casualty has been in contact with the current. The effects of electric shock will be severe pain due to muscle contraction, possible minor burns at the point of contact and the casualty may be thrown off his or her feet causing further injuries. The muscle contraction may prevent the casualty from releasing contact with the conductor. The first step to be taken when this occurs is to switch the current off, or if this is not possible, pull or push the casualty clear using a non-conductive material such as a piece of timber or a broom handle. Use anything that is handy providing it is not made of metal and is dry. At the same time call for help and if possible get someone to telephone for an ambulance as it is important that the casualty is treated as quickly as possible by a qualified person.

### Emergency resuscitation

It is recommended that a study is made of the St John's Ambulance Association's first-aid manual relating to this subject. The following will be helpful and is given for general guidance only. If the casualty is breathing but unconscious, loosen the clothing about the neck and waist and lay the casualty in the recovery position (see Fig. 1.6); this prevents the casualty from inhaling fluid or vomit. Keep the casualty warm and dry and constantly check breathing and pulse rate until help arrives.

If the casualty is *not* breathing, rapid action must be taken immediately by administering mouth-to-mouth resuscitation – *every second* counts. The main concern is (a) the continual oxygenisation of the blood by inflating the lungs and (b) restarting the heart to ensure the blood reaches the brain and other vital organs. If possible get help and proceed as follows: check the airways, the tongue or if vomit may be blocking the throat. If breathing has stopped start mouth-to-mouth procedures at once. See Fig. 1.7. If this is not possible seal the casualty's lips with your thumb and seal your lips about the casualty's nostrils to force breath into the casualty. Watch the

**Fig. 1.6**  Recovery position. After resuscitation the patient should be placed in the recovery position and kept warm until examined by a doctor or taken to hospital.

(a) Support the head so that it is tilted backwards, to ensure a good airway by checking the tongue or if vomit is not obstructing the throat

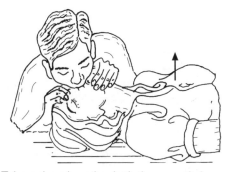

(b) Take a deep breath, pinch the casualty's nostrils together and blow into the lungs until the chest rises (see arrow)

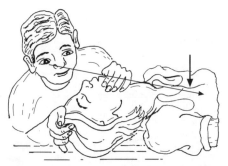

(c) Remove your mouth and watch the chest fall (see arrow). Repeat and continue inflations until the casualty is breathing or until medical help arrives

**Fig. 1.7**   Respiratory resuscitation.

chest: it should rise as the lungs are filled with air. When breathing is restored, the casualty should be placed in the recovery position. See Fig. 1.6.

If the heart has stopped beating this is usually indicated by the lips turning blue, the pupils of the eyes being dilated and absence of the carotid pulse

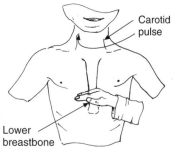

(a) A sharp blow with the side of the hand slightly to the left of the lower breastbone will sometimes restart the heart

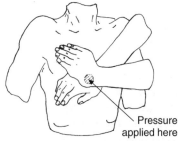

(b) Note that this must not be attempted unless the first-aider is sure the heart has stopped beating!

    (a) Kneeling by the side of the casualty feel for the lower half of the breastbone.

    (b) Place the heel of the hand on this point covering this hand with the heel of the other as shown.

    (c) With arms straight, press down repeating the pressure  at least 60 times per minute. This procedure must be continued until medical help arrives.

**Fig. 1.8**   (a) Restarting the heart; (b) external heart compression.

in the neck. The following procedure should then be adopted: see Figs 1.8(a) and (b).

Lay the casualty on his or her back and kneel alongside. Locate the lower part of the breastbone. Place the heel of one hand on the centre of the lower part of the breastbone and cover with your other hand, locking the fingers to keep them free of the rib cage. Keeping the arms straight press sharply down on the base of the breastbone and then release the pressure. This should be repeated 15 times at the rate of 60 compressions per minute. The pulse should then be checked and two more breaths of mouth-to-mouth resuscitation given. Repeat this process until the heart beat returns and continue the mouth-to-mouth resuscitation until breathing is normal. Only when heart beats and breathing are normal should the casualty be placed in the recovery position. Never leave a patient who has suffered an electric shock, or indeed any other type of accident for that matter, until expert help has arrived.

It should be mentioned here that some have expressed concern relating to the possibility of being exposed to contagious diseases when performing mouth-to-mouth resuscitation. There are commercially available plastic shields which can be used to protect anyone against this possible risk. They are cheap, fold up easily and can fit comfortably into a first-aid box. They can be obtained from Laerdal Medical Ltd, Laerdal House, Goodmead Road, Orpington, Kent, BR6 0HX. Tel.: 01689 876634.

### Accidents on site

As it is essential for the craftworker to work safely it is necessary to consider some of the more common hazards they face on site. Every year many people are killed and injured in accidents involving the use of scaffolds or ladders and the vast majority of these accidents could have been prevented by common sense and the application of a few simple precautions.

*Scaffolding*
While the erection of scaffolding is regarded as a specialist activity to be undertaken only by trained operatives, it is important for anyone using a scaffold to be able to recognise its type and whether or not it is safe to work on.

Scaffolds are basically a temporary working platform and those used for building purposes are of two main types:

(a) Scaffolds used in the construction of a new building where one side of the scaffold is supported by putlogs resting on the building fabric as the building is constructed. It is tied to the building through windows and doorways and is sometimes referred to as putlog scaffolding.

(b) Independent scaffolds are those which are more commonly used on existing buildings for maintenance or refurbishment work, and can be used without actually being attached to the structure. Obviously some method of temporary tie to the structure will be necessary on any but low-rise buildings. Figures 1.9(a) and (b) show

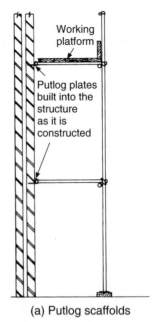

(a) Putlog scaffolds

This type of scaffold relies for its support by securing it to the structure.

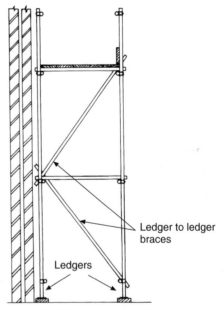

(b) Independent scaffolds

The same safety regulations apply to independent scaffolds as those of tower scaffolds, e.g. outriggers must be fitted when the working platform is three times the height of the base. The alternative is to tie the scaffold to the structure.

**Fig. 1.9**   Scaffolds.

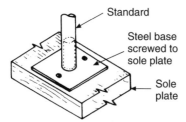

(a) Firm non-slip base for standards

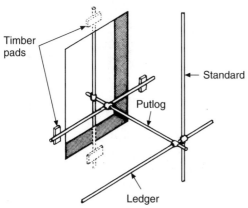

(b) Method of tying scaffolds to a building through window or door openings

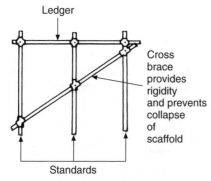

(c) Cross-bracing standards

**Fig. 1.10**   Securing scaffolding.

sections of both types of scaffold. See also mobile scaffolds.

Some of the main safety features about scaffolding are as follows (see Fig. 1.10):

(a)   All scaffolds must have a secure base as shown in Fig. 1.10(a) with a steel base plate to spread the weight of the scaffold and prevent the tube ends sinking into the ground.

(b)   Scaffolds should be securely tied to the buildings as shown in Fig. 1.10(b), this being intended to prevent the scaffold falling away from the building.

(c)   The cross brace shown in Fig. 1.10(c) forms a triangle (a recognised safe form of structure) and prevents the scaffold collapsing sideways. Note also the braces shown in Fig. 1.9(b) which serve the same purpose.

(d)   The main requirements regarding working platforms and toe boards are shown in Figs 1.11(a) and (b), the toe boards, guard rails and wire mesh screens being necessary to prevent operatives and equipment sliding or falling off the working platform.

(e)   It is necessary to space the putlogs carefully to avoid creating a 'trap', or an excessive overhang of the walk boards, as shown at B in Fig. 1.11(b). The permissible distance that a board may overhang a putlog is $4 \times$ its thickness. For example, if boards of 40 mm are used they may overhang the putlog by not more than $4 \times 40 = 160$ mm. The distance between putlogs shown at A on Fig. 1.11(c) depends on the thickness of the board. Table 1.1 gives some examples.

Figure 1.11(d) shows the recommended way to join walk boards.

The same general requirements also apply to mobile scaffolds as illustrated in Fig. 1.12, but there are also additional regulations which include the following:

(a)   They should be used only on level surfaces.

(b)   They should be moved only by pushing at the base.

(c)   Care must be taken not to foul any overhead cables or wires when the scaffold is in motion.

**Table 1.1**   Distance between putlogs.

| Thickness of plank | Distance between putlogs |
|---|---|
| 32 mm | 1 m |
| 40 mm | 1.5 m |
| 51 mm | 2.6 m |

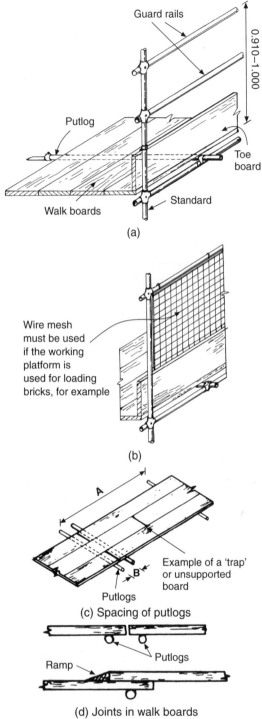

Fig. 1.11  Details of working platforms.

(a)

Guard rails

0.910–1.000

Putlog

Toe board

Standard

Walk boards

(b)

Wire mesh must be used if the working platform is used for loading bricks, for example

(c) Spacing of putlogs

A

B

Example of a 'trap' or unsupported board

Putlogs

(d) Joints in walk boards

Ramp

Putlogs

Overlapping of boards should be avoided but where they do occur a ramp must be used.

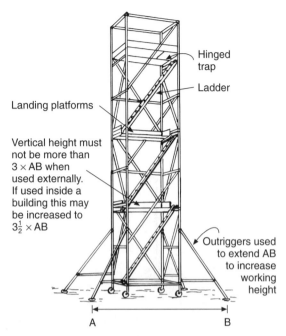

Fig. 1.12  Mobile tower scaffold.

Hinged trap

Ladder

Landing platforms

Vertical height must not be more than $3 \times AB$ when used externally. If used inside a building this may be increased to $3\frac{1}{2} \times AB$

Outriggers used to extend AB to increase working height

A          B

The legal provisions relating to scaffolds and the means of access to them are contained in detail in the Construction Regulations 1996.

*Trestle scaffolds*
Figure 1.13 illustrates two typical scaffolds of this type, both of which are useful for relatively low-level work. It is essential to use them within the guidelines shown, as they are entirely free standing and do not have the rigidity of a proper scaffold. Check the base if there is any sign of 'wobble' and if necessary a length of cord attaching the trestle to a convenient fixing will doubly ensure its safety.

*Ladders*
Ladders should be protected by the application of several coats of clear varnish; paint must not be used as it will cover any defects such as cracks that may be present. They should never be stored in such a way as to cause permanent bowing or twisting, the best method being to stand them upright or place them in racks which provide full support for the stiles. The best ladders have a steel tie rod under each rung, these rods having two functions:

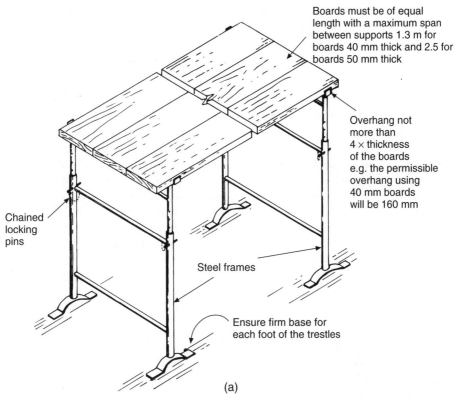

Boards must be of equal
length with a maximum span
between supports 1.3 m for
boards 40 mm thick and 2.5 for
boards 50 mm thick

Overhang not
more than
4 × thickness
of the boards
e.g. the permissible
overhang using
40 mm boards
will be 160 mm

Chained
locking
pins

Steel frames

Ensure firm base for
each foot of the trestles

(a)

This type of trestle gives a good working platform the height of which can be adjusted by the telescopic supports which
are pinned through the main trestle.

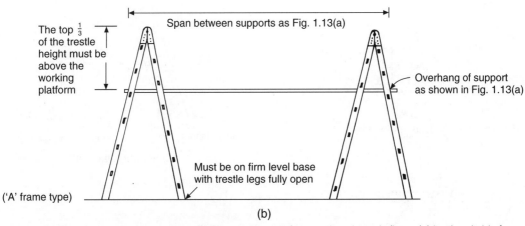

The top $\frac{1}{3}$
of the trestle
height must be
above the
working
platform

Span between supports as Fig. 1.13(a)

Overhang of support
as shown in Fig. 1.13(a)

Must be on firm level base
with trestle legs fully open

('A' frame type)

(b)

This type of scaffold does not provide such an effective working platform as that shown in figure (a) but is suitable for access.

General:
1. Trestles more than 3.6 m high must be tied to the structure, they must not be used where a person could fall more than
4.5 m. It is not permissible to use trestles on a platform scaffold unless it is firmly secured to avoid collapse.
2. Maximum height of working platforms is 2 m unless toe boards and guard rails are provided and access is by means of
a separate ladder.

**Fig. 1.13**   Trestle scaffolds.

(a) Should the rung break the rod will still support the user as a temporary measure and prevent a fall.

(b) They ensure the stiles do not fall away from the rungs.

The recommendations relating to the slope of ladders are illustrated in Fig. 1.14(a) and must be observed to ensure safety.

When in use, ladders should be well secured by lashing the top to a ledger on a scaffold or some equally good fixing, and using one of the two methods shown to ensure the bottom of the ladder does not slip. See Fig. 1.14.

Typical examples of the misuse of ladders are shown in Figs 1.14(c) and (d). Detail (c) shows a board forming a working platform, resting on a rung of the ladder. Quite apart from being potentially dangerous this will subject the rung to undue strain.

Detail (d) shows a ladder that is too short for the job and the dangerous way in which an attempt has been made to extend it. Do *not* use two short ladders tied together; *always* use a ladder long enough for the job.

*Handling ladders*

Quite apart from making ladders secure when in use, great care must be taken when moving them about and when raising them ready for use. Some very serious injuries are on record due to accidents involving the raising or lowering of ladders. Short, light ladders can, with care, be handled by one person, but long, heavy ladders will require at least two persons, one of whom must always be at the foot of the ladder, having one or both feet on the bottom rung to prevent it sliding when raised.

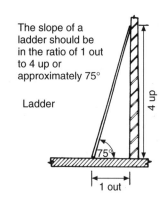

(a) Recommended slope of ladders

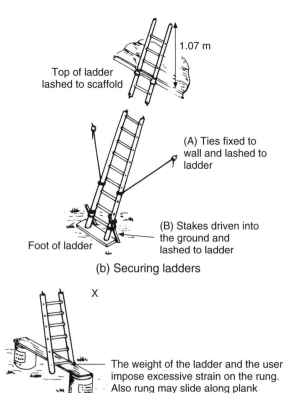

(b) Securing ladders

(c) Dangerous practice with ladders

(d) Dangerous practice with ladders

**Fig. 1.14**   Safety when using ladders.

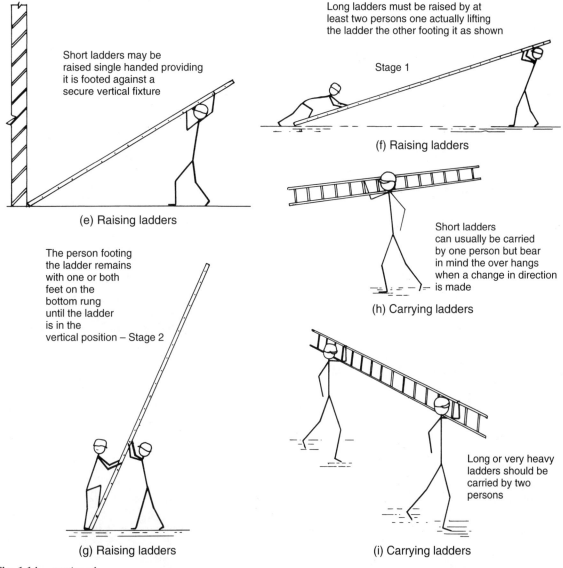

Short ladders may be raised single handed providing it is footed against a secure vertical fixture

(e) Raising ladders

Long ladders must be raised by at least two persons one actually lifting the ladder the other footing it as shown

Stage 1

(f) Raising ladders

Short ladders can usually be carried by one person but bear in mind the over hangs when a change in direction is made

(h) Carrying ladders

The person footing the ladder remains with one or both feet on the bottom rung until the ladder is in the vertical position – Stage 2

(g) Raising ladders

Long or very heavy ladders should be carried by two persons

(i) Carrying ladders

**Fig. 1.14**   *continued*

Before raising a ladder always look out for any overhead obstructions – they can be a trap for the unwary. Overhead cables are a common hazard, especially on existing buildings. When carrying ladders from one place to another, never be tempted to put the arm through the rungs to make it easier to carry. It might be a little more comfortable this way but if one stumbles or trips over it could result in a broken arm! Figures 1.14(e)–(i) show the correct ways of carrying and raising ladders. Remember,

too, the same procedures must also be carefully followed when lowering ladders.

### Health and safety when working in and near lift shafts

*Excavations*
This term relates to trenches, pits or holes below ground level. Regulations in respect of excavations

require certain precautions to be taken intended to prevent persons falling in, and also to ensure that effective support is provided to guard against collapse of the walls. Fencing must be erected around any excavation into which persons could fall more than 2 metres, but clearly it is common sense as well as legally compulsory under the HSW Act for a place of work to be made safe for both workers and the general public. Similar fencing is also necessary for lift shafts or stair wells as shown in Fig. 1.15(a).

Excavated material should be deposited at a sufficient distance back from the edges of a trench or hole to prevent excessive mass causing the sides to collapse. Suitable materials such as stout timbers or steel plates should be used to shore up the inside of the excavation, the extent to which this is necessary depending upon the nature of the soil and the depth of the hole or trench, as shown in Figs 1.15(b)–(c).

The main causes of the trench walls caving in are related to the following:

(a) The type of ground in which the excavation is made.
(b) The depth of the excavation.
(c) Whether or not the trench is near a building or moving traffic.
(d) The possibility of the ground near the edges of the trench being unable to support the weight of the soil excavated, machines and materials.

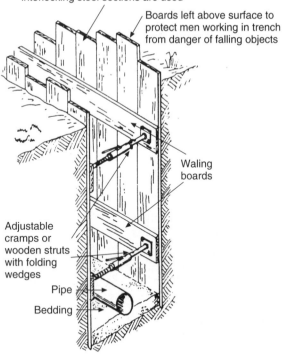

(b) Close boarding of trenches, used where there is any likelihood of trench collapse, e.g. sand or loose shale

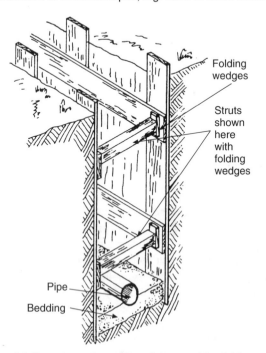

(c) Open boarding of trenches, used in stable ground for depths of up to 1.5 m

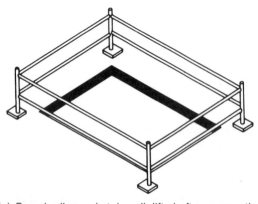

(a) Guard rail round stair well, lift shaft or excavation

**Fig. 1.15**

The Health & Safety Regulations require that **extensive and deep excavations must be inspected daily and thoroughly examined every seven days by a competent person**, as a certain amount of experience is required to detect the first indication of any danger.

Work in trenches, such as the installation of drains, must sometimes be carried out in unfavourable weather conditions and in such circumstances it is mandatory for the employer to provide protective clothing.

### Electric power tools and temporary lighting

It is sound advice to read and thoroughly understand before use the instructions provided with power tools, especially those tools with which the user is not familiar. The following six points however, are a general guide for safe use.

(a) The supply voltage should always be within the range for which the tool is designed.

(b) For work on building sites voltages of 110 should be used as this is far safer than the mains voltage of 240, a shock from the latter often proving fatal. A portable step-down transformer can be used to reduce the voltage from 240 to 110 volts, care being exercised in handling the transformer as these are fragile and easily damaged. As an alternative, or if no mains supply of electricity is available, a small portable generating plant may be used.

(c) Portable power tools are perfectly safe providing they are treated with care and used only for work for which they have been designed. It is important that they are serviced regularly by a competent electrician.

(d) All electric power tools must be earthed unless they are double insulated. Double insulated equipment is indicated by the mark shown in Fig. 1.16.

(e) Flexible electrical cables used for lighting or tools should never be allowed to trail on the floor where they are liable to be damaged; instead they should be tied or suspended at high level out of harm's way. Only tough rubber or PVC cables should be used, and all temporary connections should be both waterproof and unbreakable.

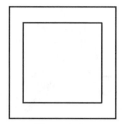

**Fig. 1.16**   Double insulation mark on power tools.

(f) Portable electric hand lamps must be adequately earthed and should not be used with a socket plugged into a light pendant which has only live and neutral connections. These lamps should have an electrically insulated handle and be fitted with a wire guard around the bulb. Most of the lamps are fitted with a clamp to enable them to be secured to a suitable point thus leaving both hands free.

*Safety with hand tools* (Figs 1.17(a)–(g))
Most building craftworkers do a great deal of their work with hand tools and many quite serious accidents are caused when these are defective or badly maintained. It is not only the user who is endangered, but also workmates who may be in the vicinity. The use of unsafe tools is an offence under the HSW Act.

A common case of using an unsafe tool is an improperly fixed hammer head flying off its handle. Other common dangers which occur with hand tools are shown in the following illustrations.

Diagram (a) shows what a plumber's hammer should *not* look like. Damaged handles should always be promptly replaced, not tied up with string or wire. The head should be securely wedged to the handle using both wooden and barbed metal wedges, as in (b). Nails are not suitable as wedges.

When using a spanner, make sure it fits the nut. *Never* use one that 'nearly' fits. Discard old spanners (c) and pipe wrenches which may slip in use and result in severe hand or wrist injuries.

Cold chisels, caulking irons, punches and socket-forming tools are all used with a hammer and continuous use causes the top of these tools to turn over, commonly called 'mushrooming' as

(a) A hammer like this is a danger to both the user and his workmates

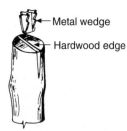

Metal wedge

Hardwood edge

(b) Fixing hammer heads correctly with wooden and metal wedges

(c) Worn and opened spanners can slip resulting in hand injuries

(d) 'Mushroom' headed tools can result in loss of sight or serious eye injury

(e) Correct form of grinding for the use of hammers on chisels, points, caulking tools and socket formers

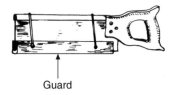

Guard

(f) Guards should be fitted to all sharp cutting tools when not in use

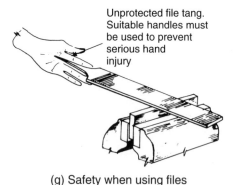

Unprotected file tang. Suitable handles must be used to prevent serious hand injury

(g) Safety when using files

**Fig. 1.17** Safety with hand tools.

illustrated in (d). The mushroomed parts often fly off when struck with a hammer and can cause severe eye injuries, so keep the heads of all tools of this nature ground as shown in (e).

The edge of cutting tools should be kept sharp and protected when not in use. Never carry sharp tools in a pocket, and always work with the hands behind the cutting edge. Although a screwdriver is not normally classified as a cutting tool, it is unwise to use one on work held in the hand, use a suitable vice instead. A simple guard on a saw (f) makes it safe when not in use and also protects the cutting edge.

The use of a file without a handle can be very dangerous as the sharp tang can pierce the hand as shown in (g). Make sure the work is securely held in the clamp or vice when filing operations are being carried out.

**General site safety**

One of the first rules of safety is tidyness. Building debris should never be left lying about. Nails protruding from pieces of timber as shown in

**Fig. 1.18**   One of the most common hazards on site. Never leave nails like this in timber.

Fig. 1.18 is one of the most common causes of foot injury; such nails should always be removed or hammered over. Pieces of pipe or tube can roll if trodden on and cause a fall which could be serious, especially if the person falling is carrying something such as a ladder or gas cylinder.

Remember also that while work is being carried out which does not necessarily constitute a serious danger to oneself, this may not be the case for people in the vicinity. Be extra careful when anyone else is working below, and in all cases display proper warning notices and cordon off possible danger areas.

Protective footwear should always be worn on site and in the workshop: boots or shoes which are strong and hard wearing, preferably with steel reinforced toe caps to protect the foot against the type of hazard shown in Fig. 1.19.

Eye injuries are painful and can result in loss of sight, so always wear the appropriate type of safety goggles or glasses when exposed to dust or flying chippings as in Fig. 1.20. Goggles are made with a variety of lenses of different impact resistance for various operations. The correct type must be worn or they may not provide the required protection.

Protection of the skin by using suitable gloves may be necessary in some cases when handling

**Fig. 1.20**   Eye protection. Goggles are essential when cutting away brickwork, concrete, etc.

abrasive materials or those materials containing strong acids or alkalis such as cement or lime mortars. The use of barrier cream can prevent skin disease such as dermatitis, and also prevents the entry of grease and dirt into the pores of the skin.

### Hearing protection

In comparatively recent years it has been found that the hearing of a person may be seriously affected if he or she is subjected to persistent noise. Many power tools now in use on site, such as cartridge firing tools, large hammer drills and percussion tools, generally have a very high noise level. Operatives using such tools or working in close proximity to them should protect their hearing by using suitable ear muffs.

### Knee pads

Plumbers spend much of their working lives on their knees, often on cold, hard or damp surfaces, which, apart from the discomfort, can be the cause of many ailments, such as rheumatism and a common complaint called 'housemaid's knee'. Many of the older generation of plumbers fabricated a form of knee pad, using pieces of discarded carpet with leather straps sewn on to secure them. It is now possible to purchase proper pads made of leather which give a good degree of flexibility and are comfortable and safe to use.

### Storage of pipes and cylinders (Figs 1.21(a)–(c))

The incorrect storage of pipes or gas cylinders can be a possible site hazard. Diagram (a) shows how

**Fig. 1.19**   Protective footwear. Falling objects are extremely dangerous, especially to the head and feet.

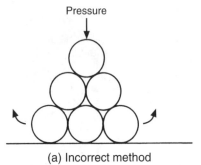

## (a) Incorrect method

Pressure on the top of a stack of pipes or cylinders can force out those below. Foot or hand injury may be the result of incorrect stacking.

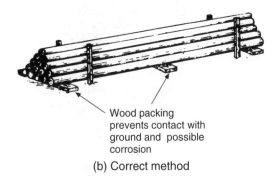

Wood packing prevents contact with ground and possible corrosion

## (b) Correct method

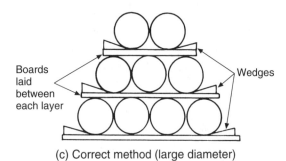

Boards laid between each layer

Wedges

## (c) Correct method (large diameter)

**Fig. 1.21** Storage of pipes and cylinders.

stored circular members can roll outwards owing to the pressure of those on top, and to prevent this occurring wedges or stakes should be used as shown in (b).

If the site is a large one and large quantities of pipe are held in store, a properly constructed tube rack is a better arrangement (c). Not only can differing sizes be separately stored, but protection from corrosion can be provided by keeping the tubes clear of the ground and covering the whole rack with waterproof sheeting.

## Safety signs

Prominent notices should be displayed on site to warn of possible hazards; to indicate what protective clothing should be worn; and to provide general information. Some typical examples are shown in Fig. 1.22.

Safety signs fall into four main categories which can be recognised by their shape and colour. In some cases additional information is provided specifying the nature of the danger.

*Prohibition signs*

| Shape | Circular |
|---|---|
| Colour | Red border and cross bar |
| | Black symbol on white background |
| Meaning | Shows what *not* be done |
| Example | No smoking |

*Mandatory signs*

| Shape | Circular |
|---|---|
| Colour | White symbol on blue background |
| Meaning | Shows what *must* be done |
| Example | Wear hand protection |

*Warning signs*

| Shape | Triangular |
|---|---|
| Colour | Yellow background |
| | Black border and symbol |
| Meaning | Warns of hazard or danger |
| Example | Caution, risk of electric shock |

*Information signs*

| Shape | Square or oblong |
|---|---|
| Colour | White symbols on green background |
| Meaning | Indicates or gives information of safety provision |
| Example | First-aid point |

*Protective clothing*

Construction work is usually associated with dirt, dust, dampness and mud, and additionally for the plumber oil and grease; therefore suitable overalls will protect the user's personal clothing. Quite apart from this the use of protective clothing in cold weather conditions enables the wearer to keep warm. It is not easy to concentrate on a job if one is feeling cold and wet. In the event of plumbers

**Fig. 1.22**  Typical notices warning of potential hazards.

handling lead, the use of overalls is essential so that at the end of the working day they can be removed and the possibility of contaminating public or personal vehicles, or indeed one's home, is minimised. It is recommended that clothes that have been in contact with lead should be washed separately to avoid contaminating other clothing. Protective clothing should be comfortable and such that one's movement is not constrained. It should be worn properly, not as is sometimes seen with boiler suits tied around the waist by the arms; this practice has been known to cause accidents.

Many employers sensibly insist on the use of safety helmets on their sites, as employees are less likely to sustain serious head injuries in the event of an accident. It is compulsory to wear a safety helmet for certain types of work which include the following:

(a)  Where overhead work is in progress.
(b)  On demolition sites.

(c) When loading or unloading vehicles by means of mechanical equipment such as cranes or lifts.

(d) In excavations more than 1.2 m deep.

(e) When piling work is in progress.

It is important that the inner head band is adjusted correctly to suit the wearer as it can cause both discomfort in use and possibly render the helmet useless in the event of an accident. For identification purposes it should be clearly marked with the wearer's name.

## Toxic and dangerous substances and materials

Plumbers frequently handle potentially dangerous substances and it has always been accepted in the trade that lead falls into this category. Sensible plumbers have always recognised the dangers of lead and taken the necessary precautions to avoid these, some of which are listed here.

Molten lead gives off fumes containing lead and operations involving lead welding or melting lead for caulked joints should always be carried out in a well-ventilated area. In extreme cases a suitable respirator should be worn. Protective clothing must be worn but should be removed for meal breaks and before leaving the site. Personal cleanliness is essential, and before commencing work a good barrier cream should be applied to the hands. Before a meal is consumed, and on completion of work, hands and finger nails must be thoroughly washed and scrubbed with a suitable nail brush.

Remember that lead can be absorbed into the body via the mouth, through inhalation when it is vaporised, and in some cases through the skin.

### Solvents and adhesives

Fumes from many of these substances are both toxic and addictive and great care should be taken to avoid inhaling them. In most cases they also constitute a very real fire hazard and great care must be taken when, for example, making solvent-welded joints on plastic pipes. When not in use all caps and stoppers on the containers of these materials should be replaced, not only for safety reasons, but also for economy. If stoppers are not replaced after use these materials will quickly become unsuitable for

the purpose for which they were intended, owing to the evaporation that takes place.

### Asbestos

This material has been used for a long time in the construction industry for a wide variety of purposes. It has been found to be extremely dangerous in both fibre and powder form and the manufacture of asbestos and asbestos products has ceased. Its principal use in mechanical services was for insulation and in the form of asbestos cement products such as corrugated roofing sheet and flue pipes. Where work on asbestos cement products has to be carried out, possibly for maintenance purposes, the principal danger is when cutting the material with a saw or drilling, thus producing asbestos dust. Providing this is done in a well-ventilated area and a suitable face mask is worn, there is minimum danger to the installer. Asbestos products are most dangerous when used for heat insulation and because of its hazardous nature it is, in many cases, being removed. This can only be done by specialist firms using the correct equipment and taking precautions to ensure no contamination of the surrounding area occurs. No one under the age of 18 is permitted to undertake work of this nature and when such material is encountered, no attempt should be made to remove it. It is the subject of the Asbestos Regulations 1969.

## Safety with compressed gases and blowlamps

### Liquid petroleum gases (LPG) (Figs 1.23(a)–(d))

The gases most commonly used by plumbers and heating fitters on site for portable heating equipment are propane and butane. Cylinders must always be used and stored in the upright position in a ventilated area. Notices indicating the storage of explosive gases and no smoking signs must be prominently displayed both inside and outside the store. Read any instructions provided with the equipment and ensure the cylinders are not placed near a source of heat. Regulators may be of the fixed or adjustable type, comply with the relevant British Standard and be suitable for the gas being used. Before connecting a regulator, snift the valve

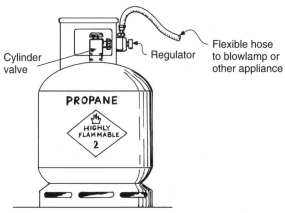

(a) Typical portable gas cylinder

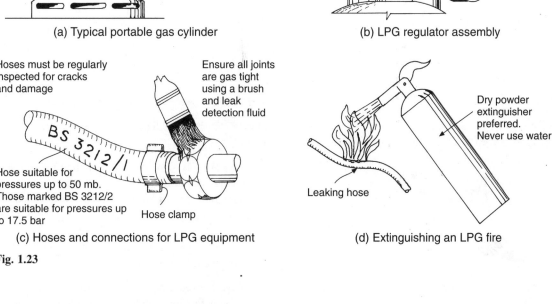

(b) LPG regulator assembly

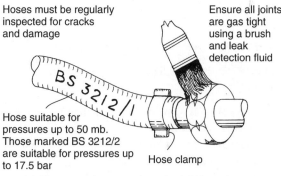

(c) Hoses and connections for LPG equipment

(d) Extinguishing an LPG fire

**Fig. 1.23**

to blow out any dust and make sure the mating surfaces are clean, dry and undamaged. All hexagon nuts are kerfed (notched) and have a left-hand thread. Use only flexible hoses carrying the BS number 3212. See Figs 1.23(a)–(d) which illustrate the equipment commonly used and further safety instructions.

*Prevention of fire risks on site*
Many materials used in the construction industry are highly flammable and care should be taken to ensure that fire risks are reduced to a minimum.

Fire caused by carelessness with the blowlamps can be dangerous and costly. Be especially careful when making soldered capillary joints in awkward corners, under suspended floors and in roof spaces. Figure 1.24(a) shows how convection currents can

draw the flame from a blowlamp through a hole in the ceiling and start a fire in a sub-floor or roof space. Certain types of pipe insulation are very flammable and such insulation should be temporarily removed to give plenty of clearance for the flame when making soldered joints (see Fig. 1.24(b)).

Another fire hazard with which plumbers need to be concerned is the solvent used for making joints on synthetic plastic materials. Many of these solvents give off heavy flammable vapour while in use, and it is dangerous to smoke or use a flame in the vicinity where these joints are being made.

Suitable fire-fighting equipment must always be available and ready for use, not only on building sites but also when small maintenance jobs are

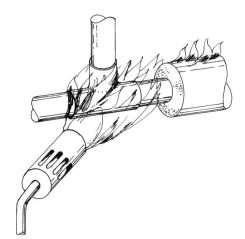

(a) Flame can be drawn through a hole in the ceiling by convection causing flammable materials to ignite

(b) Ensure any flammable insulation is removed before sweating a soldered joint

**Fig. 1.24**   Fire risks with blowlamps.

**Fig. 1.25**   The three essentials of fire. For a fire to be maintained, heat, oxygen and fuel must all be present. Therefore to put out a fire one or more of these should be removed.

Fire extinguishing is based on this very simple principle:

(a) *Starving* – removing the fuel.
(b) *Smothering* – preventing the access of oxygen.
(c) *Cooling* – removing the source of heat.

   Fire risks are classified in three categories as follows:

(a) *Class A risks* – mainly extinguished by cooling, usually with water.
(b) *Class B risks* – most efficiently extinguished by smothering and preventing oxygen joining with fuel.
(c) *Class C risks* – a non-conductive smothering agent must be used.

   Table 1.2 gives details of extinguishing agents and types of fire for which they are suitable.
   It is important to recognise that not all extinguishing agents are suitable for all types of fires. Using the wrong extinguisher can be dangerous. For example, to throw water on an electrical fire would give the fire-fighter an electric shock and in the case of a petrol or oil fire, water would easily spread the blaze. It is therefore important to know the various types of extinguishing agents and the types of fire for which they are suitable. It should be noted that BCF extinguishers are especially useful when dealing with expensive electronic equipment as BCF does little physical damage. It is, however, derived from the chemicals bromine, chlorine and fluorine and special care is necessary with its use as these substances give off dangerous fumes when these extinguishers are used.

being carried out. Detailed information on fire control will be found in the Health and Safety at Work booklets obtainable from the Stationery Office and listed at the end of this chapter, but the following information will be helpful in extinguishing or controlling fires until the arrival of the fire brigade.
   Broadly speaking there are three essentials to every fire and these can be shown as the three sides of a triangle as in Fig. 1.25. If any one part of the triangle is removed, combustion cannot continue.

**Table 1.2** Extinguisher selection chart types and uses.

| Class of fire ref. BS 4547 1970 | Extinguishing principles | A.B.C. all purpose powder | Metal powder | $CO_2$ gas | Foam | Water | BCF |
|---|---|---|---|---|---|---|---|
| *Class A.* Fires involving solid materials usually of an organic nature. Examples: wood, paper, textiles, etc. | Water cooling or exclusion of air | Yes Excellent | No | No | Yes | Yes Excellent | Yes |
| *Class B.* Fires involving liquids or liquefiable solids. Examples: oil, fat, paint, etc. | Flame inhibiting or surface blanketing and cooling | Yes Excellent | No | Yes | Yes Excellent | No | Yes |
| *Class C.* Fire involving gases | Flame inhibiting | Yes | No | Yes | Yes | No | Yes |
| *Class D.* Fires involving metals: magnesium, sodium, titanium, etc. | Exclusion of oxygen and cooling | No | Yes Excellent | No | No | No | No |
| Fires involving electrical hazards | Flame inhibiting | Yes | No | Yes Excellent | No | No | Yes |

## Lifting and handling materials

*Manual lifting and carrying*
Many thousands of accidents caused in the handling and carrying of goods are reported annually to HM Inspectors of Factories. Building is among those industries having the highest accident rate and one of the major contributory factors can clearly be identified as the lack of training provided in the 'know-how' of lifting. Brute force and ignorance are of little use in lifting and can lead to strain or injury. It is more effective to 'think out' the lifting operation and then to rely on the skilful application of the right muscles. A load can be raised with less effort by using the strong leg and thigh muscles as opposed to those in the abdomen and back. This procedure is illustrated in Fig. 1.26.

The factors to be considered in the lifting operation are:

(a) maintain the back in a straight, though not necessarily vertical, position
(b) tuck the chin well in and do not drop the head forward or backward
(c) keep the arms straight and close to the body

(a) Correct
Legs are straightened to lift the load. Back straight.

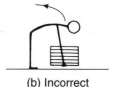

(b) Incorrect
Weight of the load taken by back.
Leg muscles not being used.

**Fig. 1.26** Correct and incorrect ways of lifting.

(d) place feet about 200–300 mm apart (no wider than the hips) with one foot slightly forward and pointing in the direction of travel
(e) grip the object firmly with the whole hand (not just the finger tips) using gloves for protection where necessary
(f) bend the knees and lift by straightening the legs

(g) lift in easy stages: floor to knee – knee to carrying position

(h) ensure the load does not obscure vision

(i) move off in the direction of the advanced foot, keeping the load close to the body

(j) when setting down the object, reverse the lifting procedure

It is impossible to specify accurately the maximum load which can be lifted by an individual as this depends on the physical characteristics and the age of the person. The HSW Act stipulates that no person shall be required to lift, carry or move any load that is so heavy as to be likely to cause injury. If an object is too heavy, too large or of an awkward shape, do not attempt to lift it alone – seek assistance or use a mechanical lifting appliance.

A lot of lifting can be avoided by the old-fashioned method of using rollers. Short pieces of steel pipe placed under a suitable support carrying the load will enable it to be moved very easily on flat hard surfaces. Use of rollers on sloping surfaces is, however, not recommended, as a load can get out of control and be a possible cause of accident.

**Levers**

A lever may be defined as a simple machine that has been used for thousands of years. Plumbers and heating fitters commonly use levers in the form of a crowbar or length of steel pipe to move heavy objects. Pipe wrenches and similar tools also employ the principle of leverage. It is seldom necessary on site to go through lengthy calculations to determine the length of a lever required to move, for example, a boiler or roll of sheet lead; from experience it is well known that the longer the lever the less effort is needed.

Levers are classified in first, second and third orders; Figs 1.27(a)–(c) show all three orders, and some practical applications. A worked example of the first order of levers is shown in Fig. 1.28, as this is a common example from a safety aspect where a responsible person has to make an accurate calculation.

In this practical example a cradle is to be used over a parapet wall. Note in Fig. 1.28 that as a safety measure E and B must be three times greater than L and A. Assuming the mass of the cradle and any load it has to carry to be 250 kg, the

Key: L = Load to be moved
F = Fulcrum or pivot point
E = Effort necessary
A = Distance between fulcrum and load
B = Distance between fulcrum and effort

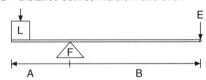

The fulcrum here is between the load and the end of the lever, the effort required to move the load depends on the position of F in relation to L and E. To reduce the effort shorten A or lengthen B. The basic formula states that where $A \times L = B \times E$ the lever will be in equilibrium.

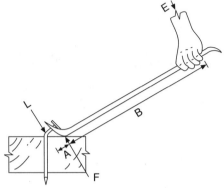

The use of a crow bar to remove a large nail from a block of wood is a typical example of the first order of levers. Float-operated valves also employ this order.

(a) First order

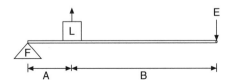

In this case the load is situated between the fulcrum and end of the lever.

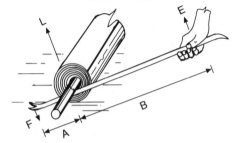

Moving a roll of lead sheet. In this case the load is situated between the fulcrum and the end of the lever. Another example of the second order of leverage is the wheelbarrow.

(b) Second order

**Fig. 1.27** Levers and leverage.

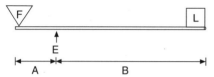

This order of levers differs from the first and second in that the effort is between the fulcrum and the load. The hand vice shown is a typical example; spring bow compasses are another.

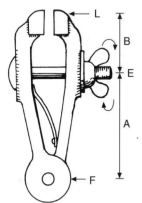

Although this order has little to do with a reduction of effort it has many applications in the engineering and manufacturing industries. It should be noted that in this case distance B is not strictly correct as the example shown employs another simple machine, the screw thread!

(c) Third order

**Fig. 1.27** *continued*

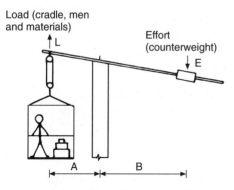

**Fig. 1.28** Application of leverage to supporting suspended scaffold or cradle.

distance A = 1.2 m and length B = 3.6 m, the effort (counterweight) needed to support it in equilibrium may be calculated by transposing the basic formula L × A = E × B in terms of E:

$$\therefore \quad E = \frac{L \times A}{B}$$

$$E = \frac{250 \times 1.2}{3.6} = 83.333 \text{ kg}$$

The effort or mass (weight) to maintain the cradle in equilibrium is 83.333 kg but remember that both E and B must be multiplied by 3 in order to comply with safety requirements. The minimum length of B will be 3.6 m × 3 = 10.8 m and the mass (counterweight) must be 83.333 kg × 3 = 249.999 kg approximately 250 kg.

**Pulleys**

The pulley, being a wheel, is in effect a continuous lever. With the single fixed wheel pulley the only mechanical advantage is a change in the direction of the effort E to a more convenient position, i.e. pulling downwards is easier than lifting upwards. In order to lift a load using a single pulley wheel, an effort of at least the same magnitude as the load must be applied. Such a machine is said to have a mechanical advantage (MA) of one, which can be expressed as a formula:

$$\text{Mechanical advantage} = \frac{\text{Load}}{\text{Effort}}$$

A gin wheel is an example of a single fixed wheel pulley and is illustrated in Figs 1.29(a)–(b).

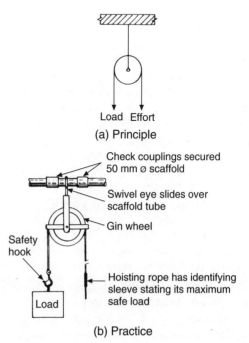

**Fig. 1.29** Simple pulley or gin wheel.

The single movable pulley shown in Fig. 1.30 will give a mechanical advantage of two. With this arrangement one end of the rope is fixed and the pulley itself is cradled in the rope. The load is suspended from the pulley block and the effort is used to pull downward on the free end of the rope. Because the load is suspended from two ropes each taking half the strain, the effort required is only 50 per cent of the load.

The common pulley tackle shown in Fig. 1.30 has only single wheels in each block, whereas Fig. 1.31

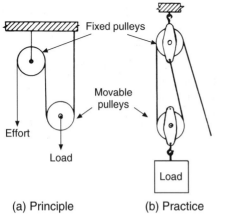

(a) Principle     (b) Practice

Sketch shows the working action of a single movable pulley.

**Fig. 1.30** Double pulley hoists.

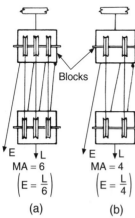

By pulling rope 'E' on the first pulley a distance of 1 m, the load will be raised by approximately 166 mm. In the case of pulley (b) a greater effort is required but the load will be raised by 250 mm.

**Fig. 1.31** Multi-pulley systems enable heavy loads to be lifted with little effort.

illustrates pulley blocks having more than one wheel, the number of wheels determining the mechanical advantage. It may appear at first sight that using a multiple wheel pulley or a lever is a method of obtaining something for nothing, i.e. a small effort to raise a greater load, but this conclusion is not correct. These machines cannot create additional energy and the work done by the effort is never less than the work done on the load. What in fact happens is that the load is moved through a smaller distance than the effort, i.e. by pulling the rope a distance of 1 metre, the load is lifted perhaps 100 mm.

With the arrangement shown in Fig. 1.30 the effort moves twice as far as the load, so the pulley is said to have a velocity ratio (VR) of two. Velocity ratio can be expressed as follows:

$$\text{Velocity ratio} = \frac{\text{Distance moved by effort}}{\text{Distance moved by load}}$$

If a machine were 100 per cent efficient then the velocity ratio would be identical to the mechanical advantage. In practice, friction losses must be considered in addition to the mass of the pulley blocks, while the actual mechanical advantage requires the efficiency (%) of the machine to be taken into account.

Mechanical advantage can also be expressed as the formula

$$\text{Mechanical advantage} = \\ \text{\% Efficiency} \times \text{Velocity ratio}$$

### Slings

Mechanical handling involving lifting is usually accompanied by means of a sling made of chain or rope of steel wire, fibre or composition construction. Reference to any catalogue of lifting appliances will indicate a wide range of slings made for specific lifting operations, some common types which may be used on the building site being shown in Figs 1.32(a)–(d).

Adequate strength for a specific lifting task is of vital importance. This can be found by referring to the Safe Working Load (SWL) that is either indicated on the sling or is recorded in the test report which must be provided for every lifting

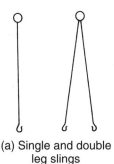

(a) Single and double leg slings

(b) Arrow sling

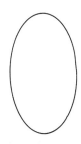

(c) Endless chain sling

(d) Canvas sling used for materials on which a chain sling might slip

**Fig. 1.32**   Types of sling.

**Table 1.3**   Two-legged slings at different angles.

| *Approx. angles only* | | | | | |
|---|---|---|---|---|---|
| Angle of sling | 0° | 30° | 60° | 90° | 120° |
| % of SWL (approx.) | 100 | 96 | 86 | 69 | 50 |

appliance used on site or in the workshop. An SWL of 1,500 kg means the sling should not be used for loads in excess of that figure.

Even if the lifting capacity of the sling is satisfactory, careless slinging can cause the SWL to be exceeded. The lifting capacity of any sling *decreases* as the angle between the legs *increases* – the effect can be seen from Table 1.3.

For example, a two-legged sling with an SWL of 1 tonne (1,000 kg) used at an angle of 30° can carry a maximum load of

$$1,000 \times \frac{96}{100} = 960 \text{ kg}$$

Used at 90° the same sling will safely carry only

$$1,000 \times \frac{69}{100} = 690 \text{ kg}$$

The following are typical examples of lifting operations encountered by the plumber which could be hazardous.

(a) Awkward objects such as boiler sections will swing about if the sling is 'off centre': lift slowly until the load is off the ground.
(b) Compressed gas cylinders should not be lifted with chain slings. A fibre or composition rope sling may be used to lift one cylinder – never more than one at a time – provided it is correctly adjusted to prevent slipping.
(c) Long lengths of large-diameter pipes may bend (deflect) or even slip from the sling. Slings should be positioned at one-quarter of the pipe length from each end, see Fig. 1.33.
(d) Pulleys must never be attached to any element of the building, e.g. the roof truss, without checking that it will take the load.

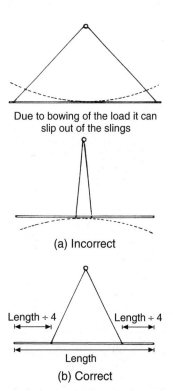

Due to bowing of the load it can slip out of the slings

(a) Incorrect

Length ÷ 4      Length ÷ 4

Length

(b) Correct

**Fig. 1.33**   Correct and incorrect methods of using slings.

## Further reading

British Red Cross Society, 9 Grosvenor Crescent, London, SW1X 7EJ. Tel.: 0207 235 5454.

Construction Site Safety (Safety Notes), CITB, Bircham Newton, nr King's Lynn, Norfolk, PE31 6RH. Tel.: 01485 577 577.

Fire Protection Association, Malrose Avenue, Boreham Wood, Herts, WD6 2BJ. Tel.: 0208 207 2345.

*Hazards at Work: T.U.C. Guide to Health and Safety*, obtainable from T.U.C. Publications Dept, Congress House, Russell Street, London, WC1B 3LS. Tel.: 0207 636 4030.

Health & Safety Publications, PO Box 1999, Sudbury, Suffolk, CO10 6FS. Tel.: 01787 881165. For specific enquiries: HSE Information Centre, Broad Lane, Sheffield, S3 7HQ.

HMSO, PO Box 276, London, SW8 5DT. Tel.: 0870 600 5522.

Royal Society for the Prevention of Accidents, Edgbaston Park, 353 Bristol Road, Birmingham, B5 7ST. Tel.: 0121 248 2000.

## Self-testing questions

1. Describe briefly the purpose of (a) a guard rail, and (b) toe boards, when used on scaffolding.
2. State the reason why ladders should not be painted and what methods are used for their protection and storage.
3. State the ratio of vertical and horizontal distances relating to the safe slope of ladders.
4. List the precautions to be taken when using a blowlamp in a roof space or under suspended wooden floors.
5. State the three essential conditions which must be present before a fire can be started, and name the extinguishing agent recommended in the following cases:
   (a) a petrol fire
   (b) burning timber
   (c) an electrical fire
6. State the recommended supply voltage for the safe use of electrical power tools and lighting on site.
7. Describe briefly the main purposes of the Health and Safety at Work etc. Act and state the main responsibilities of employers and employees under the Act.
8. What type of sling is recommended to hoist gas cylinders? What is the maximum number of cylinders which can be lifted in a sling?
9. What is the meaning of the abbreviation 'SWL' and what is its importance?
10. If a load of 6 kg is to be lifted using a crowbar 900 mm long and the fulcrum is placed 100 mm from the load, what effort will be required to raise it using the formula

$$E = \frac{L \times A}{B} ?$$

11. Make a simple sketch showing the symbol for double insulated power tools.
12. List the correct first-aid procedures for treating a person suffering from electric shock.
13. A prohibition order may be issued if unsafe or dangerous working practices are discovered in a workplace. State who issues such an order and its implications.
14. List the safety requirements relating to the use of LPG.
15. List five possible accidents or dangerous occurrences where the HSE must be notified immediately.

# 2 Hand tools

After completing this chapter the reader should be able to:

1. Recognise and name the common tools used by the plumber.
2. Name the tools required to be supplied by the plumber.
3. State the use of each tool.
4. Identify faults in hand tools.
5. State maintenance requirements for hand tools.
6. Select and list appropriate tools for a given job.

## Introduction

A saying much favoured by plumbers is that 'Good tools are required to do a good job'. Good tools are, however, very expensive and in order to keep costs to a minimum the following points must be considered.

### Selection
It must be realised that low-cost items are not always a bargain. To obtain the best value for money, high-quality tools which will last should be purchased.

### Maintenance
To ensure a long, effective and safe working life, it is essential to keep tools in a good state of repair.

### Basic tool kit
By agreement between employers and employees the plumber is required to provide a basic tool kit which is sufficient to allow the employee to undertake the normal range of plumbing work.

When the apprentice or trainee achieves the status of a trained plumber he or she is paid a tool allowance by the employer. This enables the tools to be maintained in good condition and to replace those which have become worn.

The employee is not normally expected to provide expensive tools such as, for example, stocks and dies, pipe vices and large bending machines and power tools.

## Classification of hand tools

It is not practicable to describe within the scope of this book every tool the plumber uses, but it is necessary to consider briefly the more common items so they can be recognised and their uses understood. The application of the various tools to particular activities is more fully detailed in later chapters.

The hand tools used by the plumber can best be grouped under the following general practical operations.

### Measuring and setting out
*Compasses*, *dividers*, *try squares*, *straight edge*, *flexible steel tape* (marked in metric), *chalk line*, *spirit level* (those shaped like a canoe are termed 'boat levels') and the *plumb bob* are tools so widely used as to require no further explanation.

### Cutting (sheet and pipe)
For sheet materials such as lead, copper and aluminium, *tinman's snips* having either straight or curved blades are used. The recommended handle is

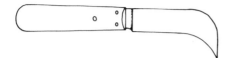

**Fig. 2.1** Lead knife used for cutting sheet lead.

the 'open-ended' or 'no nip' pattern which avoids pinching the hand when they are closed.

Sheet lead, being a soft material, can also be cut using a knife (Fig. 2.1) that 'scores' the metal, which is then separated along the scored line.

Holes are cut in metal vessels, e.g. water storage tanks, by *hole saws* or *hole cutters* (see Fig. 2.2). Both tools are rotated in a clockwise direction, the saw by hand, breast or power drill. Different diameter holes can be produced by changing the cutters. If power tools are used, only those having low speeds are suitable for these cutters to avoid over-heating the blades, especially when cutting metal. Suitable lubricants on the blades will also prolong their life.

A variety of tools are available for cutting pipe, the choice being determined by the material to be cut. The *hacksaw* is the most common tool, the best pattern having a tubular frame. Three factors are important when buying hacksaw blades: the nominal length, the blade steel and the number of teeth per 25 mm. These factors are explained in Table 2.1 and Fig. 2.3.

When hand cutting, the blade teeth point forward and at least three consecutive teeth must be in contact with the section being cut (this avoids broken teeth). Sparing application of a lubricant will prolong the life of the blade.

It should be noted that the teeth on all saw blades are 'set' so that the actual cut is wider than the saw blade. This ensures the saw blade does not bind in the cut and gives adequate clearance round the blade to allow the cuttings to fall clear. Saws for cutting timber have larger teeth than those used for cutting metal and can be reset using a saw set. Unless the reader has had some instruction on saw sharpening and setting, this job is best left to a specialist, or make friends with a good carpenter. Relatively cheap wood saws are now available having very hard teeth but sharpening and resetting are not possible with these and they must be replaced when they become worn.

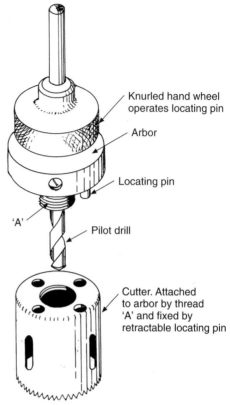

(a) Heavy-duty pattern
with interchangeable cutters

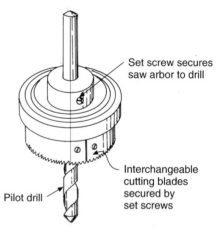

(b) Light pattern suitable for only thin
metal sheet, wood or plastic

**Fig. 2.2** Tools for cutting circular holes in various materials.

**Table 2.1** Hacksaw blades.

| Blade steel | |
| --- | --- |
| *Tungsten steel* | *High-speed steel* |
| General purpose work | Maximum blade life and speed of cutting. For hard material. |

Both types available as follows:

Flexible – hard on cutting edge only; unbreakable in use; for unskilled operators.

All hard – hardened (brittle) over whole blade: rigid; gives accurate cutting for skilled operator.

| Metal thickness being cut | Teeth per 25 mm | |
| --- | --- | --- |
| | *Hard material* | *Soft material* |
| Up to 3 mm | 32 | 32 |
| 3–6 mm | 32 | 24 |
| 6–12 mm | 24 | 18 |
| 12–24 mm | 18 | 14 |

Arrow on blade indicates direction of cut. Care must be taken to ensure blade teeth are pointing forward, towards front of hacksaw. In use pressure is exerted on forward stroke but not on backward stroke. (See also Table 2.1.)

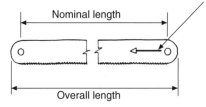

**Fig. 2.3** Hacksaw blades.

## Pad saws

*Pad saws* (see Fig. 2.4) are sometimes called 'keyhole' saws and were indeed used for that purpose in the days when keys were very large. Nowadays these saws are mainly used for cutting large circular or irregular shapes in timber boards and sheet material such as plywood, plasterboard or hardboard. The blades can be obtained separately from the handles and range from fine to coarse cutting. Use a fine-toothed blade if cutting thin wooden sheets, as this will prevent persistent bending or snapping of the blade.

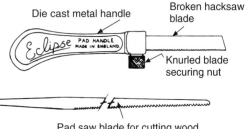

Pad saw blade for cutting wood. Obtainable in various lengths and teeth form from fine to coarse

**Fig. 2.4** Pad saw.

Handles are made in both wood and metal. Those made of wood have the advantage that the blade length can be varied but if this type is used it is important to ensure the blade is firmly secured by the set screws. Should the blade work loose it may slip and penetrate the wrist. Those with a metal handle lack the flexibility of blade length but are safer in use and are made to permit broken hacksaw blades to be fitted. These are very useful to a plumber where a pipe must be cut in a position where it is impossible to use even a junior hacksaw.

*Pipe cutters* are of three basic types: wheel, roller (Figs 2.5(b) and (c)) and link. A wide range of multi-wheel and roller cutters are available for use on the various pipe materials (with the exception of cast iron for which the link cutter only is used). Reference to a manufacturer's catalogue is advised where there is doubt as to the best cutter to be used for particular pipe materials.

The three- or four-wheel cutter is of most use where a full turning circle cannot be achieved. These need only be rotated through 120°–130° and this is a real advantage where space is limited. The formation of both external and internal burrs produced by these cutters is a disadvantage as the burrs have to be removed, but the latter can be avoided if the cut is completed using a hacksaw.

As roller cutters require to be turned through 360°, they have a limited use. The rollers prevent an external burr being formed, and also position the cutter at 90° to the pipe preventing the 'spiral cut' that is common with the multi-wheel type.

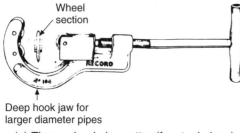

Wheel section

Deep hook jaw for
larger diameter pipes

(a) Three-wheel pipe cutter (for steel pipes)

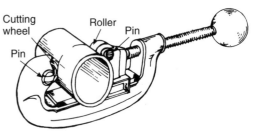

Cutting wheel    Roller    Pin

Pin

(b) General purpose single wheel roller cutter
for copper tubes

The rollers steady the tool and enable a thin fast, cutting
wheel to be used. The ends of a cut should meet when
cutter is rotated round the pipe if the pins in the roller or
cutting wheel become worn, two or three cuts will be
made on the pipe – this being known as 'tracking'.

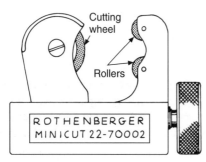

Cutting
wheel

Rollers

ROTHENBERGER
MINICUT 22-70002

(c) Mini tube cutter for copper tubes

These small cutters are very useful in situations where those
of a larger size are difficult to operate, i.e. under floors or in
corners. The type shown will cut diameters of from 8 mm to
22 mm copper tubes.

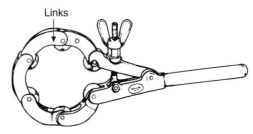

Links

(d) Link pipe cutter

Used for cutting cast iron pipes. Adjustable in that links
may be removed for smaller pipe.

Pipe wall

(e) Action of link pipe cutter

The hard outer skin of the casting is penetrated by the wheels
and parted by splitting due to the wedging action of the wheels.

**Fig. 2.5**   Pipe cutters for copper and steel tubes.

Link cutters are used mainly for cast iron drain,
discharge and flue pipes. Cutting does not actually
occur with this brittle pipe material, the break being
achieved by the even pressure exerted around the
pipe by the wheel (see Fig. 2.5(e)).

*Burrs* form inside a pipe when cutters are used,
and as these restrict the bore they must be removed
using either a pipe *burring reamer* or a *round/half
round file*. The reamer illustrated in Fig. 2.6(a) is

for use in a carpenter's brace but other patterns can
be obtained.

*Plastic pipe cutters*
Figure 2.6(b) shows the cutters that are available
for making a clean cut on polythene, polypropylene
and polybutylene pipes, avoiding the woolly fibres
which have to be removed when such pipes are cut
with a saw.

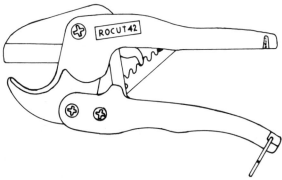

(a) Pipe deburring reamer

Reamers are made in a variety of sizes, each size being used for three or four pipe sizes, to remove burrs which have formed on the inside of the pipe when cutting.

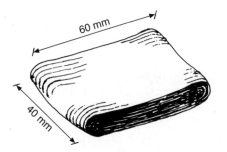

(b) Plastic pipe cutters for polythene, polybutylene and cross-linked polythene

**Fig. 2.6**

Fig. 2.7  Wiping cloth. Suitable for patching and wiping solder dots on lead sheet. Sizes shown are approximate.

### Tools for laying and fixing sheet roof materials

*Wiping cloths*

These are made from a hard-wearing heat-resistant fabric known as 'moleskin', a member of the 'fustian' group of fabrics. Close inspection of this material will reveal a grain which should run along the length of the cloth. A cloth having dimensions as shown in Fig. 2.7 is a good general purpose tool

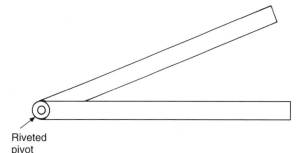

Fig. 2.8  Bevel. A useful tool for measuring angles. The illustration shows a bevel made from two pieces of bright low-carbon steel, 20 mm wide × 3 mm thick and approximately 300 mm in length.

suitable for wiped solder patches or solder dots on sheet lead fixings. Wiping cloths may still be obtained from specialist plumbing tool stockists, or offcuts are sometimes available from manufacturers of clothing made from moleskin. See Chapter 5 for the type of solder used.

*Bevel*

This is a simple tool similar to a square but having a movable blade to enable angles other than those of 90° to be set out. Typical examples of its use are for setting out step flashing and pipe bending. A 600 mm steel folding rule is a good substitute and is probably more useful to a plumber than bevels obtained from tool stores, which are mainly used by carpenters.

In the absence of a folding rule a bevel can be simply made out of two convenient lengths of steel strips. The ends are rounded off and secured with a rivet or nut and bolt as shown in Fig. 2.8.

To prevent damaging roof-covering materials, many tools are made from a non-metallic substance. Plumber's wood tools were traditionally manufactured from boxwood, beechwood, hornbeam or lignum vitae, but due to the high cost of these timbers, manufacturers now produce some of them from high-density polythene.

The more common tools are as follows.

*Lead dresser* (Fig. 2.9)   This is used for flattening sheet lead. Its face must be kept smooth to avoid damage to the lead.

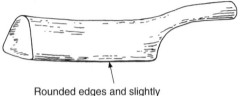

Rounded edges and slightly
bowed face reduce tool marks on the sheet

**Fig. 2.9**  Dresser.

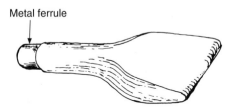

Metal ferrule

**Fig. 2.10**  Chase wedge. Used to move sheet lead into a
corner and for straightening corners after bossing is
completed. In the latter case great care is needed to avoid
stretching the lead which would cause cracking.

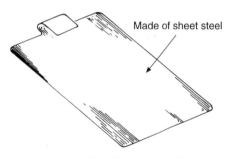

Made of sheet steel

**Fig. 2.11**  Drip plate. The drip plate is laid between two
sheets of lead when the top sheet is to be driven into a
corner by a chase wedge. A useful size is 100 mm × 150 mm.

*Chase wedge* (Fig. 2.10)   This is used for chasing
in angles, upstands and drips, and is used in
conjunction with a drip plate. A *drip plate* is shown
in Fig. 2.11 and is made from approximately 1 mm
thick steel sheet (100 mm × 150 mm in size). It is
positioned between the overcloak and undercloak
when forming details such as lead drips, to allow
the sheets to slide freely over each other. (See also
Fig. 11.46(a) for another use of the drip plate.)

*Mallets* (Fig. 2.12)   These are made of boxwood,
lignum vitae or hard rubber and are used mainly on
tools with wooden handles having no metal ferrule.

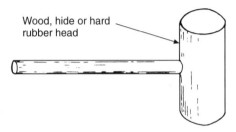

Wood, hide or hard
rubber head

**Fig. 2.12**  Flat-face tinsmith's mallet for general purpose
work on metal sheets.

Section

**Fig. 2.13**  Bossing stick. Used for bossing corners and
rolls in sheet lead.

Section

**Fig. 2.14**  Bending dresser. This is made of boxwood
and now mainly as an alternative to a bossing stick.

*Bossing stick* (Fig. 2.13)   This is a tool for sheet
lead work and can be used with advantage where
it is difficult to work with a bossing mallet, as for
example on roll ends and overcloaks. Sometimes
used for bossing external corners in conjunction
with a bossing mallet.

*Bending dresser* (Fig. 2.14)   The name of this tool
is derived from the period when large-diameter
lead pipes were bent by hand. They are similar in
appearance to the bossing sticks but have a broader
face. They are mainly used on lead sheet for
bossing chimney aprons and back gutters or internal
corners where their broad rounding face avoids
excessive tool marks on the lead.

*Bossing mallet* (Fig. 2.15)   Also a tool for sheet
lead work, this is used in forming various roofwork
details such as bossing corners, roll ends and drips.
To reduce fatigue in use, the handle is made flexible
by using malacca cane.

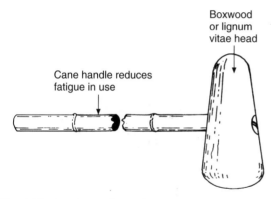

Boxwood or lignum vitae head

Cane handle reduces fatigue in use

**Fig. 2.15** Bossing mallet. An all-purpose bossing tool for sheet lead. Some plumbers also find them useful for copper and aluminium sheet work. Mallets used with these latter two materials lose their smooth surfaces and should not be used on sheet lead.

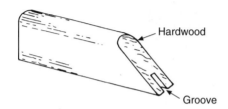

Hardwood

Groove

**Fig. 2.16** Step turner. Used for folding or turning the edge of sheet lead for flashing.

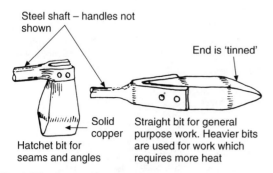

Steel shaft – handles not shown

End is 'tinned'

Solid copper

Hatchet bit for seams and angles

Straight bit for general purpose work. Heavier bits are used for work which requires more heat

**Fig. 2.17** Copper bits or soldering irons.

*Step turner* (Fig. 2.16)   This is used to form the 'turn-in' on step flashings. These can be bought or made from a piece of hardwood.

*Copper bit or soldering iron* (Fig. 2.17)   Copper soldering bits have a wooden handle and are usually

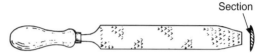

Section

**Fig. 2.18**   Rasp. Used for shaping soft materials such as lead or wood and has large teeth which will not clog easily. The teeth are cleaned from time to time using card wire.

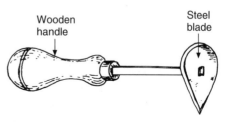

Wooden handle

Steel blade

**Fig. 2.19**   Shavehook. Used for removing the lead oxide surface film on sheet lead and lead pipe prior to soldering or lead burning.

specified by the mass of the copper head. Soldering irons are used for soft soldering of certain roofwork details, particularly zinc sheet. Hatchet-type bits are useful when soldering angles.

*Rasp* (Fig. 2.18)   Similar in shape to a file but has much coarser teeth enabling it to be used for soft materials such as lead or shaping wood. Although originally a tool principally used for lead pipe jointing it is still a very useful addition to the jobbing plumber's tool kit.

*The shavehook* (Fig. 2.19)   This is used to clean the surface of lead sheet prior to soldering or welding. Various shapes of blade are available – that shown is heart shaped.

**Pipework-jointing tools**

Sanitation and water services require the use of a wide variety of pipework materials and, in order to identify suitable tools, it is necessary to consider briefly the method of joining each pipe material. More detailed information on jointing procedures can be found in Chapter 5 which deals with pipework-working processes.

*Steel pipe threading tools*

*Chaser die stocks*   The tools used for cutting threads on steel pipes are known as stocks and dies, the stocks being the body and handle of the tool, the die being the actual thread cutter. The pattern shown in Fig. 2.20(a) has interchangeable sets of dies (chasers), allowing threads to be cut on a range of different diameter pipes. It should be noted that chasers which are interchangeable in a common stock are numbered 1 to 4. It is important that these numbers correspond to those marked on the stock when they are changed as failure to do this will result in a defective or stripped thread. Dies can be obtained to cut parallel or taper British Standard Pipe (BSP) threads.

Die stocks that cut a thread on only one size of pipe are also available (Fig. 2.20(b)), separate die heads being required for every pipe diameter. They are very useful for threading pipes in position, and by reversing the dies, close-to-wall threading can be achieved.

*Receder dies*   Chaser dies are drawn on to the pipe by means of the thread which is being cut. This action places great strain on the pipe and can, in certain circumstances, result in the pipe being crushed. To obviate this, receder die stocks as illustrated in Fig. 2.20(c) are designed with a screw feed that is firmly clamped on to the pipe. When the die is rotated it is pulled on to the pipe along the screw feed in a similar way to a nut pulling itself on to a bolt. This operating principle removes the strain on the pipe which is encountered with chaser dies. The threads on receder dies are parallel, but taper threads can be cut by adapting the stock and causing the chasers to move outwards from the pipe (to recede) as threading is in progress (see Fig. 2.20(c)).

Other patterns of stocks and dies are available, such as solid or block, circular split and two piece. These are, however, used mainly for bolt threading, although the solid die is still in use for lightweight tube such as electrical conduit.

Lubrication is essential when cutting threads to (a) avoid excessive friction and (b) prevent 'stripping' of the thread. Cutting oils and pastes are specially formulated for this purpose and can be applied with an oil can, a brush or in the form of an aerosol.

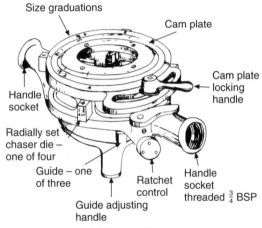

(a) Chaser die stock

Chaser die stocks are adjustable and can be used for threading pipes of different sizes.

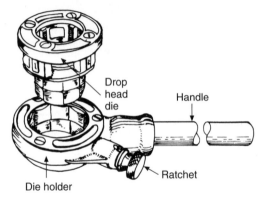

(b) Ratchet drop head die stock

Interchangeable heads save time when threading pipes of different diameters in that the dies for different sizes of pipe will fit into the same die holder.

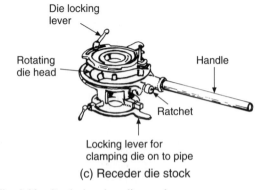

(c) Receder die stock

**Fig. 2.20**   Steel pipe threading tools.

**Fig. 2.21** Hexagon die nut. The die nut is used to clean used or badly cut threads. It is made of very hard tool steel.

**Fig. 2.22** Rethreading file.

*Hexagon die nut and rethreading files* The tools shown in Figs 2.21 and 2.22 are used for repairing damaged or rusty threads – they do not replace stocks and dies.

BSP threads are always used for pipework in the plumbing industry, but for general threading a variety of thread forms are available, ranging from coarse to fine, the principal threads being British Association (BA), used mainly for electrical work on diameters of up to 6 mm, and Metric (ISO).

### Copper tube jointing tools

Certain basic tools are required for jointing copper pipes such as tube cutters, reamers and smooth-cut files, depending on the particular requirements of the type of fitting being used.

### Swaging tools

Many special tools are available for capillary jointing, see Chapter 3. A common tool widely used, however, is the *socket former* (Fig. 2.23) which provides an economic method of making soldered or brazed capillary joints. Figure 2.23(b) shows a swaging tool for manipulative compression fittings.

*Bent bolt* (Fig. 2.24) This is used in conjunction with a hammer to open branch holes in a pipe. Since the demise of lead pipes a bent bolt is now used only by those plumbers who specialise in welding or brazing copper tubes. The technique is shown in Book 2, Chapter 1.

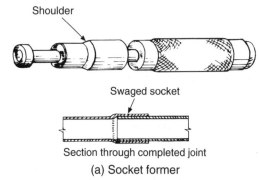

**(a) Socket former**

Tool is driven into tube which opens as it passes over the shoulder. Tube is annealed before being opened. Provides an economic capillary joint.

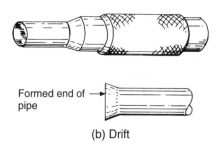

**(b) Drift**

Used for forming a bell mouth when preparing a manipulative compression joint or bronze welded joint on copper tubes.

**Fig. 2.23** Swaging tools for copper pipes.

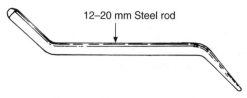

**Fig. 2.24** Bent bolt. The bent bolt has many uses for the plumber including its use when opening holes for branch joints in copper pipes.

## Pipe-bending tools

The various bending *machines* that are available for copper and steel pipes are described in Chapter 5.

### Bending springs

Small-diameter copper tube can be easily bent by the use of a bending spring (Fig. 2.25) made from square section polished steel wire. They are positioned inside the pipe before bending is commenced and provide internal support to prevent

**Fig. 2.25**  Bending spring. The spring is placed inside the pipe at a point where a bend is to be made to maintain the true bore of the pipe during the bending process.

collapse of the pipe wall. Springs should be lightly oiled before use as they are then more easily removed after the bend has been made.

## Tools for fitting and fixing pipework

In this section various general tools will be considered which enable plumbing work to be fixed and fitted to the building fabric.

### Wrenches and spanners

There are three basic wrenches, illustrated in Fig. 2.26, used for rotating pipes when jointing, e.g. when tightening screwed joints.

*Straight pipe wrench* (Fig. 2.26(a))  These are commonly referred to as 'Stilsons' because the original pattern was produced by an American company called Stilson. A range of sizes (150–1,500 mm long) are available, the length to be used being dependent on the pipe diameter or the leverage required. Never extend the handle with pipe as this can cause the tool to be strained and damaged.

*Footprint pipe grip or wrench* (Fig. 2.26(b)) 'Footprints' is the common name given to this type of wrench because the trade mark of the manufacturer is a footprint. They can be obtained in lengths from 150 mm to 350 mm to suit pipe size and leverage.

*Chain wrench* (Fig. 2.26(c))  These are obtainable in lengths of 500 mm to 1,283 mm, and for nominal pipe diameters of from 3 mm to 200 mm.

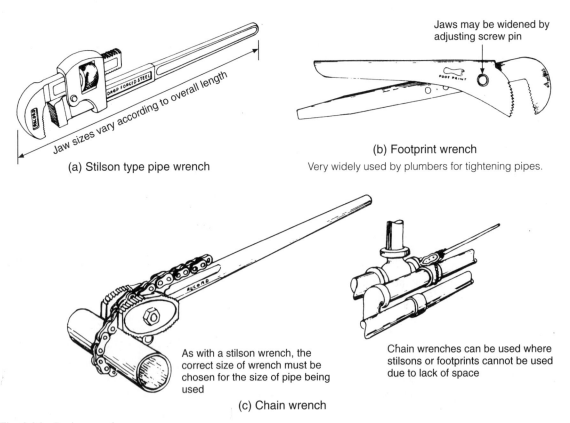

(a) Stilson type pipe wrench

Jaw sizes vary according to overall length

(b) Footprint wrench

Very widely used by plumbers for tightening pipes.

As with a stilson wrench, the correct size of wrench must be chosen for the size of pipe being used

(c) Chain wrench

Chain wrenches can be used where stilsons or footprints cannot be used due to lack of space

**Fig. 2.26**  Basic wrenches.

The hardened teeth of the tools described above make them suitable only for steel pipe, and any application to soft materials such as copper and brass must be avoided since the hardened steel will damage the soft metal. Use of these tools must also be limited to circular sections such as pipe, and they must never be used on square-faced coupling nuts. Teeth marks left on pipe and fittings must be removed by file, as the sharp edges can be a hazard.

*Spanners*   The correct tool to use on square, hexagonal or octagonal nuts and couplings is a spanner. Three types are in common use in the plumbing industry: open-ended, ring and adjustable, and these are illustrated in Fig. 2.27.

*Ring* spanners should be used where possible as they completely encircle the nut and prevent slipping. Obviously they cannot be used for tightening the nuts on compression-type fittings as the copper tube would not permit access of the spanner to the nut. They are very useful, however, for carrying out maintenance work on boilers and other mechanical appliances.

The *open-ended* pattern is used where it is only possible to apply the tool from the side, e.g. when tightening the coupling nut of a pipe fitting.

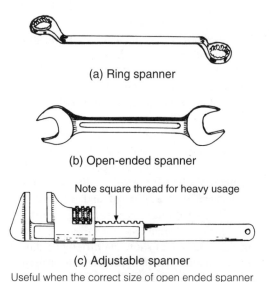

(a) Ring spanner

(b) Open-ended spanner

Note square thread for heavy usage

(c) Adjustable spanner

Useful when the correct size of open ended spanner is not available.

**Fig. 2.27**   Common types of spanner.

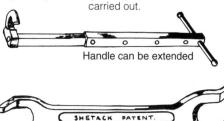

Swivel head made in two sizes for ½″ and ¾″ BSP Nuts: heads are reversible to enable both tightening and untightening operations to be carried out.

Handle can be extended

SHETACK  PATENT.

**Fig. 2.28**   Basin wrenches. Both these spanners are useful when hot and cold water connections are made to baths, basins and sink units.

An *adjustable* spanner can be used instead of the open-ended pattern and, depending on its size, can be adjusted to suit a range of nuts.

*Basin wrenches* (Fig. 2.28)   These are for use in tightening nuts where an ordinary spanner will not reach, such as the nuts on lavatory basin taps, bath taps and taps fitted to sink units. These wrenches are available in a range of sizes to fit bath and lavatory basin wastes and tap back nuts.

*Waste fitting holder* (home made)   A useful tool that can easily be made by a plumber for screwing up bath overflows and preventing waste fittings turning while the back nut is being rotated, is shown in Fig. 2.29. Generally a short 25 mm nominal bore steel pipe is the most suitable material to use.

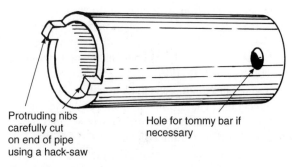

Protruding nibs carefully cut on end of pipe using a hack-saw

Hole for tommy bar if necessary

**Fig. 2.29**   Waste and overflow spanner. It is easily made from a short length of 25 mm nominal bore steel pipe.

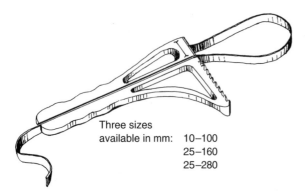

Three sizes
available in mm:   10–100
                    25–160
                    25–280

**Fig. 2.30**   Strap wrench (marketed under the name of Boa Constrictor).

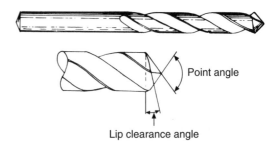

Point angle

Lip clearance angle

**Fig. 2.31**   Twist drill. This is used for drilling holes in almost any material except for bricks and concrete, where a masonry drill is needed.

*Strap wrench*   Figure 2.30 shows this type of wrench which is very similar in both appearance and action to an oil filter wrench used by car owners. Its use is confined to gripping polished surfaces such as the easy-clean shield on a tap, or fittings made of plastic. The tool is made of very tough plastic and unlike steel tools will not damage the work.

*Leverage*   It should be noted that all spanners and wrenches employ the principles of leverage and it is important to remember to apply this principle carefully and ensure the correct size of wrench is selected for a specific job. For example, one that is too small would not exert sufficient leverage, whereas one that is too large may damage the work on which it is used. Never extend a wrench with a piece of tube as this is likely to damage it.

### Drills

*Twist drills* (see Fig. 2.31) are used for making holes in conjunction with hand or power drills. Two main types of steel, *high-speed steel* and *high-carbon steel*, are used in twist drill manufacture.

High-speed steel (HSS) contains under 1 per cent carbon to which is added alloying agents such as tungsten, chromium, vanadium, cobalt and molybdenum. These drills are able to cut even at a dull red heat without loss of hardness or rapid blunting of their cutting edge.

High-carbon steel contains 1–2 per cent carbon to which is added alloying agents such as tungsten or chromium. If used at high speeds, i.e. in a power drill, and the operating temperatures exceed 250 °C, the drill becomes blunt owing to loss of hardness.

Correct cutting lubricants must be used to prolong drill life and improve the quality of drilling, the following lubricants being recommended with differing materials:

| | |
|---|---|
| Low-carbon steel | Soluble oil |
| Copper and brass | Dry, paraffin or soluble oil |
| Aluminium | Paraffin or soluble oil |
| Cast iron | None required |
| Plastics | None required |

Twist drills for general use have a point angle of 118° and a lip clearance angle of 10°–12°. Other materials require different angles for effective cutting, e.g. for plastic, the point and lip clearance angles are 75° and 20° respectively.

The *masonry drill* is similar to the twist drill but has a tungsten carbide tip which is very hard and wear resistant. These drills are suitable for brick, tile, stone and concrete, and can only be resharpened on a special grinding wheel.

### Hammers

A hammer consists of a steel head and a wooden shaft, the face of the head and pein being hardened and tempered. The wood used for the shaft must be well seasoned, straight grained, and is usually ash or hickory. The head must be securely fixed to the shaft by means of wedges.

Figures 2.32(a)–(c) show the different patterns of pein available. The plumber uses all of these, although the straight pein is traditionally referred to as a 'plumber's hammer' as it is useful for forming hard sheet metal.

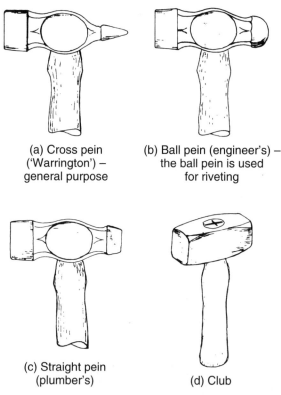

(a) Cross pein
('Warrington') –
general purpose

(b) Ball pein (engineer's) –
the ball pein is used
for riveting

(c) Straight pein
(plumber's)

(d) Club

**Fig. 2.32**  Hammers. Hammer shafts are usually made of hickory wood which endures the impact of hammering without breaking.

(a) Cold or hard chisel

A general-purpose cutting tool for use on most materials.

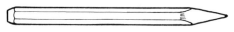

(b) Brick point

Useful for cutting away hard bricks or concrete.

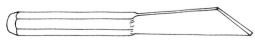

(c) Plugging chisel or joint raker

Used to remove the jointing media between bricks so that a wood-fixing plug or lead wedge may be inserted.

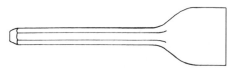

(d) Bolster or floorboard chisel

Can be used to cut the tongues of floorboard to facilitate lifting.

**Fig. 2.33**  Types of chisel.

For cutting brickwork or concrete the use of a 'club' or 'lump' hammer is recommended (see Fig. 2.32(d)) which has a heavy head and two striking faces. The short handle reduces fatigue in use yet enables a heavy blow to be delivered.

Hammers are classified by the shape of the head and the mass of steel used.

### Chisels

A selection of 'cold' or 'hard' chisels (see Fig. 2.33(a)) are required by the craftworker to cut holes in brick and concrete. These are specified by length and diameter, a useful size being 15–20 mm diameter and 250–350 mm long.

*Pointed chisel* (Fig. 2.33(b))  This type of chisel is also of use when concrete or hard brickwork has to be removed. The point tends to split rather than

cut very dense materials. They are also specified by diameter and length.

*Plugging chisel* (Fig. 2.33(c))  This type of chisel is used for removing the mortar joint between bricks in order to insert the edges of flashings or to position wooden plugs for fixings.

*Bolsters*  These belong to the same family as cold chisels, the difference being in the shape of the blade which is wider and thinner. These tools are not generally used by the plumber, the exception being the 'electrician's' or 'floorboard' chisel (see Fig. 2.33(d)) which is of the same shape as a bolster with a thinner blade. This is required for cutting through the joint between tongued and grooved flooring in order that a saw can be inserted to cut away the tongue thus enabling the board to be lifted.

Chisels, bolsters and points are all manufactured from high-quality steel often alloyed with a small percentage of chromium or tungsten to increase their toughness.

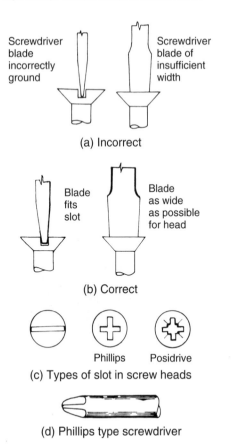

(a) Incorrect

(b) Correct

Phillips    Posidrive
(c) Types of slot in screw heads

(d) Phillips type screwdriver

**Fig. 2.34** Screwdrivers. A screwdriver of the correct size should always be used, i.e. the end of the screwdriver should fit snugly into the slot at the end of the screw.

**Table 2.2** Classification of files.

| File type | Teeth per 25 mm |
| --- | --- |
| Rough | 20 |
| Bastard | 20–25 |
| Second cut | 30–40 |
| Smooth | 50–60 |

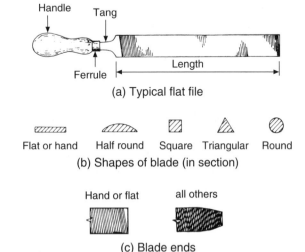

(a) Typical flat file

Flat or hand    Half round    Square    Triangular    Round
(b) Shapes of blade (in section)

Hand or flat    all others
(c) Blade ends

**Fig. 2.35** Files.

### Screwdrivers

It is essential to select the correct screwdriver from the many patterns available (see Fig. 2.34). For slotted screws they are identified by length and width or diameter of the blade, the point size being used only to specify those for recessed-headed screws. The blade of the screwdriver is of cast or alloy steel with a wood or plastic handle which may be electrically insulated. When using slotted screws the blade should fit the width and length of slot.

### Files

These tools are manufactured from high carbon steel and are classified according to shape of section, length, and spacing of the teeth as shown in Table 2.2 and Fig. 2.35.

The choice of file type depends on the operation being undertaken and the amount of metal that has to be removed. For general filing the most popular lengths are 250 and 304 mm. It is essential for safe use that a good-quality handle is fitted to the file tang. Apart from the different shapes, files are also made in various cuts for use on different materials (see Table 2.2).

### Thread taps

A thread tap as illustrated in Fig. 2.36(a) cuts an internal (female) thread, either right or left hand, being made to produce the different types of thread listed previously in the section dealing with external (male) threads.

To form an internal thread three taps are available as illustrated in Fig. 2.36(b) a taper tap, (c) a second, and (d) a plug or bottoming tap, plus the tap wrench (e) which is required to hold the tap. Details of the use of taps is described in the chapter

(a) Typical tap for cutting female threads

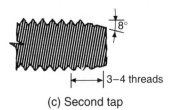

8–10 threads

(b) The taper tap is first screwed through the hole and is followed by the second tap

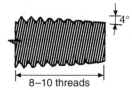

8°

3–4 threads

(c) Second tap

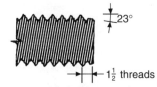

23°

1½ threads

(d) The plug, or bottoming tap, provides the completed thread

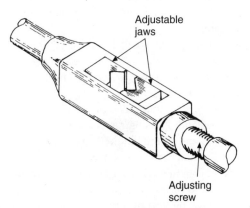

Adjustable jaws

Adjusting screw

(e) Tap wrench for holding taps

**Fig. 2.36** Thread taps. When a threaded hole is required in a thick sheet material, first a hole is drilled and then the taps are screwed into the hole to form a thread. Taps are also used for cleaning old or damaged female threads. To provide a good thread, taps (b), (c) and (d) are used.

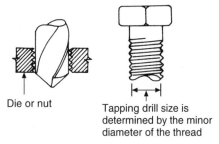

Die or nut

Tapping drill size is determined by the minor diameter of the thread

**Fig. 2.37** Approximate method of determining a suitable tapping drill.

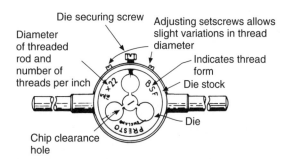

Die securing screw

Adjusting setscrews allows slight variations in thread diameter

Diameter of threaded rod and number of threads per inch

Indicates thread form

Die stock

Die

Chip clearance hole

**Fig. 2.38** Illustration of split circular die.

on working processes. Figure 2.37 illustrates the method of selecting a suitable tapping drill.

*Split circular dies*
The dies illustrated in Fig. 2.38 are made specially for cutting male threads on solid round rods. The diameter of the die varies to that of the thread and it is important that the die stock is of suitable size for the die. Although for a 'one-off' job both die holders and dies can be obtained separately, they are usually sold in sets covering a range of sizes and thread forms.

*Screwdrivers*    Many modern plumbing appliances require an electrical supply and it is not unusual for the plumber to make the electrical connections to the appliance from a socket outlet. The following tools will be a useful addition to the plumber who has to make electrical connections.

Smaller screwdrivers than those normally used will be required so that damage to the small setscrews used in electrical appliances is avoided. These screws mainly range from 2, 4, 6, 8 BA

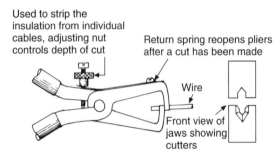

Used to strip the insulation from individual cables, adjusting nut controls depth of cut

Return spring reopens pliers after a cut has been made

Wire

Front view of jaws showing cutters

**Fig. 2.39**   Wire-stripping pliers. To avoid damage to the conductor it is suggested that the correct setting of the cutting jaws is checked on a piece of scrap cable prior to use.

threads and it is worth noting that an 8 BA set screw has an approximate diameter of only 1.5 mm.

Two other useful tools will also be necessary.

*Electrical tools*

*Cable-stripping pliers*   Figure 2.39 shows one type of these pliers which are adjustable to allow for various diameters of cable. It is important that prior to use the jaws are correctly set to avoid damage to the cable. Always check the setting on a piece of scrap cable before use. The cable to be stripped is then placed in the jaws, the pliers tightened and rotated through 90°; the insulation may then be simply pulled off.

Non-adjustable pliers are also available which have a series of cutters in the jaws, which can accommodate various cable diameters, the main disadvantage of these being a direct pull cannot be obtained when the insulation is drawn off.

*Side-cutting pliers*   These are used for cutting cables to the required lengths. A typical example is shown in Fig. 2.40.

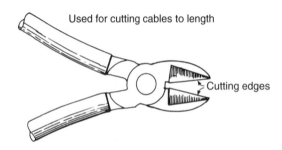

Used for cutting cables to length

Cutting edges

**Fig. 2.40**   Side-cutting pliers.

**Care and maintenance of tools**

To ensure tools have a long life and remain effective in use, it is essential that they are cared for and correctly maintained. It is bad practice to wait until a tool is not suitable for use before maintenance is carried out. Routine maintenance, and this includes cleaning, must be undertaken regularly.

It will be useful to consider the more important aspects of care and maintenance which apply to common hand tools.

The working parts of all tools must be lubricated, steel tools lightly coated with oil to prevent rusting and the edges of cutting tools maintained in a sharp condition.

*Wood tools*
(a)  Apply raw linseed oil to prevent drying and cracking.
(b)  Maintain a smooth working surface by treating with sandpaper.
(c)  Ensure mallet handles are securely wedged and not broken.

*Cold or hard chisels*
(a)  The 'mushrooming' on these and other cutting tools must be removed at the earliest stage of formation by grinding.
(b)  Sharp cutting edges must be maintained by grinding.
(c)  Repeated grinding will leave the cutting end too 'stubby' or 'thick'. When this stage is reached the chisel should be 'drawn out' to its correct shape as follows.
(i)  Heat the cutting end of the chisel to bright red and by using hammer blows 'draw out' the end to the required shape as shown in Fig. 2.41 using a new chisel as a pattern.
  Hammering must only be carried out while the chisel is at the red heat or cracking will occur.
(ii)  When the correct shape is obtained, the chisel end must be heated again to bright red and allowed to cool slowly. The blade is then in a soft condition (annealed) and the final shaping can be carried out using a file or grinding wheel.

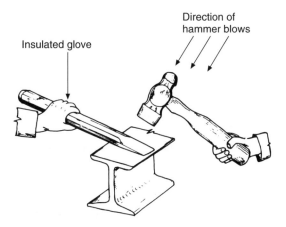

**Fig. 2.41** Drawing out a cold chisel.

**Table 2.3** Oxide colour and corresponding temperature.

| Oxide colour | Temperature (°C) |
|---|---|
| Pale straw | 230 |
| Dark straw | 240 |
| Yellow brown | 250 |
| Purple/brown | 260 |
| Purple | 270 |
| Deep purple | 280 |
| Deep blue | 300 |

(iii) After final shaping of the blade it is again heated to a cherry red and cooled quickly (quenched) in oil or water depending on the type of steel, i.e. water or oil hardening. This process leaves the steel in a hard brittle condition (hardened).

(iv) To obtain the correct hardness, the chisel must be tempered. The blade is cleaned to bare metal and heat is applied to a point roughly 50 mm from the cutting edge. Coloured oxide films will appear where the heat is applied, these colours indicating the temperature of the steel (see Table 2.3) and as the heat is conducted so colour changes occur.

(v) When the tip colour is purple (270 °C) the chisel is quenched in water. The eventual hardness of the chisel depends upon the temperature indicated by the colour at which it is quenched. At certain specific temperatures it is known that the iron/carbon relationship will provide steel of a certain hardness, the colour purple indicating a suitable hardness for chisels used in general plumbing work. Rapid quenching then 'freezes' or 'sets' the steel at the desired state of hardness.

*Dies*

(a) Worn or damaged chasers must be discarded or imperfect threads will be cut.

(b) Chasers are manufactured as a set of four and they must never be interchanged with those of another 'set'.

(c) When placing chasers in the die stock it is essential that the number on the die corresponds with that of the slot in the stock.

*Files*

(a) Clean file teeth using card wire.

(b) Store files away from other tools.

(c) Wrap files individually in rag when not in use.

(d) Do not use split handles.

*Hammers*

(a) Ensure the shaft is of the correct size for the head and secured by wedges.

(b) If the shaft dries out the head will become loose – soak the hammer head in water from time to time. Never use a hammer with a loose head.

(c) Replace damaged shafts.

*Pipe cutters*

(a) Replace blunt cutter wheels making sure the correct wheel is used – wheels are available for plastic, cast iron, copper and stainless steel.

(b) Replace worn rollers.

(c) Roller and wheel pins (spindles) must be removed when worn or spiral cutting will occur.

*Pipe grips and wrenches*

(a) Ensure the teeth are not 'clogged' with jointing paste or other matter or the tool will slip when in use.

(b) When teeth become worn, the defective part or the whole tool must be replaced.

### Screwdrivers

Never grind the end of a blade for slotted screws to a knife edge. The blade end must be the same thickness as the slot in the screw head.

### Spanners

Open-ended spanners which are 'splayed' (strained open) must be discarded as they are dangerous. Ring or open-ended spanners which are worn must also be replaced for safety reasons.

### Twist drills

(a) Maintain correct point angle and sharp cutting edge.
(b) If accurate drilling is required it is inadvisable to sharpen by hand – use a grinding jig or have the drill sharpened by an expert.
(c) Approximate angles and edges can be achieved by grinding, a special wheel being needed for tungsten-tipped drills.

## Further reading

Information relating to tools may be obtained from the following sources:

Clarkson Osborne International, PO Box 37, Penistone Road, Sheffield, S6 3AH. Tel.: 0114 276 8622.

Monument Tools Ltd, Restmoor Way, Hackbridge Road, Hackbridge, Wallington, Surrey, SM6 7AM. Tel.: 0181 288 1100.

Record Tools, Oscar Works, Meadow Street, Sheffield, S3 7BQ. Tel.: 0114 272 6662.

Rems (UK) Ltd, 1a Greenleaf Road, Walthamstow, London, E17 6QQ. Tel.: 0208 521 9168.

Ridge Tools, Arden Press Way, Pixmore Avenue, Letchworth, Herts, SG6 1LH. Tel.: 01462 485335.

Rothenberger (UK) Ltd, 14–18 Tenter Road, Moulton Park Industrial Estate, Northampton, NN3 1PS. Tel.: 01604 646231.

S.K.S. Dormer Ltd (drills and threading equipment), Rother Valley Way, Holbrook, Sheffield S20 3RW. Tel.: 0114 251 2614.

## Self-testing questions

1. State the fault in pipe cutters which leads to tracking or spiral cutting.
2. List the information necessary when selecting hacksaw blades.
3. State the maintenance requirements for wood tools.
4. List the tools that would be required to form and install the lead flashings round a chimney stack.
5. Select a suitable tool for removing the head work of a tap.
6. State the type of wrench that is generally used for the installation of LCS (low-carbon steel) pipework systems.
7. Name the three types of thread taps available.
8. Select a suitable tool for cutting a male thread on a small circular bar.
9. Explain the terms hardening and tempering when related to tool steels.
10. Name the tool shown in Fig. 2.42 and describe its use.

**Fig. 2.42** Question 10.

11. Describe why it is necessary to use a drip plate and roll overcloaks in sheet lead.
12. Why is it necessary to use a lubricant when cutting threads on LCS pipes?

# 3 Mechanical and special tools

After completing this chapter the reader should be able to:

1. Name and state the correct uses and potential dangers of the main fuel gases used by mechanical service engineers.
2. State and describe the dangers of using defective and damaged gas heating equipment.
3. Recognise the need for care in the use of compressed gases.
4. State the correct procedures for assembling portable welding equipment and commissioning it for use.
5. List the main powered workshop and site equipment used by the plumber and have a knowledge of its use.
6. Understand and list the safe working procedures with machine tools.
7. Describe the merits of pipe-freezing equipment and state its uses and advantages.

## Gas heating equipment used by the plumber

*Portable combustible gas equipment*
Soldering is one of the techniques with which plumbing has been identified, certainly since the Roman era and possibly even before that. Until the invention of the first paraffin and petrol blowlamps, solder for wiping joints on lead pipes was melted in a cast iron pot over a convenient fire, this process being accomplished by first pouring the solder over the joint until sufficient heat was conducted by the solder into the metal to be joined. A joint was then made by shaping and wiping the semi-molten metal with a moleskin cloth before the solder solidified. Joints on sheet materials could be made using a copper bit or soldering iron which again was heated in a suitable fire, quite often the householder's cooking range. The introduction of paraffin- and petrol-fuelled blowlamps enabled plumbers to carry their own source of heat and quickly became indispensable for small repair jobs on lead pipe or lead roofs. The introduction of blowlamps fuelled by compressed gas was another step forward, and this source of heat, being reasonably economical,

clean and very convenient, is now used almost exclusively for soft soldering and light brazing work.

The flame temperature and structure of any blowlamp or torch depend on whether a combustible gas is burned in air, where only approximately 21 per cent oxygen is available, or in an atmosphere of pure oxygen. Where a small concentrated flame having a high temperature is required, as, for example, in fusion welding processes, the use of pure oxygen is essential. Various combinations of oxygen and fuel gases are available, but the most commonly used is that of oxygen and acetylene. In most instances both gases are compressed in cylinders to high pressures, acetylene to 15.5 bar, oxygen to 137 bar, and in this form they are easily portable. One of the main advantages of this combination of gases is the very high temperatures they produce. Although not generally used for mechanical services work owing to its lack of portability, it is possible to use low-pressure acetylene where the gas is generated using special equipment. Low-pressure acetylene is usually confined to industrial processes where it is found to be more economic than the use of dissolved acetylene contained in cylinders.

**Table 3.1**  Flame temperatures of fuel gases when used with pure oxygen.

| Fuel gas | Flame temperature (°C) | Chemical symbol |
|---|---|---|
| Acetylene | 3,250 | $C_2H_2$ |
| Hydrogen | 2,950 | $H_2$ |
| Natural gas (methane) | 2,800 | $CH_4$ |
| Propane | 2,850 | $C_3H_8$ |
| Butane | 2,820 | $C_4H_{10}$ |

*Fuel gas*  The term *dissolved acetylene* relates to the compression of acetylene in cylinders where, because of the unstable nature of this gas, special precautions are necessary. This will be discussed later in this chapter. The first gas welding equipment was invented in 1837 by a French engineer who used hydrogen to weld lead with a reasonable degree of success. Hydrogen is a very clean fuel gas, but is expensive to produce and for this reason its use in plumbing has declined. It is, however, still used by jewellers for welding or brazing precious metals. Table 3.1 shows the flame temperature of various fuel gases burned with pure oxygen.

*Liquid petroleum gases (LPG)*  These gases are mainly derived from crude oil, from which, during the refining process, they are given off and retained in special containers for further processing. Of these gases the two most familiar are propane and butane. Like all fuel gases, except hydrogen, they are chemical combinations of hydrogen and carbon, both of which burn freely in oxygen. Like natural gas with which they have much in common, all being hydrocarbons, they are naturally odourless, but during their production a chemical is added to give the distinctive smell associated with these gases which serves as a warning should a leak occur in a cylinder or the equipment with which it is used. Both of these gases are heavier than air and any leaked gas tends to lie in pockets in low-lying areas, for which reason special care is necessary when working in confined spaces or basements. Like all gases they can be compressed easily and liquefy at comparatively low pressures.

When the valve on an LPG cylinder is opened the pressure is released, causing the liquid to revert to a gas. In order to convert water to steam, heat is required: in the same way liquid gases need heat to gasify. This heat is absorbed from the air surrounding the cylinder outlet and at low air temperatures the heat absorbed from the atmosphere surrounding the regulators may cause the gas passing through it to freeze. This applies to all compressed gases and it is not unusual for frost to appear on the regulator during very cold weather when large quantities of gas are abstracted from a cylinder. The recommended procedure in these circumstances is to wrap the regulator in a cloth soaked in very hot water. (This applies to all regulators used with high-pressure gas cylinders.) If possible the equipment should be taken to a warmer area, but it *must not* be subjected to direct heat.

Propane and butane cylinders are marked in a variety of ways and marketed under various trade names depending on the supplier, a typical example being Calor gas, which is in fact butane. The small throw-away cylinders used with small blowlamps favoured by 'do-it-yourself' enthusiasts are too expensive for use by the professional plumbers who generally use larger containers. The most suitable sizes for portability are those containing 3.9 kg of propane or 4.5 kg of butane. Cylinders containing larger quantities of gas are available but they are too heavy to be carried easily. Note that these gases are sold by weight and one can ascertain how much gas is left in the cylinder by weighing it and deducting the weight of the cylinder. The physical properties of these gases are shown in Table 3.2.

**Table 3.2**  Physical properties of liquid petroleum gases.

| Property | | Butane | Propane |
|---|---|---|---|
| Freezing point at atmospheric pressure | | −140 °C | −186 °C |
| Boiling point at atmospheric pressure | | −10 °C | −42 °C |
| Specific gravity (taking air as 1) | | 2 | 1.5 |
| Volume of air required to burn 1 unit volume of gas | | 30 | 23 |
| Ignition temperature | | 480–540 °C | 480–540 °C |
| Limits of flammability (% of gas in air mixture) | upper | 8.5 | 11 |
| | lower | 1.9 | 2.0 |

The cylinders of fuel gases are normally red in colour, with the exception of acetylene, which is maroon, and butane, which is generally blue or grey. Propane cylinders are not only red but also labelled 'propane, highly flammable'. All fuel gas unions and nuts have left-hand threads to distinguish them from those of non-combustible gases.

Propane and butane gases are normally burned in air, although propane is sometimes used with compressed oxygen for flame cutting steel as it is less expensive than acetylene. When fuel gases are burned with pure oxygen, a more concentrated flame having a higher temperature can be achieved than with the same gas burned in air. The reason for this is that air contains approximately 21 per cent oxygen, the remainder being mainly nitrogen, carbon dioxide and a group of what are called *inert* gases. The nitrogen, carbon dioxide and inert gases will not burn and, unlike oxygen, do not support combustion but are heated during the burning process only to be discharged into the atmosphere with the products of combustion.

Table 3.3 indicates the composition of the atmosphere and includes the inert gases previously referred to. The atomic structure of these gases is such that, unlike oxygen and nitrogen, for example, they form compounds with other atoms only very rarely. Most of these gases have some commercial application. Argon, for instance, is used in electric light bulbs, while neon is used in some types of fluorescent lighting. Argon is also used in a process known as gas-shielded arc welding. In this case it is used to prevent the oxygen in the air gaining access to the area of the weld and it acts in a similar way to that of a flux.

**Table 3.3** Composition of the atmosphere.

| Gases | Percentage by volume |
|---|---|
| Oxygen | 20.95 |
| Nitrogen | 78.08 |
| Carbon dioxide | 0.03 |
| Argon | 0.93 |
| Neon ⎫ | |
| Xenon ⎬ | Less than |
| Helium ⎪ | 0.003 |
| Krypton ⎭ | |

Oxygen and acetylene form a useful combination for welding purposes. Acetylene gives a higher flame temperature than other fuel gases and, when burned with pure oxygen, produces a small highly concentrated flame which can be used for all types of fusion welding and brazing.

*Acetylene* Acetylene is heavier than air and is a compound of hydrogen and carbon, and it is produced by the reaction of water and calcium carbide. Calcium carbide has the appearance of small grey stones, is prepared from quicklime and coke and is a compound of carbon and calcium. When calcium carbide is brought into contact with water, acetylene gas is given off freely. Unlike other fuel gases, acetylene is an *unstable* gas which, if compressed to pressures of approximately 2 bar, is prone to self-detonation, i.e. it will explode. For this reason acetylene must be *dissolved* in acetone, a spirit with an affinity for acetylene. Acetone will absorb approximately 25 times its own volume of acetylene under normal temperatures at atmospheric pressure. Figure 3.1 shows this pictorially. By dissolving acetylene in acetone, it is stabilised and

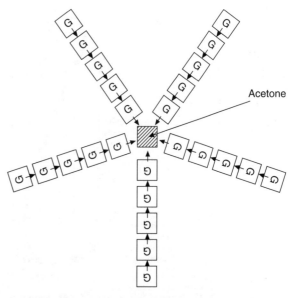

**Fig. 3.1** Volumes of acetylene into acetone. Each box labelled 'G' indicates one volume of acetylene. Acetone possesses the unique capability of absorbing 25 times its own volume of acetylene.

can then be compressed without the danger of explosion. To avoid disturbing the acetone when the cylinder is moved, it is absorbed by the charcoal or kapok filling. To prevent the acetone entering the gas regulator, acetylene cylinders have a flat base and they *must* always be used in the upright position. The pressure relief valve in the base of the cylinder has now been discontinued; one only is fitted opposite the valve.

As with equipment for all fuel gases the threads are left-handed, and hexagon union nuts are kerfed or notched for clear identification in poor light or darkness.

*Oxygen*   The commercial method of producing oxygen is to abstract it from the atmosphere by a process of distillation. In brief, the air is compressed, causing it to liquefy; in this state the oxygen separates out of the nitrogen and can be filtered off.

Oxygen cylinders are painted black and have right-handed threads. *Oxygen does not burn* but supports combustion, i.e. it is essential that it is present for combustion to take place.

### Oxy-acetylene equipment

*Regulators*   The main features of oxy-acetylene equipment are shown in Fig. 3.2. The function of the regulator is to reduce the high pressure of the gas contained in the cylinder to a lower working pressure, which is determined by the size of nozzle used. The larger the nozzle the higher the outlet gas pressure should be. Types of regulator vary – a single-stage regulator is normally used on a manifold system where several welding points are

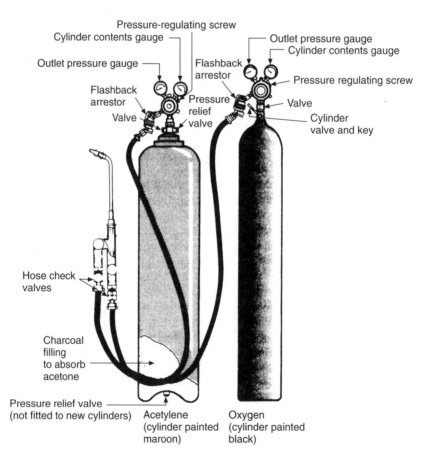

**Fig. 3.2**   High-pressure welding equipment.

served from a central store and both oxygen and acetylene are controlled by a master regulator. For portable equipment the two-stage type is more effective as it reduces any possible fluctuations in the gas flow.

Most regulators are fitted with two pressure gauges, one of which indicates the gas pressure in the cylinder, the other showing the pressure at the blowpipe. This pressure is varied to meet the requirements of the nozzle in use by turning the regulator screw in a clockwise direction to increase the pressure, or anticlockwise to decrease it. When the equipment is out of use the cylinder valve must be turned off; any pressure on the hoses is relieved by opening the valves on the blowpipe until both gauges register zero, after which the regulator screw must be screwed outward to release the pressure on the rubber diaphragm inside. Failure to do this will result in eventual distortion of the diaphragm, giving false readings on the gauges. A pressure relief valve is built into most types of regulator to protect them against excessive pressure build-up.

*Hoses and flashback arrestors*  Hoses for acetylene are red in colour, those for oxygen are blue. They are reinforced with canvas to enable them to withstand the working pressures to which they are subjected. Hose unions must be attached to the hose by special clips; pieces of wire or tape must not be used as they could be highly dangerous. The unions to which the blowpipe is attached should incorporate non-return valves called hose protectors. These prevent the flame, in the event of a backfire, passing into the hoses. It is also recommended that flashback arrestors are fitted to the regulator outlets. These fittings ensure that if the hose accidentally catches fire the gas supply is automatically shut off, protecting both the regulator and the cylinder.

A *flashback* may be defined as the ignition of the gases in the nozzle, or in some cases the blowpipe itself. In most cases it causes a bang or series of bangs which sound like rapid rifle fire. If this happens the procedure is to turn off the fuel gas and plunge the nozzle into a bucket of water to cool it down. The oxygen valve should be left open as this prevents the ingress of water to the blowpipe. If a flashback causes the gases to ignite in the mixing chamber of the blowpipe, it is usually identified by

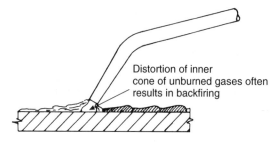

**Fig. 3.3**  Cone of unburned gases too close to work.

a loud squealing noise. The same extinguishing procedure should be followed, but this form of flashback, known as *backfire*, can be more serious than those previously mentioned and can damage the blowpipe – which should be returned to the manufacturers for an overhaul.

The usual causes of flashback are:

(a)  Holding the nozzle so that the cone of unburned gases is too close to the work (see Fig. 3.3).
(b)  The nozzle is not properly tightened into the blowpipe allowing air to enter the mixing chamber.
(c)  Nozzle overheating causing pre-ignition of the gases. This often happens when one is welding in a corner or allowing slag to build up on the tip of the nozzle when welding steel. Slag must be removed at once as it insulates the tip of the nozzle, causing it to overheat and bringing about the conditions causing backfire. Most suppliers of welding equipment supply nozzle cleaning fluid which is simply a dilute acid, but it will be found that soaking the nozzle in vinegar is quite as effective.

*Nozzles*  Most nozzles are made of copper to which is added a small percentage of titanium which increases its hardness. Copper is used because it is a good conductor of heat; and heat, unless dissipated quickly, results in persistent backfire. If the nozzle becomes blocked, never attempt to clear it with steel wire or needles, etc. Only copper wire or proper nozzle cleaners supplied by the manufacturer should be used. If the top of the nozzle is out of square, as shown in Fig. 3.4, it will almost certainly cause persistent backfiring. It is possible to square it

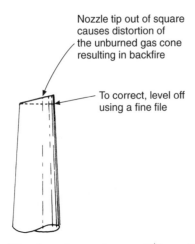

Nozzle tip out of square causes distortion of the unburned gas cone resulting in backfire

To correct, level off using a fine file

**Fig. 3.4** Tip of nozzle out of square. This condition is brought about by scraping the nozzle on a hard surface to remove slag build-ups on the tip. If approved methods are used this situation will not occur.

up with a fine file, but the number of times this can be done is limited. Most nozzles are made in such a way that approximately 3 mm can be removed before the nozzle hole begins to become enlarged, at which stage it should be discarded. The size of nozzle used will depend on the material and its thickness. Table 3.4 indicates which nozzle would be suitable for a specified thickness of material and the recommended gas pressure for steel welding. When welding copper, larger nozzles than those indicated will be necessary owing to the high thermal conductivity of this metal.

*Blowpipes* These vary considerably, depending on the manufacturer and the purpose for which they are used. The small blowpipe illustrated in Fig. 3.5 is typical of most types and can be used for welding lead and light sheet metal.

**Table 3.4** Welding nozzle sizes for low-carbon steel up to 8 mm thick.

| Thickness of material (mm) | Nozzle size | Operating pressure in bars: oxygen and acetylene | Gas consumption | |
|---|---|---|---|---|
| | | | Acetylene (litres per hour) | Oxygen (litres per hour) |
| 0.9 | 1 | 0.14 | 28 | 28 |
| 1.2 | 2 | 0.14 | 57 | 57 |
| 2.0 | 3 | 0.14 | 86 | 86 |
| 2.6 | 5 | 0.14 | 140 | 140 |
| 3.2 | 7 | 0.14 | 200 | 200 |
| 4.0 | 10 | 0.21 | 280 | 280 |
| 5.0 | 13 | 0.28 | 370 | 370 |
| 6.5 | 18 | 0.28 | 520 | 520 |
| 8.2 | 25 | 0.42 | 710 | 710 |

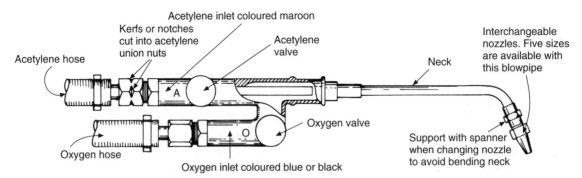

Acetylene inlet coloured maroon

Kerfs or notches cut into acetylene union nuts

Acetylene valve

Interchangeable nozzles. Five sizes are available with this blowpipe

Acetylene hose

Neck

Oxygen valve

Oxygen hose

Oxygen inlet coloured blue or black

Support with spanner when changing nozzle to avoid bending neck

**Fig. 3.5** Welding blowpipe. This is the BOC model 'O'-type blowpipe and is one of the most suitable for lead welding and for welding other light-section materials. Larger blowpipes are very similar in construction.

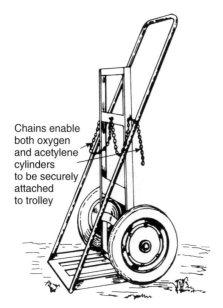

Chains enable both oxygen and acetylene cylinders to be securely attached to trolley

**Fig. 3.6** Cylinder trolley. A trolley enables compressed gas cylinders to be safely used and easily transported on site.

*Spanners* Most manufacturers supply a set of open-ended spanners with new sets of welding equipment. These should be kept with the equipment to avoid loss. Their use ensures that the union nuts are not damaged or overtightened. Remember that welding equipment is expensive to buy or replace and that using damaged equipment can be dangerous. The cylinder key should always be left in the acetylene cylinder valve allowing it to be shut off quickly in the case of an accident.

*Cylinder trolleys* A common type of trolley is illustrated in Fig. 3.6. A trolley allows welding equipment to be moved on site and ensures that the cylinders are always used in the upright position. A chain is supplied with the trolley to secure the cylinders because if a cylinder falls over when in use the regulator will almost certainly be seriously damaged.

*Assembly of equipment* The cylinders must be secure and standing in an upright position. Before the regulators are fitted, each cylinder must be 'snifted' by opening the valve momentarily. The gas released during this operation blows any grit or

debris from the metal seating to which the regulator is fitted. This will avoid damage to the carefully machined metal surfaces and maintain a gas-tight joint. The regulator can now be fitted to the cylinder. After it has been checked that no dust or dirt is present in the hoses – by blowing through them – they can be attached to the regulators or, if provided, the flashback arrestors. The blowpipe should then be attached to the hoses and fitted with a suitable nozzle. *No oil or grease must be allowed to come into contact with oxygen* as it will emulsify to become a dangerous fire risk if near a flame. One should also be aware that one of the characteristics of chemical change is that heat is given off and the reaction of oxygen and oil is so violent that spontaneous combustion may occur.

*Lighting-up procedure* Always open the cylinder valves slowly. One turn should be sufficient. They could then be closed quickly should there be an emergency. A sudden surge of gas can damage the regulators and the gauges. The fuel gas valve on the blowpipe should then be opened and the regulator adjusted to the working pressure required. This ensures that any air in the hose is purged which might otherwise cause a flashback. The same procedure is repeated with the oxygen equipment. Both valves on the blowpipe should then be turned off and an inspection for leaks conducted using soapy water on the joints. Once it is established that the equipment is not leaking, the acetylene valve is opened and ignited with a spark lighter. Avoid matches as they are often discarded before they are extinguished and could cause a fire. The acetylene valve on the blowpipe should be adjusted so that the gas burns without the formation of black smoke. Then the oxygen is turned on until the desired flame is achieved.

The metals to be welded will influence the type of flame required and this is dealt with in Book 2.

*Shutting-down procedure* The valves on the blowpipe should be shut down (the acetylene valve first), then the cylinder valves closed and those on the blowpipe reopened. This process releases the pressure on the regulators, allowing the pressure gauges to show a zero reading. The regulator adjusting screws should then be untightened until they rotate freely in order to prevent damage to the regulator diaphragm.

*Portable electric drills*

These are an indispensable tool for all operatives in the mechanical services industry, and are so commonly used as to need little description. As with all electrical tools they are made for use with a 240 volt mains supply or 110 volts using a step-down transformer, the latter being preferable for site work. For light work rechargeable-battery-operated drills are useful where no power supply is available. When using powerful drills and large drill bits always check the correct speed is selected and use the side handle, because if the drill snatches it can throw the operator off his or her feet. All electric drills should be regularly maintained by a competant person and frequent checks made on the cable for broken conductors and damage insulation.

## Machine tools used in plumbers' workshops

*Drilling machines*

Owing to the fact that the nature of plumbers' work quite often necessitates the fabrication of special brackets and supports, and other types of work that could almost be described as engineering or blacksmithing, equipment such as power drills is not uncommon in plumbers' workshops.

Some manufacturers of portable electric drills make a special stand to which the drill can be clamped for workshop use while allowing it to be moved about on site. However, this arrangement lacks the flexibility and advantages of extensive use offered by a fixed power drill. Such a drill is shown in Fig. 3.7 and is designed for bench fitting. A similar machine which stands on a column is called a pillar drill; it can be fixed directly to the floor. This brings us to the first important point relating to power tools: they must be firmly bolted down to a suitable support to ensure their safety in use.

The chuck speed on the drill illustrated can be varied by changing the belt on the stepped pulley. Such a facility is invaluable as the use of a large drill on soft metals or thin sheets can often result

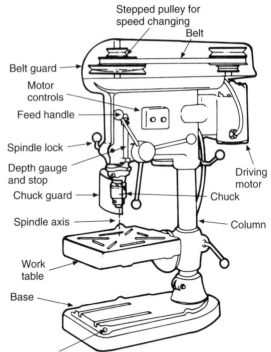

Fig. 3.7 Bench drilling machine.

in the drill binding in the hole, causing the work to spin round. Two more very important points emerge here:

(a) The correct drill speed should be selected to suit both the drill size and the material being drilled – when in doubt choose a low speed.
(b) Make sure the work is securely clamped – *do not* hold it by hand.

Figure 3.8 illustrates a tool called a machine vice which can be bolted down to the drill table. Large pieces of work may be clamped to the table as shown in Fig. 3.9. For convenience, a selection of bolts and clamps should be retained in a suitable box near the drill.

Lubrication of the work during drilling is important for two reasons:

(a) It reduces friction and thus prevents overheating of the drill.
(b) It lessens the possibility of breakage, especially when small drills are used.

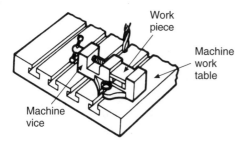

**Fig. 3.8** Machine vice. Machine vices should be bolted to work table to ensure absolute safety.

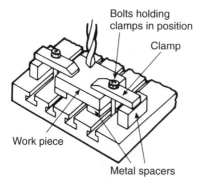

**Fig. 3.9** Clamping work to the work table. Method used for securing large work pieces to machine work table.

It should be noted that lubrication is essential to most cutting or threading operations; a good-quality cutting oil or paste should be used. When drilling copper, bronze, soft brass or aluminium, a little paraffin is a useful lubricant. Brass rods or strip obtained from an engineering supplier usually contain a small percentage of lead which gives it free-cutting qualities. This material, and cast iron which also cuts freely, does not require the use of a lubricant and should be machined dry.

The use of a chuck guard (see Fig. 3.7) is essential with fast-rotating machines such as power drills. It prevent one's hair or clothing becoming entangled with the chuck, which would result in a serious accident. It is also recommended that suitable eye protection is worn when using drilling machines as the metal (or swarf) removed by the drill can be thrown far and wide.

### Grinding tools

While the plumber generally does not use machine tools, it is quite common to find grinding equipment provided in plumbing workshops for the sharpening of cutting tools, and if these are misused they can be very dangerous. The worst accident that can occur with grinding equipment is probably the explosion or shattering of the abrasive wheel, with disastrous consequences to anyone in the vicinity. Wheel explosion is usually due to incorrect fitting of the wheel, which should only be changed by a trained and qualified person.

To ensure safety in the use of grinders, the following observations may be helpful to those with limited experience of these tools, but it must be stressed again that the fitting or changing of wheels should only be done by a person properly trained to the requirements of the standards of the Abrasive Wheel Regulations.

The wheel should be fitted on the spindle with two recessed metal discs, which take a bearing against two cardboard washers. A check is made before using the grinder to ensure these cardboard washers are in place, and the wheel is secure on the spindle (see Fig. 3.10(a)).

The wheel must be provided with a suitable guard, but some variation of the amount of the exposure of the wheel is permissible, depending upon its use. The wheel exposure shown in Fig. 3.10(b) is suitable for general purpose work. The sides of the wheel should not be used for grinding as stresses may be set up causing it to explode. When the face of the wheel becomes worn it should be redressed (to remove any irregularities) by a trained operative. See Fig. 3.10(c). Do not forget that dressing will slightly reduce the diameter of the wheel and it will be necessary to re-adjust the tool rest to the correct gap. See Fig. 3.10(c).

Eye protection is essential when grinding is being carried out and some machines are fitted with a clear shatter-proof glass window which must always be in position when the machine is in use, as in Fig. 3.10(d). Safety goggles of the correct specification must also be worn, whether or not the machine is equipped with a safety window and ensure the operator can stand comfortably on a clean dry floor.

### Angle grinder

These are used for many purposes in the construction industry, the various grinding blades

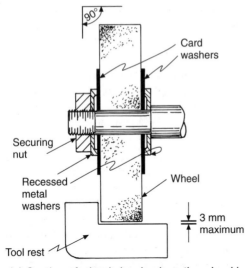

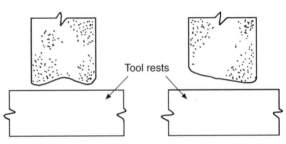

(c) The worn irregular faces of these wheels would be dangerous to use. Redress or replace!

(a) Section of wheel showing how the wheel is fitted on the shaft and tolerances for tool rest

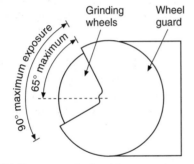

(b) Exposure of wheel for normal operation

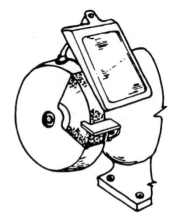

(d) Eye protection on grinding wheels

Shatter-proof glass window must be in position and goggles worn by the operator when wheel is in use.

Cutting disc. Special discs are available for cutting a wide variety of materials

(e) Angle grinder

**Fig. 3.10**  Grinding wheels.

used permit the cutting and shaping of both metals and masonry. The recommendations relating to safety are very similar to those governing the use of bench grinders, especially regarding eye protection, e.g. goggles must be worn. When cutting masonry a lot of fine dust is created and under these circumstances some form of face mask should be used to filter out the dust. When using these tools above ground level work on a secure staging or scaffold because if the blade jams it can throw the operater off balance. A typical angle grinder is shown in Fig. 3.10(e).

*Power saws* These vary considerably and the type used depends largely on the work they are required to do. Bandsaws, similar to those used for wood working, are often found in engineering workshops but are rarely used by the plumber. Probably the most robust and useful for general purpose work is the power hacksaw shown in Fig. 3.11. It works in a similar way to a hand hacksaw, the only difference being that with most types the blade cuts on the back stroke, not the forward stroke as with hand saws. Care is necessary when replacing the blade to make sure it is the right way round. Correct tensioning of the blade is important if breakage is to be avoided, and the work must be securely clamped in the vice because if it moves this will also result in a broken blade. Power saw blades are very expensive. All saws of this type are equipped with an adjustable screw which automatically operates a switch when the saw reaches the end of the cut. Once correctly adjusted it should not be interfered with. Circular-type saws have also been developed for cutting metals; a bench-mounted type is shown in Fig. 3.11. It will perform all the operations done by a power hacksaw, but it has the added advantage of being able to cut bevels very accurately owing to the fact that it can be turned and clamped in any position within 45° to the right or left of centre. As the speed of the cut is controlled by the operator, these machines are considered by many to be superior to power hacksaws for cutting light-gauge tubes and non-ferrous metals. Another type of rotary saw for cutting pipes is shown in Fig. 3.13. It is a hand-operated tool useful on site for cutting steel pipes quickly and accurately. The pipe is held in a special pipe vice supplied with the saw while the saw is rotated round it. When cutting long lengths of pipe with this tool it is important to support both ends of the pipe to avoid damaging the saw blade.

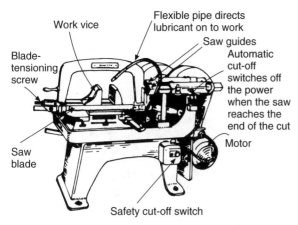

**Fig. 3.11** Power hacksaw.

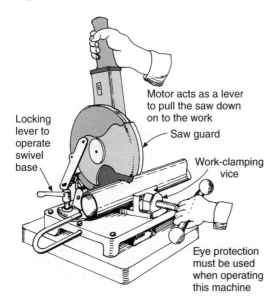

**Fig. 3.12** Rotary metal saw.

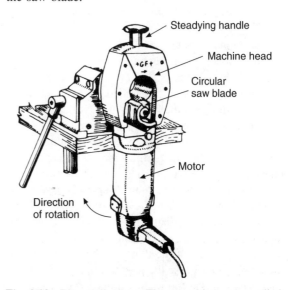

**Fig. 3.13** Pipe-cutting saw. These machines are supplied with a self-centring vice which holds the pipe rigid while the machine head carrying the circular saw is rotated by hand around the stationary pipe.

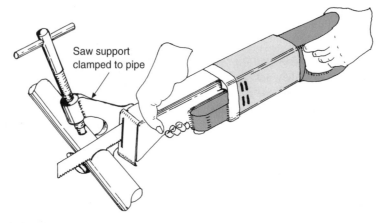

**Fig. 3.14** Hand-held electric power saw.

*Hand-held power hacksaws*

Figure 3.14 shows a typical saw of this type. Many are supplied with a clamp-on attachment to ensure a square cut, but they can also be used freehand in awkward corners. During use the work must be properly secured, eye protection worn, any guards provided by the manufacturer must be in place and the operator working on a firm steady base.

As with drilling operations, the use of power saws, of all types, on metal requires an adequate supply of the correct lubricant to prevent excessive wear and overheating of the cutting edges.

*Power threading machines*

A typical machine of this type is shown in Fig. 3.15. Portable threading machines have been in use for many years on construction sites where they save both time and the hard manual work involved in cutting threads on steel pipes by hand. This is especially true on sites where a large quantity of steel pipe is to be installed. By means of a special attachment, pipe nipples (short lengths of pipe) can be threaded to any length. Nipples can be purchased, but stock lengths are limited and the traditional method of cutting lengths of pipe held in a connector (long-screw) is both laborious and time consuming. Some of these machines have a rotating die head, others cause the pipe to rotate through a stationary die; the former has an advantage in that pipes can be threaded after they have been bent. The chaser dies are changed and adjusted in a

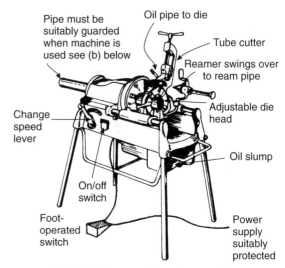

(a) Portable power threading machine

(b) Telescopic pipe guard used to conform to safety regulations for power threading machines

**Fig. 3.15** Power threading machine.

similar way to those of hand-threading equipment, care being taken to ensure they are fitted in correct sequence in the die head. Special dies are also available for bolt threading which is especially useful if pipe brackets or adjustable hangers are to be made. The type shown also incorporates a tube cutter and a deburring tool (reamer) so that all the necessary operations can be carried out without removal of the work from the machine.

To conform with the safety regulations, any rotating pipe protruding from the machine must be guarded, and in no circumstances should any attempt be made to thread a pipe containing a bend. To do so would endanger those in the vicinity, and could also lift the machine from the floor causing it to fall over, thus incurring a large repair bill. A further safety requirement is the ability to stop the machine immediately should an accident occur. Some manufacturers provide a foot switch for this purpose, other machines are equipped with an automatic cut-off switch mounted on the machine itself.

Many of these machines, especially those suitable for threading a wide range of pipe diameters, are equipped with a gearbox giving slower cutting speeds for larger pipes. They also incorporate an oil reservoir from which coolant is pumped over the chasers when the machine is in operation, and which is returned to the oil bath through a filter for reuse. The principal maintenance requirements of these machines are to top up the oil bath, clean the filter and change the oil at periodic intervals. Only oil recommended by the manufacturers should be used.

### Safety switches

All fixed workshop machines should be equipped with not only on/off switches but also a safety switch which will enable the machine to be switched off quickly and easily in the event of an accident. These switches are identified by the large red mushroom-shaped 'off button' (see Fig. 3.16).

## Socket and branch-forming tools

These tools have been developed over a period of years for forming sockets and branches on

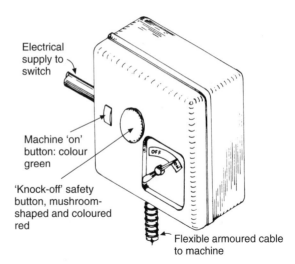

Electrical supply to switch

Machine 'on' button: colour green

'Knock-off' safety button, mushroom-shaped and coloured red

Flexible armoured cable to machine

**Fig. 3.16** Safety 'knock-off' switch. Health & Safety Regulations recommend that these switches are fitted to all fixed workshop machinery so that in the event of an accident any machine can easily be isolated.

light-gauge pipes and tubes. They are especially suitable for copper pipe and are used extensively by European plumbers and pipe fitters. They are designed for hard-soldered (brazed) joints and are not suitable for soft soldering as the socket and branch entries made with these tools are of insufficient depth to make a strong joint.

### Socket-expanding tools

Various types are available to suit all pipe diameters, but most work on the same principle: that of causing a series of segments to expand on the inside of the pipe, thus forcing out the walls to form a socket. The type shown in Fig. 3.17 is suitable for copper pipes of up to 28 mm outside diameter. It will be seen that the head of the tool that forms the socket is made of six separate segments through which a tapering pin is forced, causing them to open when the handles are pulled together, thereby expanding the tube end. The segments are held together by a circular spring which, when pressure on the handle is released, pulls the segments together and holds them in place. This particular tool is especially useful as it can be used to form a socket on a pipe close to a wall or corner.

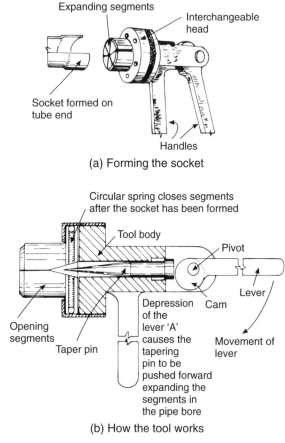

(a) Forming the socket

(b) How the tool works

**Fig. 3.17** Socket expander.

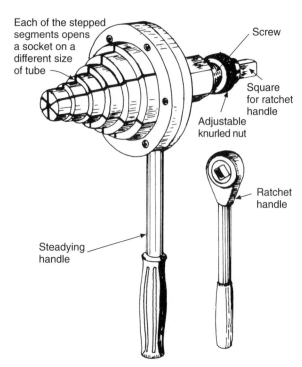

**Fig. 3.18** Socket expander for opening tubes up to 108 mm nominal bore. This tool operates in a similar way to the lever type. But as it is used for larger pipe sizes more effort is necessary to open the segments. In this case the taper pin is forced inward by rotating the screw. The adjusting nut can be set to limit the socket expansion and, where a series of sockets are to be formed, ensure consistency of the expanded diameter.

The tool shown in Fig. 3.18 works on the same principle, but is designed for opening sockets on large pipes; in this case the tapering pin forcing the segments outward is operated by a threaded screw.

Considerable savings on fittings can be effected by using these tools, especially with the larger sizes of tubes. They also enable short lengths of tube to be joined together and used which might otherwise be scrapped. The tube end on which the socket is formed should be annealed, as this reduces excessive wear and strain on the tools and less effort is required by the operator – it also lessens the likelihood of splitting the pipe. A practical hint worth remembering to ensure a truly circular socket is to first partly open the socket, then to release pressure on the tool and rotate it a few degrees before expanding the socket to its full size. When it

is formed in one operation the section of the socket looks like a hexagon nut with rounding edges which may result in a defective joint. If it is found that the socket is eccentric to the pipe bore, the cause is usually owing to unequal heating during the annealing operation. It is essential that the tube end on which the socket is to be formed is heated *evenly* to a dull red colour.

*Branch-opening tools*
As with socket-forming tools, branch-forming tools can rapidly recoup their initial cost, especially on a contract where copper tubes having a diameter upward of 35 mm are to be installed. Tees, in these sizes, are very expensive. There are several types of tool for this purpose, some forming a saddle on the branch pipe which is brazed over a hole in the main

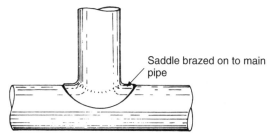

**Fig. 3.19** Saddle-type branch. This type of branch simply requires a hole in the main pipe. The saddle is swaged out with a special tool.

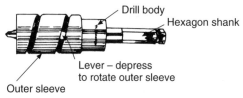

(a) Drill closed to cut smallest diameter hole within its range

The drill shown will bore a hole having a maximum diameter of 28 mm which is suitable for a 42 mm branch extractor

(b) Tapering drill fully exposed

**Fig. 3.20** Expanding drill. When the lever in the spiral groove is depressed the outer sleeve can be rotated to expose a larger section of the tapering drill permitting a larger diameter hole to be made.

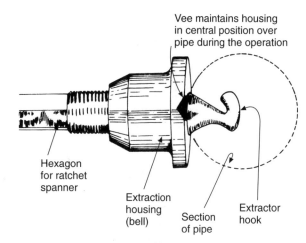

**Fig. 3.21** Branch hole extractor. As the extractor hook is rotated it raises the edge of the hole sufficiently to enable a brazed joint to be made.

pipe. For example, see Fig. 3.19. Others are made to pull up the sides of the branch hole in a similar way to that of using traditional methods with a hammer and bent bolt. It is the latter type of tool that is described as follows:

The first prerequisite is an expanding drill. This is used in conjunction with a branch hole extractor tool. The expanding drill is simply a tapering drill over which is mounted a sleeve which can be adjusted to govern the size of the hole actually made in the pipe. By depressing a lever on the side of the sleeve, which is made with a spiral groove, it can be raised to expose more of the taper drill, permitting it to cut a larger hole. The sleeve is graduated to a series of sizes, one of which is selected to enable the correct size of hole to be drilled for the branch diameter required. Figures 3.20(a)–(b) illustrate a typical drill of this type. The edges of the hole are drawn up by means of a special extractor (see Fig. 3.21). It has a hook-like end which enables it to be 'joggled' into the hole. The extractor is carried in a guide which, owing to its shape, is called a *bell*. This guide has a vee groove in its base which locates it square on the pipe. The ratchet lever is placed over the hexagon on the extractor which, as it is rotated, draws the hooked end through the hole, forcing the edges upward to form a raised socket. Figure 3.22 shows the operation necessary to form a branch hole using this tool. The one slight defect with branch openings formed in this way is that the top edge of the branch hole is not level, which results in lack of cover on the flanks of the branch. This can be overcome if the bottom of the branch pipe is filed away as shown in Fig. 3.23. Great care must be taken to ensure it does not penetrate into the main pipe, as this would have the serious effect of obstructing the main flow of water, one of the unforgivable sins a plumber must never commit. To ensure this does not happen the branch entry in the opening is limited by raising two indents on the tube with the special pliers shown in Fig. 3.24.

(a) First drill with Unidrill automatic

(c) Insert extractor hook into drill hole and turn bell against drill hole

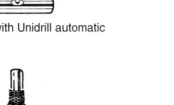

(b) Turn extractor to the left with ratchet. Take care that bell fits firmly

(d) Mark input depth of branch and form dimple as shown in Fig. 3.24

**Fig. 3.22**   Stages of branch formation using extractor tools.

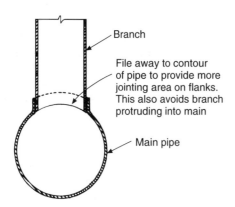

Branch

File away to contour of pipe to provide more jointing area on flanks. This also avoids branch protruding into main

Main pipe

**Fig. 3.23**   Section through branch made with forming tools.

Raised dimples formed in pipe

**Fig. 3.24**   Dimple pliers. Used to make an external indent on the branch pipe governing its entry into the main.

The equipment discussed here is capable of forming branches of up to 42 mm on copper tubes. It is a generally recommended rule that copper should be worked only in its annealed state; however, opening branches using these extractors is one of the few cases where this rule does not apply. If it has been annealed it will be found that as the extractor is pulled upward, the softened pipe is unable to resist the downward pressure of the bell, and distorts. When using these tools it is helpful to apply a little grease or tallow to the extractor each time a branch is opened as less effort will be required to draw it through the hole. When opening holes for large-diameter branches the same principle is used, but special clamps are necessary to secure the extractor unit firmly on to the pipe.

It is also possible to operate a similar extractor by means of what is in effect a geared-down electric drill. Further information on these and other tools can be obtained from suppliers mentioned in the further reading section at the end of this chapter.

### Electrical methods of soldering

The use of a blowlamp has been traditional in the plumbing industry for many years not only for making joints on pipelines, but also for such

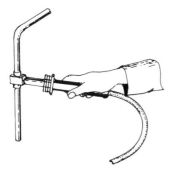

(a) Shows electrodes clamped on integral solder ring fitting. Can also be used with end feed types

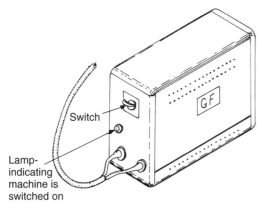

Switch

Lamp-
indicating
machine is
switched on

(b) Transformer and controls

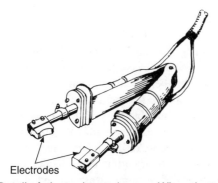

Electrodes

(c) Detail of electrodes and tongs. When the tongs are pressed round the copper fitting an automatic switch activates the electrodes

**Fig. 3.25**  Electric soldering machine.

purposes as melting lead for running joints, thawing out frozen pipes and many other situations where a portable source of heat is required. One of the dangers of using blowlamps is the fire risk

associated with their use, especially when working in roof spaces and under suspended floors where careless use can be a serious fire risk. These risks can be reduced by the use of the electrical resistance soldering equipment shown in Fig. 3.25 which is capable of making satisfactory joints on copper tubes up to 54 mm in diameter. The equipment consists basically of a portable transformer which operates from either a 240 or 110 volt supply producing a current which can be varied between 90 and 190 amps at very low voltages of between 5 and 8 volts. Heat is applied through the tongs which clamp round the fitting, automatically switching on the flow of current. The use of this equipment allows joints to be made on pre-insulated pipes or pipes close to decorated surfaces and easily damaged materials such as wood and plastic, an invaluable asset to the plumber or heating fitter when working in occupied houses.

*Core drills*

Eye protection, a face mask and suitable protective clothing must be worn when operating this equipment, and if drilling above ground level, a good working platform must be used. Most readers are familiar with the small tungsten-carbide-tipped masonry drills for drilling masonry. Larger holes are traditionally cut with a hammer and chisel or point. An alternative is to use a technique where a series of small holes are drilled around the perimeter of the masonry to be removed, which can then be cut out with a suitable chisel. Where the hole is circular, special drilling equipment is available for drilling holes of up to 150 mm in diameter. The main advantages of using this method are saving of time, less possibility of damage to the building fabric and the holes are more easily made good. Figure 3.26 illustrates the core drill and its associated components. The drill itself is hollow, rather like a short length of steel pipe in appearance, but having a series of cutting teeth on one end made of tungsten carbide or tungsten diamond. These drills and the extension pieces are of varying length to accommodate drilling holes of differing depths.

The hole must be started with a pilot drill to a depth of approximately 50 mm. Keep an eye on the drill to ensure it is at right angles to the surface of

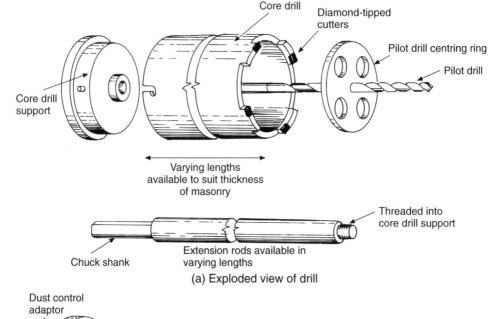

(a) Exploded view of drill

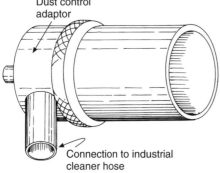

(b) Special adaptor used with dry core
drills to avoid masonry dust

**Fig. 3.26** Conventional dry core drilling. These drills enable large circular holes to be cut very accurately thus avoiding damage and making good of masonry. Typical uses include hole for discharge pipework and gas flues.

the masonry so that the hole is not cut on the skew. The core drill is now fitted over the pilot drill and drilling is commenced until the cutting segments start to cut into the masonry. The pilot drill can then be removed leaving the core drill to cut, in effect, an ever deepening circular groove in the masonry, the core of which passes into the hollow centre of the drill leaving an accurate hole.

Masonry drilling produces a lot of fine dust and special adaptors can be obtained which can be connected to an industrial type of vacuum cleaner

to remove the dust as it is produced. This is very useful when working in occupied properties. It is essential that a suitable power drill is used with these drills and should be of the rotary percussion type, rated at between 1,000 and 1,500 watts. The speed range will depend on whether the tool in use is tipped with tungsten carbide, which requires speeds of about 250 revs per minute, or diamond-tipped tools which require higher speeds of between 1,500 and 3,000 revs per minute. It is obviously sensible to check if an existing power drill is suitable before investing in expensive drilling equipment. It should also be noted that these drills are suitable for brick- or blockwork, but should not be used on hard natural stone, e.g. granite or flintstone, or dense concrete. When these tools lose their cutting edge they can be resharpened, and in some cases retipped. This operation is best left to companies who specialise in this work. These drills will give a long and trouble-free life if they are well cared for and used only for the work for which they are designed.

## Drain and discharge pipe cleaners

Although there are many firms which specialise in clearing blocked pipes, many plumbers also take

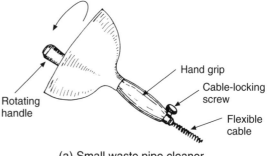

(a) Small waste pipe cleaner

Turning the handle causes ejection of the flexible cable simultaneously causing it to rotate.

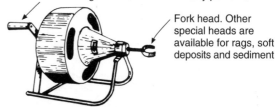

(b) Tool capable of cleaning drains to a distance of up to 15 m

This works on the same principles as the waste pipe cleaner but is more robust being equipped with a longer and stronger cable.

**Fig. 3.28**   Waste pipe and drain cleaners.

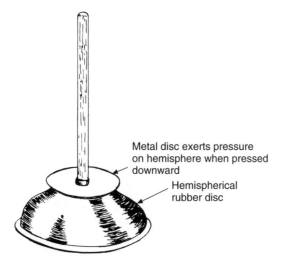

**Fig. 3.27**   Plunger. This is a traditional plumbers' tool for cleaning wastes. A rapid up and down movement of the handle alternately pressurises and depressurises the air trapped under the hemisphere. This has the effect of moving the blockage to and fro until it breaks up and can be flushed away. The success of this tool depends largely on the severity and cause of the blockage. Note that any integral overflows must be sealed to obtain any effect with this tool.

on this work. The old-fashioned plunger shown in Fig. 3.27 is effective, although its success is limited to some degree by the cause of blockage and how effectively integral overflows can be blocked while it is in use. Such a plunger works by the application of a positive pressure on the blockage when the plunger is depressed, and a reduction of pressure when it is pulled up. This has the effect of moving the obstruction to and fro in the pipe which in some cases causes it to break up, when it can then be flushed out by opening a tap. More effective tools for clearing stubborn obstructions are shown in Fig. 3.28. The small hand-held type (Fig. 3.28(a)) is operated by turning the handle which causes a spiral-wound stainless steel rotating cable to be fed into the pipe; it is able to pass round traps and bends to break up blockages. The larger type (Fig. 3.28(b)) is very similar in operation but has a stronger cable and is capable of cleaning drains to a distance of up to 15 m. For the company wishing to specialise in this type of work, electrically powered and motorised equipment is also available with a full range of accessories.

### Freezing equipment for water services

Pipe-freezing equipment has proved to be a useful aid to plumbers and pipe fitters in cases where repairs are to be effected in systems of pipework which would take a long time to drain down and refill. There is the added possibility in some cases, especially heating systems, that air locks may occur on refilling, often taking a long time to clear. It is not always convenient to shut down a main completely in, say, industrial premises which may rely on water for production purposes. It is in such cases that pipe-freezing equipment may be used to advantage. All types of equipment have two common features: a jacket, that may be rigid or flexible, which is strapped on to the pipe to be frozen, and a hose connected to the jacket into which is injected the refrigerant. This has the effect of cooling the water in the pipe until it freezes. The time taken to freeze the water will depend on two factors: the diameter of the pipe and the temperature

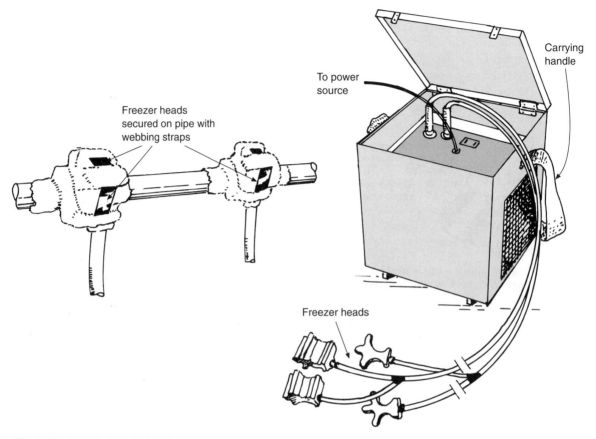

**Fig. 3.29**  Electrical method of freezing pipes.

of its contents at the beginning of the operation, which should, of course, be as low as possible.

There are four pipe-freezing systems available, electric, aerosols, carbon dioxide and liquid nitrogen. The last is beyond the scope of the plumber as it requires trained operators using specialist equipment, and is only economical on pipe diameters in excess of 75 mm.

*Electric pipe-freezing machine* (Fig. 3.29)
An alternative to freezing by gases is an electrical freezer which works on a similar principle to that of a refrigerator. It will freeze pipes of 12–15 mm nominal bore for an indefinite period, unlike those methods using gases which have a very limited freeze time. They also have an additional advantage in that there is no problem with the dispersal of the gases as is the case with other methods. These machines are normally made to operate on 240

volts, but if required for site work 110 volt types can be made to order. It is quite likely that when fully developed for larger pipe sizes, this method of freezing may supersede those using gases.

*Gas freezing equipment*
Aerosol freezing employs hydrafluorocarbon as a freezing agent. Freezing using this method is limited to pipes of up to 25 mm in diameter for it to be economical. A freeze can be effected on a pipe having a nominal diameter of 20 mm, containing water at 10–15 °C, in about 10–12 minutes. This is a fairly cheap method and the canisters can be obtained at most plumbers' merchants. Manufacturers of freezing equipment employing gases must use those which are environmentally friendly, but it is very important to read carefully any instructions regarding the safe use of such products.

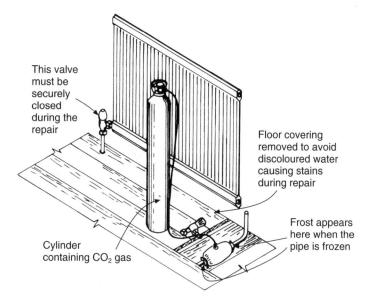

**Fig. 3.30**  Replacing a defective radiator valve using $CO_2$ pipe-freezing equipment.

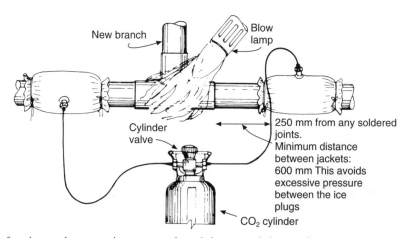

**Fig. 3.31**  Using freezing equipment to insert a new branch into an existing service.

Carbon dioxide ($CO_2$) freezing equipment operates on the principle that liquid $CO_2$ at high pressure, when released into the atmosphere, becomes a solid and forms what is known as dry ice. This solid has a temperature of minus 78.5 °C which rapidly abstracts heat from the pipe. The $CO_2$ is released into a strong, reinforced nylon jacket which is tied round the pipe to hold the dry ice in position while freezing takes place.

A successful freeze is usually indicated by the appearance of frost at each end of the jacket and the pipe itself will feel very cold. A blowlamp may be used for the repair work providing it is kept at least 250 mm away from the jacket. Figures 3.30 and 3.31 show two typical applications of this method of isolating a section of pipe to enable repair work to be carried out.

*Safety precautions with $CO_2$ gas*
$CO_2$ is non-toxic and non-flammable. It is, however, heavier than air and care should be taken to disperse any gas that escapes from the ends of the jacket before repairs are commenced. Although non-toxic, it has an affinity for oxygen which it will extract

from the surrounding air very quickly; therefore plenty of ventilation is essential. It is also intensely cold and can cause cold burns or frostbite – the protective gloves supplied with the equipment should be worn at all times. Cylinders should be stored indoors and never exposed to a source of heat. They are fitted with a bursting disc safety device which will rupture if the cylinder is overfilled or heated above 50 °C. It is recommended that the cylinder should be transported outside a vehicle in a specially made cradle. This is not because $CO_2$ is any more dangerous than other compressed gases, but should the bursting disc operate inside the vehicle it might distract the driver's attention. Suppliers of freezer equipment are mentioned at the end of this chapter under the heading of further reading. It is essential, to avoid accidents and obtain a successful freeze, that any operating instructions provided by the manufacturer of the equipment are carefully read and complied with.

## Further reading

Much useful information can be obtained from the following sources:

Arctic Products Ltd, Baird Road, Corby, Northamptonshire, NN17 5ZA. Tel.: 01536 264000.

*BOC Hand Book of Operating Instructions for Gas Welding and Cutting and Safe Under Pressure.* British Oxygen Company (BOC), Customer Services, Priestley Road, Worsly, Manchester, M28 2WT.

*Pipe Cutting Threading Machines. Electric Soldering Equipment.* George Fischer Sales Ltd, Paradise Way, Coventry, CV2 2ST. Tel.: 02476 535535.

Rems, Rothenburger, Rigid Tools. Address as in Chapter 2.

Specialist cutting of masonry and stonework. Duradiamond Ltd, Unit 7, Murnin Road Industrial Estate, Bonnybridge, Stirlingshire, FKA 2BP. Tel.: 01324 814036.

Mr Szyplinski, No. 1, Skeats Wharf, Pennyland, Milton Keynes, MK15 8AY.

## Self-testing questions

1.  (a)  Explain the meaning of the term 'dissolved acetylene',
    (b)  Why is it necessary to use acetylene cylinders in the upright position?
2.  State why the regulators on compressed gas cylinders sometimes freeze in cold weather and explain what action should be taken when it occurs.
3.  State three means of identifying an acetylene cylinder and two methods of recognising union nuts used with acetylene equipment.
4.  Describe in detail the correct procedure for assembling and commissioning oxy-acetylene welding equipment.
5.  (a)  State the main advantage of a two-stage regulator.
    (b)  Using Table 3.4 state the size of welding nozzle and the correct operating gas pressure for use with 3.2 mm thick low-carbon steel.
    (c)  Why is it necessary to use a larger welding nozzle for welding copper than for steel?
6.  (a)  Explain three ways in which a flashback can occur in a welding blowpipe.
    (b)  Describe the course of action to be taken if the blowpipe flashes back and a loud squealing noise is heard.
7.  Describe two methods of securing the work piece to the machine table when using a drilling machine.
8.  Describe why it is essential to apply a suitable lubricant or coolant when using thread-cutting equipment or machines, such as power drills and saws.
9.  Identify the safety switching arrangements that should be fitted to all fixed power tools in order to conform with the Health & Safety Regulations.
10.  (a)  State the safety recommendations that must be observed when using pipe-freezing equipment.
    (b)  Describe two situations in which pipe-freezing equipment may be used with advantage.
11.  List the faults commonly found on bench grinders.
12.  State two reasons why it is essential that goggles *must* be worn when using grinding tools.

# 4 Materials and fittings for gas, water services and sanitary pipework

After completing this chapter the reader should be able to:

1. Name the main materials used for pipes, fittings and components.
2. Understand the purpose of British and ISO standards in relation to building construction.
3. Describe the constituents of the main alloys used in the manufacture of plumbing materials.
4. Recognise and describe the types and grades of steel and copper pipework and principles of jointing.
5. Select materials, fittings and components for differing types of work.
6. Identify the characteristics and limitations of pipework used for sanitation.
7. State the advantages and disadvantages of various pipework materials.
8. Describe simple site tests to identify the main thermoplastic materials used for sanitation.
9. Demonstrate knowledge of the physical and chemical properties of thermoplastics used for pipework services.

## Introduction

The installation of pipework systems for the supply of hot and cold water – and to a lesser extent gas – forms the major part of the work of the plumber. This means that it is essential for the plumber to have a thorough understanding of the performance of pipework materials in current use.

The earliest recorded use of metallic pipes for water supplies is about 2750 BC when lead and copper piping was used by the Egyptians. Lead piping was also used by the Greeks and Romans, and apart from terracotta (clay), wood (bored trunks of oak, elder and elm) and the introduction of steel, it was the principal pipe material used by plumbers until after the Second World War. Since then there was a gradual decline in the use of lead pipe owing to its cost and weight, but mainly because of the danger it has been found to pose to health. It was, however, still in use for repair and maintenance work until the 1986 Water Bylaws prohibited its use for potable water supplies.

The principal materials used nowadays for plumbing and gas services are steel, copper, lead, brass and plastics. To appreciate the qualities of each material, their physical and working properties must be understood, also their suitability for the type of work for which they are employed. Guidance relating to the foregoing is given by various standards listed as follows.

## Standards and legislation relating to the construction industry

### Codes of practice (CP)
These comprise recommendations for a minimum standard of work and materials. CPs deal with various practices in industry generally and although they are not a legal document the recommendation of a relevant code is often required by specifying authorities. Many of these codes when updated are incorporated into British Standards. BS 6700 for example deals with both hot and cold water and supersedes the CPs relating to the subject.

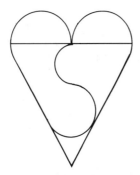

**Fig. 4.1** The BSI kite mark symbol. This can only be used on materials and equipment that meet the standard of BSI.

**Fig. 4.2** British Board of Agrément logo. (Courtesy of the British Board of Agrément.)

*British Standards Institution (BSI)* Incorporated by Royal Charter, BSI is the independent body for the preparation of British Standards. It is the UK member of the International Organization for Standardization (ISO) and the UK sponsor of the British National Committee of the International Electrotechnical Commission (IEC). As with CPs the relevant BS sets out minimum standards and recommendations for many aspects of construction work, and providing these standards are met will satisfy most regulations, bearing in mind that regulations are enforceable by law. Various items of equipment and appliances can be tested by BSI at a manufacturer's request, and if the standard is met such items may carry the BSI kite mark, see Fig. 4.1.

*International Organization for Standardization (ISO)* With the entry of the UK into the EEC, some standardisation of components and appliances is necessary for the interchange of goods and services between the participating countries. The purpose of ISO is to ensure that various national standards are 'harmonised' in order for them to be acceptable to other countries within the European Union.

*European Committee Standardization (EN)* Many of the components used in plumbing and heating services are marked BS EN accompanied by a relevant number. For example, copper tube previously conforming to BS 2871 Part 1 is now designated as BS EN 1057. This indicates that copper tubes conforming to this standard are acceptable throughout the European Union.

*DIN Standards* These are the German equivalent to British Standards and are generally accepted in this country.

*British Board of Agrément* The function of this body is to test and certify that any new materials or systems used in the construction industry are suitable for this purpose. Approval by this body is indicated by the logo shown in Fig. 4.2, which is attached to the component or appliance.

**Working properties of materials**

Consideration of mechanical and other properties will inform the plumber how a material will behave when being fixed and when in use.

*Malleability*
This term relates to the ability of a metal to be hammered or rolled into permanent shape without breaking. This property is required of metals that are to be formed by 'bossing'.

*Ductility*
The ability of a metal to withstand distortion without fracturing is indicated by the elongation that occurs while a sample is being tested to find its tensile strength. The term 'ductility' is used to denote this property. Neither malleability nor ductility can be measured accurately, the only test being to 'work' metals and then compare the results. Table 4.1 shows the order of malleability and ductility for cold-worked metals.

**Table 4.1**  Order of malleability and ductility.

| Malleability | Ductility |
| --- | --- |
| Lead | Copper |
| Aluminium | Aluminium |
| Copper | Zinc |
| Zinc | Lead |

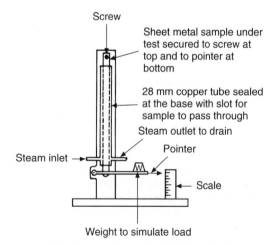

**Fig. 4.4**  Simple experiment showing how heating and loading a sample of material can increase the effect of 'creep'.

### Tensile strength

The tensile strength of a material is its ability to resist being torn apart. The method of applying the test is to subject the sample to increasing loads until it breaks, the loading that causes breakage being the ultimate tensile stress.

### Creep

The tendency of metals to stretch slowly over a period of time if under constant loading is termed 'creep' and it is an important consideration in sheet metal weathering. Creep is a form of plastic deformation and is illustrated in Fig. 4.3 where two different metal samples are seen to have lengthened owing to the application of a load. When the loads are removed, metal 'A' has returned to its original length, i.e. the elongation to load was elastic deformation. Sample 'B', however, has suffered a permanent increase in length, i.e. plastic deformation, known as 'creep'.

An experiment to compare the 'creeping' action of metals when heated is shown in Fig. 4.4. Samples of sheet metals 6 mm wide are cut and

fixed at the top by a screw and at the bottom to a pointer. The initial position of the pointer is marked and steam is passed into the tube causing the strip to become hot and expand. The expansion causes the pointer to move down the scale and, upon cooling to its original temperature, to move up again. The difference between first and final positions of the pointer indicates the amount of creep.

The effects of heat and mass will increase the tendency of sheet roofing metals to 'creep'. Hot metal has a lower tensile strength and the mass of metal will provide the load causing elongation, especially in the case of sheet metal fixed vertically or placed on steeply pitched roofs.

### Hardness

The hardness of a metal is its resistance to indentation, abrasion and deformation. Several tests can be used to determine hardness but they are all similar in principle and require that either a steel ball (Brinell) or a diamond point (Vickers and Rockwell) is pressed on to the metal surface and the indentation produced gives an indication of hardness.

A simple workshop test for hardness uses a steel ball bearing and a glass tube, and is illustrated in Fig. 4.5. When the ball is dropped into the tube it will hit the metal sample and then bounce upwards,

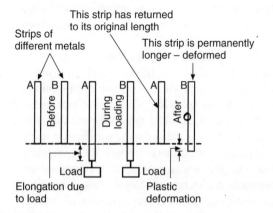

**Fig. 4.3**  Illustration showing the effects of 'creep' on materials.

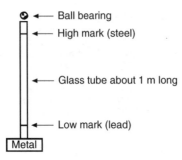

**Fig. 4.5** Simple workshop test to compare the hardness of various metals. When the ball bearing is dropped in the glass tube it bounces on the metal below. The higher the bounce the harder is the metal.

the higher the bounce the harder the metal. By using a soft metal such as lead to provide a low mark and cast iron or steel to give a high mark the whole range of plumbing materials can be tested to find how they relate to each other.

*Temper*
The temper of a metal is the amount or level of hardness and can vary between dead hard and dead soft. An example of the range of temper can be seen with copper where it would be of maximum hardness in a hammered or cold-worked condition, the dead soft state being when fully annealed.

The working properties of hardness, tensile strength and ductility are related so that if one property is changed, then changes will occur in the others. An example of this relationship can be seen when hardness is increased by work hardening: tensile strength also increases but ductility decreases. Similarly, as a metal is heated it becomes soft, its ductility increases but the tensile strength is lowered.

*Work hardening*
This relates to an increase in hardness caused by cold working, e.g. bossing or hammering. All metals are formed of many grains or crystals which become deformed during 'working'; the effect is shown in Fig. 4.6. The deformation of the crystals reduces ductility which will prevent further cold working and could lead to fracture. This is often called fatigue cracking and is referred to in more detail in Chapter 11.

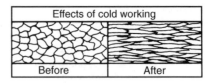

**Fig. 4.6** The effects of cold-working metals.

*Annealing*
This is the name of the heat treatment used to relieve the condition of work hardening. The metal is raised to a specified temperature; during the heating process recrystallisation takes place and the crystal grains return to normal. This treatment softens the metal, relieves internal stresses and allows further cold working to be carried out. Any attempt to anneal metal adjacent to a wooden structure will present a serious fire risk, and must be avoided or undertaken with great caution.

**Pipework materials**

There is no perfect pipework material that is suitable for all applications; each must be considered in relation to such factors as type of water, pressure, cost, bending and jointing methods, corrosion resistance, compatibility with other materials, expansion and appearance.

*Alloys*
Alloy is a term which is commonly used when describing metals, and it is important to understand the meaning of the term before discussing pipework materials and fittings.

Alloys can be produced either by mixing different metals or by mixing metals with non-metallic elements, such as carbon or phosphorus, to form a new solid having different characteristics to the parent metal. The reason for the formation of these alloys is that in many instances a metal in its pure form is unsatisfactory for a particular function. When it is blended with certain other materials it acquires additional properties such as hardness, toughness and corrosion resistance. Alloys commonly used by plumbers are shown in Table 4.2.

**Table 4.2**  Composition of alloys.

| Alloy | Main elements |
|---|---|
| Brass | Copper and zinc |
| Bronze | Copper and tin |
| Steel | Iron and carbon |
| Soft solder | Lead and tin |
| Lead-free solders | Copper and tin |
| Lead-free solders | Silver and tin |

**Table 4.3**  Percentage of carbon in tube steel.

| Type of steel | % Carbon |
|---|---|
| High carbon | 0.50–1.40 |
| Medium carbon | 0.25–0.50 |
| Low carbon | 0.15–0.25 |

### Low-carbon steel

One of the alloys which has a wide use is commonly known as *mild steel* but the more correct term *low-carbon steel* (LCS) will be used. From Table 4.2 it can be seen that steel is an alloy of iron and carbon. The typical carbon content for various steels is indicated in Table 4.3.

High-carbon and medium-carbon steels are generally used for tools and specialist engineering purposes. LCS is the usual material for steel tubes used by plumbers. It is manufactured to BS 1387 in three grades of weight: light, medium and heavy. The outside diameter of a tube of each grade is similar, the difference being the tube wall thickness, as can be seen from the examples given in Table 4.4 which also specifies the painted 50 mm wide colour band used to identify each grade.

Only medium or heavy tubes are used for gas and water services, the choice being dependent on the internal operating pressure to which it is subjected. Heavy tube is normally required for water or heating services while medium quality is specified for supplies fed from a storage cistern within the building. Tubes to BS 1387 can be obtained either painted black or coated with zinc (galvanised).

Galvanised tube must always be used for domestic water supplies as the zinc coating prevents corrosion and the process of oxidation causing the corrosion product we call rust. Water contains oxygen; the rust discolours the water and eventually the process leads to the pipe being virtually 'eaten away'.

Medium- and heavy-grade tubes are obtainable in 6 m lengths of 6 mm to 150 mm nominal bore – the bore is known as *nominal* as the *actual* internal diameter is dependent on the tube wall thickness (see Table 4.4). It is very important when storing construction materials on site that floors are not overloaded or they will collapse. This means the quantity of tube stored in one position must be carefully controlled and it is necessary to calculate the mass involved. The average mass of screwed and socketed tube of more common sizes is given in Table 4.5.

*Joints and fittings*  Steel tube for water and gas services is usually joined by means of screwed joints, the threads conforming to BS 21. Black tube (i.e. not galvanised) can also be joined by welding which is described in Book 2 of this series. Galvanised tube, however, must not be fusion welded as the heat required would remove the zinc coating and leave the steel unprotected against

**Table 4.4**  Wall thickness and diameters of LCS tubes.

| Nominal bore | Outside diameter (average) in mm | | Wall thickness in mm | | |
|---|---|---|---|---|---|
| | L | M&H | L | M | H |
| 25 mm | 33.5 | 33.7 | 2.65 | 3.25 | 4.05 |
| 50 mm | 59.9 | 60.2 | 2.9 | 3.65 | 4.5 |
| 100 mm | 113.4 | 114 | 3.65 | 4.5 | 5.4 |
| Colour band | | | Brown | Blue | Red |

**Table 4.5**  Mass/kg of LCS tubes.

| Nominal bore (mm) | kg per metre run | |
|---|---|---|
| | Medium | Heavy |
| 15 | 1.23 | 1.46 |
| 20 | 1.59 | 1.91 |
| 25 | 2.46 | 2.99 |
| 32 | 3.17 | 3.87 |
| 40 | 3.65 | 4.47 |
| 50 | 5.17 | 6.24 |

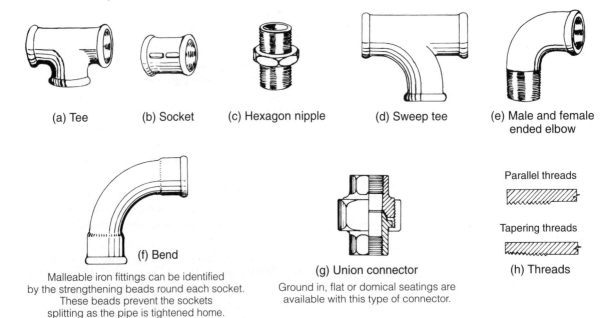

(a) Tee  (b) Socket  (c) Hexagon nipple  (d) Sweep tee  (e) Male and female ended elbow

(f) Bend

Malleable iron fittings can be identified by the strengthening beads round each socket. These beads prevent the sockets splitting as the pipe is tightened home.

(g) Union connector

Ground in, flat or domical seatings are available with this type of connector.

Parallel threads

Tapering threads

(h) Threads

**Fig. 4.7**  Typical fittings, screwed fittings and threads used for steel pipework to BS 21.

corrosion attack. A comprehensive range of pipe fittings is available both for screwed and welded joints, the latter type having no threads but the outer edge bevelled to provide the necessary joint preparation. Some of the more common screwed fittings are illustrated in Fig. 4.7.

Pipe fittings are manufactured to four British Standards. Fittings made from white cast iron and having parallel threads on the female end with taper threads on the male end are covered by BS 1256. Fittings made from white cast iron and having taper threads on both male and female ends are covered by BS 143. As cast iron is brittle, fittings to these standards are heat treated to give some ductility, i.e. the treatment for BS 1256 decomposes the hard carbides at the surface of the fitting into ductile iron and soft graphite, while that used for BS 143 anneals the surface by burning out a large proportion of the carbon. After heat treatment the fittings are referred to as malleable cast iron or malleable iron.

Steel is used to make fittings to BS 1740, these having parallel female threads and tapered male threads. Although more expensive than those made from cast iron, these fittings are able to withstand higher internal pressures. Another group of fittings

often encountered are known as *tubulars* and are formed from steel tube to BS 1387.

Malleable cast iron pipe fittings are usually provided with a beaded end, while steel fittings have a plain end. A beaded end (see Fig. 4.7(f)) provides a reinforcement on malleable fittings which strengthens the outer end and prevents splitting when it engages with a taper pipe thread.

*Compression joints for LCS tubes*  These are made of malleable iron, but instead of threads, they have special locking rings and seals which are tightened on to the pipe. They are suitable for use with gas, water and compressed air. The same type of joint using different locking rings may also be used for polythene pipes. This feature enables joints between steel and polythene to be made quickly and easily. They are extensively used by the gas and water utility companies for underground services because of their flexibility for both new and repair work. Figure 4.8 shows a straight coupling with one end having a locking ring for polythene pipe, the other threaded for steel. A full range of tees, elbows, reducers, etc., is manufactured for steel pipelines above ground for heavy-duty work, i.e. industrial and commercial work.

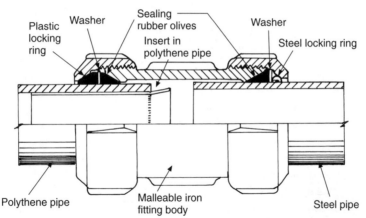

This fitting is designed to enable steel pipes to be jointed without threading. By using different olives and locking rings they can also be used for polythene services. They save time and are especially useful for emergency repairs in public buildings. They are not suitable for steel pipes that are severely corroded externally.

**Fig. 4.8**  Patent compression coupling for jointing steel and polythene pipes.

*Stainless steel*

This is the most recently developed metallic pipe material for domestic water services. The alloy used is an austenitic type, made to BS 4127 and having a chemical composition of 18 per cent chromium, 10 per cent nickel, 1.25 per cent manganese, 0.6 per cent silicon, a maximum carbon content of 0.08 per cent, the remainder being iron with small amounts of sulphur and phosphorus. The chromium and nickel provide the tube with its shiny appearance and its resistance to corrosion, protection being due to a microscopic film of chromium oxide which rapidly forms on the tube surface and prevents further oxidation. This tube is available in nominal bores of 6 mm to 35 mm having an average wall thickness of 0.7 mm. The outside diameters are similar to those of copper tubes to R250 (Table X in Table 4.6). Tubes of 15 mm to 35 mm diameter are supplied in 6 m lengths, with smaller diameters normally in 3 m lengths with longer lengths available to special order. Jointing methods are similar to those of copper tubes but special fluxes are necessary for soldered joints. For special purposes the tubes can be brazed using suitable fluxes and silver solders.

The most common method of jointing employed in the building industry is the type 'A' compression joint (non-manipulative), They are often chromium plated to match the colour of the tube.

One of the more common uses of this material is for exposed pipework in sanitary apartments, e.g. the supply of water from an automatic flushing cistern to urinals, where it offers more resistance to vandalism than copper.

It is also used extensively for discharge pipework by companies specialising in the installation of commercial kitchens where the principal method of jointing is push joints with rubber seals similar to those used with plastic pipework. When used for this purpose a special chamfer must be made if the pipe has to be cut; a special tool for this purpose is obtainable for purchase or hire from the material supplier.

The main advantages of using stainless steel are as follows: it is very strong, will accept a high polish and is easily cleaned, a very desirable characteristic when used in food preparation processes. Because of these qualities it is also used extensively for the manufacture of sanitary appliances, e.g. sinks. It is not generally used for gas work except by manufacturers of gas appliances where any compression joints used are accessible.

*Lead*

Where connections from existing lead services to other materials have to be made, the best and most economic method is to use one of the special adaptors which are available, two of which are illustrated in Chapter 5, Fig. 5.4.

Lead has also been superseded for sanitary and rainwater pipes, mainly because of its high cost and the special skills necessary for its installation. Its use is now limited to pipework in old historic buildings, where it is installed and maintained by plumbers specialising in this work. It is, however, used extensively in sheet form for weatherings. See Chapter 11.

*Copper*

Copper tube has been used in many public buildings since as early at 1900 although in those days connection was by screwed and socketed joints which required a pipe wall thick enough to thread. The first British Standard was issued in 1936 to standardise the various sizes of tube on the basis of their outside diameter. When a method of jointing copper tube without threading was introduced, the tube walls were made much thinner and were identified by the designation of 'light-gauge tubes'. Wall thickness of light-gauge tube has been reduced twice since 1936 in order to use less copper and thus lower the costs. A typical example of this can be seen in 15 mm tube which in 1936 was 1.2 mm thick, reduced in 1954 to 1 mm, and subsequently to 0.7 mm in 1964.

Copper tube in the construction industry is manufactured to EN 1057: 1996 (BS 2871 Part 1). Parts 2 and 3 of this British Standard relate to tubes for general purposes and are not relevant to building practice.

There are three specifications covering copper tube normally used in construction work and these are identified in the first two columns of Table 4.6. R250 (Table X) is the light-gauge half-hard tube which is commonly used above ground for most plumbing and heating installations. It is comparatively rigid and may be bent using a bending machine – or a bending spring if a large radius is acceptable. R290 (Table Z) refers to tubes of hard temper, the increased hardness allowing for significant reductions in wall

thickness and weight but making it unsuitable for bending.

Tubes complying with these two specifications are not intended to be buried underground. R220 (Table Y) covers tube of either half-hard or fully annealed temper, i.e. tube to be laid underground for the conveyance of gas or water. The basic differences between these tubes are their temper and wall thickness, the latter of course affecting the mass. A comparison of 15 mm diameter tube made to the three different specifications is given in Table 4.6.

Tube to R220 (Table Y) specification, often called soft copper, is supplied in sizes from 6 mm to 54 mm in 20 m coils and from 6 mm to 108 mm in up to 6 m straight lengths. The other two grades are only available in 6 m lengths in sizes from 6 mm to 159 mm. The outside diameter is the same for each type of pipe but there are differences in the internal bore due to the variations in wall thickness. It may be used underground but should be protected if laid in aggressive soils.

**Table 4.6** Comparison of 15 mm diameter copper tube to EN 1057: 1996 (BS 2871 Part 1).

| *Specification* | *Mass (kg/m)* | *Temper annealed* | *Wall thickness (mm)* |
|---|---|---|---|
| BS 2871  EN 1057 | | | |
| *Table X*  (R250) | 0.2796 | Half hard | 0.7 |
| *Table Y*  (R220) | 0.3923 | Fully annealed | 1.0 |
| *Table Z*  (R290) | 0.2031 | Hard | 0.5 |

**Joints and fittings**

Jointing methods for three types of copper tubes are shown in Table 4.7, the most commonly used being capillary or compression fittings.

The design and manufacture of capillary and compression fittings of copper and copper alloy (brass and gunmetal) are covered by BS EN 1254 Parts 1–5. Capillary fittings are of two main types, the difference being in the method by which solder is introduced into the joint, these being (a) at the end of the fitting, or (b) from a reservoir of solder in the form of a ring incorporated inside the fitting

**Table 4.7**   Jointing of copper tubes.

| Jointing method | Type of tube | | |
|---|---|---|---|
| | X | Y | Z |
| Capillary fittings | ✓ | ✓ | ✓ |
| Compression fittings | | | |
|   Type A | ✓ | — | ✓ |
|   Type B | ✓ | ✓ | — |
| Bronze welding (not normally employed or recommended for water services) | ✓ | ✓ | — |

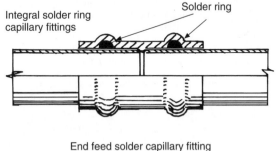

**Fig. 4.9**   Solder capillary fittings.

during manufacture. These fittings are illustrated in Fig. 4.9.

It should be noted that to conform with the water regulations only lead-free solders may be used on water services. Several alloys are available including tin/silver, tin/antimony and tin/copper. Tin/silver has very good 'wetting' or tinning characteristics but is expensive, tin/copper solders are more usual owing to lower costs. Higher soldering temperatures than those for lead/tin alloys are required, but in practice are negligible. Joints made with lead-free solder are stronger, especially at high temperatures. Traditional fluxes used for tin lead alloys can also be used for tin/silver and tin/copper solders but special fluxes have been developed to withstand better the higher soldering temperature necessary. See also page 90.

A recent development for jointing copper tubes is shown in Fig. 4.10. These fittings are made of

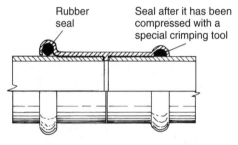

**Fig. 4.10**   Synthetic rubber seal joint for hard and half-hard temper copper tubes.

copper and look similar to a traditional capillary fitting, but instead of solder being used to make the joint, the rubber seals are compressed against the pipe by means of a special tool which fits over the raised part of the fitting. These tools may be hand or electrically operated and are loaned or bought from the fitting manufacturer. The fittings are relatively cheap and can be used where a flame is not permissible but the appearance of a traditional soldered fitting is required. Brass or plastic fittings using the 'grab ring' principle, shown in Fig. 4.11 are also alternatives to copper tube jointing.

Compression fittings are of two basic types, manipulative and non-manipulative, these names merely indicating the preparation of the tube end. The non-manipulative pattern type A requires no special preparation but for the manipulative type B, the tube end has to be formed – manipulated – using special tools. An example of each type is given in Fig. 4.12.

A comprehensive range of both capillary and compression fittings such as elbows, bends, tees and unions plus the necessary adaptors for joining metric to imperial sizes are available. Some of the more usual types are illustrated in Figs 4.13(a)–(j). Those shown are of the capillary type but most are available as compression joints. Generally all fittings have a catalogue number which should be quoted with the size of fitting required. if no catalogue is available it is necessary to indicate the branch and end sizes of tees in the following order:

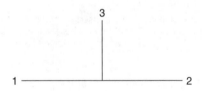

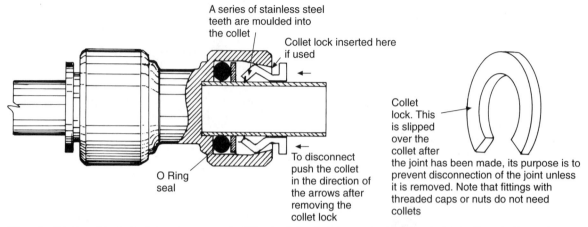

A series of stainless steel teeth are moulded into the collet

Collet lock inserted here if used

O Ring seal

To disconnect push the collet in the direction of the arrows after removing the collet lock

Collet lock. This is slipped over the collet after the joint has been made, its purpose is to prevent disconnection of the joint unless it is removed. Note that fittings with threaded caps or nuts do not need collets

When the joint is subjected to internal pressure the fitting body tends to be pushed off the pipe, but, due to its internal taper, the stainless steel teeth in the collet are forced more firmly into the pipe thus preventing its withdrawal.

**Fig. 4.11** Grab-ring-type joint – suitable for copper tubes, cross-linked polythenes and polybutylene pipes up to 28 mm nominal bore.

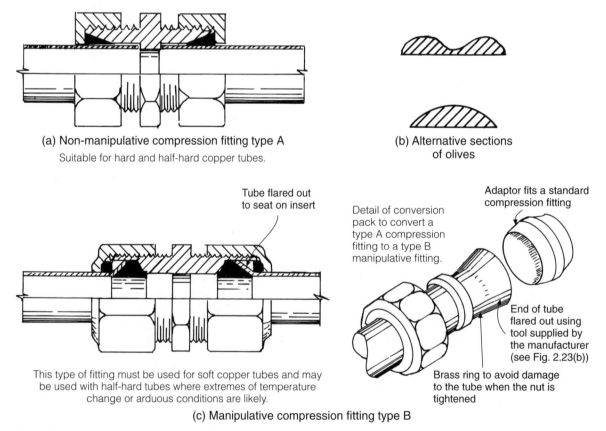

(a) Non-manipulative compression fitting type A

Suitable for hard and half-hard copper tubes.

(b) Alternative sections of olives

Tube flared out to seat on insert

Detail of conversion pack to convert a type A compression fitting to a type B manipulative fitting.

Adaptor fits a standard compression fitting

End of tube flared out using tool supplied by the manufacturer (see Fig. 2.23(b))

Brass ring to avoid damage to the tube when the nut is tightened

This type of fitting must be used for soft copper tubes and may be used with half-hard tubes where extremes of temperature change or arduous conditions are likely.

(c) Manipulative compression fitting type B

**Fig. 4.12** Copper compression fittings.

(a)

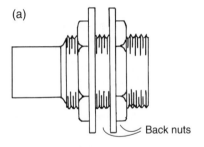

Back nuts

(b)

Back nut

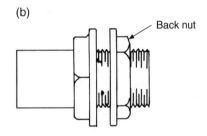

The thread here is one size larger than the nominal pipe diameter, e.g. Cu 22 mm × 1 in BSP. This provides a bell-mouth inlet giving good flow rates.

### (a), (b) Tank connectors

Back plate for screw fixing these fittings can be supplied with a built-in attachment where extra rigidity is required

Female BSP thread

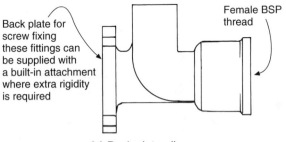

### (c) Back plate elbow

Suitable for connecting bib taps.

Female BSP end. Male thread type available

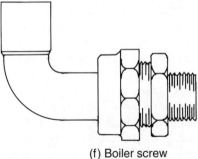

### (d) Drain cock tee

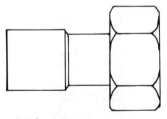

### (e) Straight tap connector

Also available with bent copper connection.

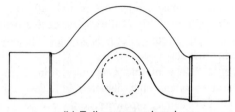

### (f) Boiler screw

Also available with female end. The union can be used on its own and those of 22 mm × 1 in BSP are useful as copper cylinder connections.

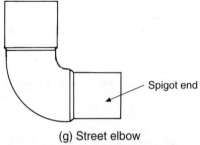

Spigot end

### (g) Street elbow

These are useful when a series of bends are required in a limited space. The spigot end, having the same OD as copper tube, can be used with any other fitting of the same size.

### (h) Full passover bend

These are time saving if the pipe to be passed over is of the same diameter of the pipe crossing it. If this is not the case the passover must be made using a bending machine.

**Fig. 4.13**   Copper fittings.

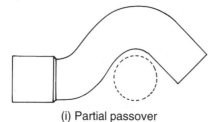

**(i) Partial passover**

As with the full passover this fitting is only suitable where the pipe shown as a broken line is of the same diameter as the pipe crossing it.

**Fig. 4.13** *continued*

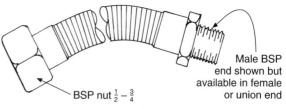

Male BSP end shown but available in female or union end

BSP nut $\frac{1}{2} - \frac{3}{4}$

**(j) Flexible union connector**

These are expensive but are time saving when used to connect pillar taps on baths, pedestal wash basins and bidets.

Therefore a 25 mm tee with the end reduced to 22 mm with a 15 mm branch would be ordered thus: 1 Tee, 25 × 22 × 15. The same information and method of ordering applies to tees for steel pipes.

Copper tubes to EN 1057 (BS 2871) are often used for sanitary pipework in public and commercial buildings where a long working life is essential. Copper is an expensive material, but its resistance to corrosion and its smooth internal bore make it very suitable for discharge pipework. Its rigidity and comparative light weight reduce the number of fixings necessary and also enable it to be fabricated off site and fitted as a complete unit. Individual sections of large units are then fitted on site using straight couplings.

A wide choice of methods for jointing large-diameter copper tubes is available. The capillary and compression joints illustrated are obtainable for pipes of up to 54 mm diameter. Compression fittings are not normally used as a means of making large-diameter joints owing to their high cost, but they are useful in cases where it is necessary to dismantle sections of pipework periodically for cleaning purposes, a typical example of this being discharge pipes fitted to urinals which frequently become obstructed.

A special form of solder capillary fitting is made for services having pipe diameters in excess of 54 mm. For sanitary pipework, joints can be fabricated by hand or by special forming tools using bronze welding or brazing to make the joints. An alternative is the use of large solder capillary fittings made of copper or cast bronze with brazed joints shown in Fig. 4.14.

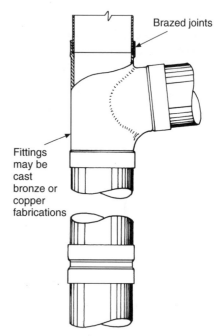

Brazed joints

Fittings may be cast bronze or copper fabrications

**Fig. 4.14** Typical brazing fittings for large-diameter (54 mm upwards) copper tubes. These fittings are sometimes used in public and commercial buildings where a long maintenance-free life is required. The type shown are for sanitary pipework, but similar fittings are available for use with large-diameter hot and cold water services.

**Cast iron**

For many years cast iron has been used as a material for both underground and overground sanitary pipework. Its use for sanitary pipework is confined to main discharge and ventilating pipes, the branch pipes being of copper or synthetic plastic materials. Pipes of less than 50 mm diameter are not normally produced, as cast iron is insufficiently flexible for direct connection to sanitary fittings other than WCs.

Cast iron is an alloy of carbon and iron, the carbon content being approximately 3%. Although it is brittle in comparison with other iron alloys, it is capable of taking some very hard knocks and will stand up to exceptionally hard wear. Its chief drawback is its weight, being a very heavy material to handle.

The standard length for overground use is 1.8 m, but shorter lengths can also be obtained, stock sizes being 1.2 m, 900 mm and 600 mm. Sanitary pipework is fabricated on site using the appropriate lengths of pipe and the choice of a wide variety of fittings.

Cast iron has mostly been superseded by PVC for overground sanitary pipework for small dwellings, but it is still extensively used on public and commercial buildings where its strength and rigidity are an advantage. Cast iron products are also used where alterations or repair work is necessary on listed buildings where the existing pipework is cast iron.

*Casting*   Cast iron pipes are made by sand or centrifugal casting. Sand cast pipes are cast with the socket in the downward position to give the metal greater density, enabling it to take the imposed stresses when it is caulked. Pipes made in this way are identified by the mould marks on the back and front of the casting, and a slight bow is often perceptible throughout its length owing to shrinkage of the metal on cooling. Centrifugally cast iron, sometimes called 'spun iron', pipes are made in a revolving mould, which as it turns throws the molten metal outward by centrifugal force, thus giving the casting as a whole greater density and a more even thickness to the pipe wall. Spun pipes are recognised by their more even texture and the absence of fixing ears on the sockets.

Traditionally cast iron pipes were treated with a bituminous solution by the manufacturers to protect against corrosion. Epoxy resin linings are now generally applied to the inside of all pipes to protect against the aggressive chemicals used in many cleaning agents. Pipes for use overground are painted externally with an anti-rust primer – usually red in colour. Underground pipework is protected externally with a coating of metallic zinc, which is then painted with acrylic paint – black or grey in colour.

When an iron casting has been made, internal stresses are set up in the metallic structure as it cools. These stresses are relieved by a method called normalising during which the casting is reheated to approximately 880 °C and allowed to cool slowly. During this process, the grain structure of the casting adjusts to a stress-free condition.

While it is very rare, a split or crack can occur in castings, either from a manufacturing defect or by careless handling. It is good practice to test castings before they are fixed, by holding them clear of the ground and delivering a light blow with a hammer. A clear ringing note will indicate that the casting is sound.

*Fixing*   Cast iron pipes are fixed to the structure of a building by ears cast on the socket, or where no ears are provided, a holderbat bracket may be built into or screwed to the wall. Special pipe nails are available for fixing, but the materials of which modern buildings are constructed are sometimes so hard that stout screws and plugs are used for fixing. The pipe should be fixed in such a way that it is spaced off the wall to permit the back to be painted. Special spacers can be obtained for this purpose but unless they are specified, short pieces of steel pipe can be used as a substitute.

*Jointing*   The traditional method of making joints in cast iron pipe is yarning and caulking with molten lead or lead wool (see Fig. 4.15(a)). Yarn or gaskin used in making these joints is a soft stranded hemp rope which may be used in its natural condition or may be impregnated with tar. (The tarred variety is best, the tar acting as a preservative.)

Lead-caulked joints are no longer permitted underground as they lack the flexibility required by the Building Regulations. Although they are permitted for above-ground drainage and sanitary systems, modern jointing methods have made the use of lead caulking obsolete.

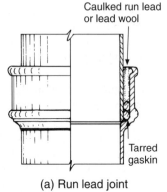

**(a) Run lead joint**

Note that this method of jointing has largely been superseded by flexible joints having rubber seals.

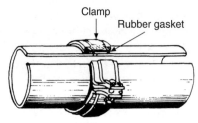

**(b) Glenwed 'time saver' joint**

This method of jointing cast iron pipe employs the use of plain ended pipes and fittings which are jointed with a rubber gasket secured by a cast iron clamp bolted in position.

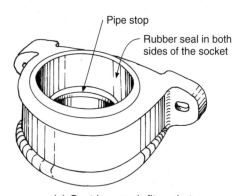

**(c) Cast iron push fit socket**

This is useful on alterations to discharge pipework in listed buildings as it has the same external appearance to a traditional BS 146 socket.

**Fig. 4.15**  Jointing cast iron pipes.

To achieve the flexibility required by the Building Regulations for underground drainage, various systems of synthetic rubber gasket jointing using plain-ended pipes have been developed.

Figure 4.15(b) shows a typical joint of this type which is also commonly used above ground.

One obvious advantage of this method of jointing is the ease with which new connections can be made to existing installations, where a complete section can be removed to facilitate the additional junctions required.

Figure 4.15(c) illustrates a joint employing synthetic rubber seals which has the same external appearance to that of the traditional socket used with caulked lead.

**Synthetic plastics**

*General characteristics*  The plastics industry has made a big impact over the last four decades in the supply of materials for the mechanical services industry. The advantages of these plastic materials are their light weight and the simple methods used in the jointing process.

The term 'plastics' is a very loose one, and can be applied to almost any material that can be manipulated. Synthetic plastics is a better term to use but even this can be confusing as only a few of these materials are widely used in plumbing. It is important for the plumber to be able to identify the type of plastic being used, as the jointing technique suitable for one type may be unsuitable for another. For this reason it is essential that the plumber is aware of a few simple tests and observations which will enable correct identification of a particular material. These simple tests will be described as each of the principal materials are discussed.

Plastic materials used in plumbing have many physical characteristics in common. Those used for pipes are invariably thermoplastics, all of which have a high resistance to corrosion and acid attack, a useful attribute especially where chemical or laboratory wastes are concerned. They have a low specific heat, a term which implies that they do not absorb the same heat quantity as metals. They are also poor conductors of heat.

One of the biggest drawbacks in the use of synthetic plastics is their high rate of expansion. Figure 4.16 illustrates the approximate comparisons of linear expansion between metal and plastic pipe. Manufacturers make allowances for this in the

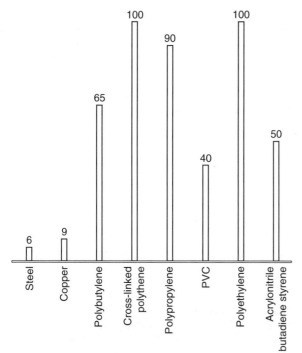

**Fig. 4.16** Expansion of commonly used plastic and metal pipes, showing the expansion in mm of a 10 m length of pipe raised through 50 °C.

design of their products, and providing the instructions for use of the material are followed, no problems from this source should be encountered during its service life.

Plastics are not such stable materials as metals and are all, to some degree, affected by the ultraviolet rays of sunlight. The long-term effect of these rays is to cause embrittlement, usually referred to as 'degradation'. Most plastics in their natural state are clear and colourless, and to offset the effects of degradation, manufacturers add colouring agents.

There are two main categories of synthetic plastics:

(a) thermosetting
(b) thermoplastics

Thermosetting plastics are generally used for mouldings, i.e. some types of flushing cistern shells and electrical equipment are typical examples of thermosetting plastics. Thermoplastics are those

which can be resoftened when heated. Most pipework materials used by the plumber fall into this category.

Most plastics are unaffected by strong acids or alkalis but they may be softened and permeated by hydrocarbons. This is not surprising as the basic elements used in the manufacture of these materials are in the hydrocarbon family, i.e. materials composed chiefly of the atoms of hydrogen and carbon. Petrol, oil, paint strippers and cellulose thinners are all hydrogen and carbon compounds which will soften and degrade thermoplastic materials.

The reader should note there are certain exceptions to this, e.g. fuel oils may be stored in vessels made of polythene. This is because there are many grades of polythene, these grades varying as to the basic material or 'polymer' of which a particular grade is made. Because of this vessels made of plastic should be clearly labelled with instructions for their use. Generally, however, unless there is no doubt about their suitability, components made of synthetic plastics should not be used with materials derived from hydrocarbons.

In a practical context cooking fats and oils are basically hydrocarbons, but while a domestic sink discharge pipe made of plastic will remain serviceable for many years, it would be an unsuitable material to use in a large commercial kitchen where vast amounts of hot water and grease would quickly lead to its degradation.

All synthetic plastics used for sanitary pipework are marketed as complete systems and it is not advisable to mix the components produced by one manufacturer with those of another, even if the basic material is the same.

Traditionally, synthetic plastics have not been used for hot water supplies owing to their relatively low softening point. Research and development have, however, enabled some types of plastics, namely cross-linked polythene (xLPE) and polybutylene (PB) to be used successfully for this purpose.

The method of making joints on plastic materials varies as to its type. Table 4.8 is not exhaustive but gives a good general guide to the types of joint suitable for the most common plastics used by the plumbing industry.

**Table 4.8** Jointing methods for synthetic plastic pipes in building services.

| Types of plastic | Mechanical joints* | Solvent welding | Fusion welding | Push fit 'O' ring (discharge pipes/overflows only) |
|---|---|---|---|---|
| Polythene | ✓ | ✗ | ✓ | ✗ |
| Polypropylene | ✓ | ✗ | ✓ | ✓ |
| Polybutylene and cross-linked polythene | ✓ | ✗ | ✗ | ✗ |
| Polyvinyl chloride | ✓ | ✓ | ✗ | ✓ |
| Acrylonitrile butadiene styrene | ✓ | ✓ | ✗ | ✓ |
| Cross-linked polythene | ✓ | ✗ | ✗ | ✗ |

*Note that mechanical joints may be combinations of 'O' ring seals, grab rings, rubber or metal olives

## Plastics commonly used for water services and sanitary pipework

*Polythene*  This is possibly the most commonly used plastic in the polyolefin group of plastics. Polythene pipes up to 150 mm nominal bore are made to BS 3284 and are classified as 'high' or 'medium' density. High-density polythene is capable of withstanding slightly higher temperatures and is more rigid than that of medium density, which is comparatively flexible and has a slightly lower melting point. It is available in straight lengths and is mainly used for chemical discharge pipework and cold storage cisterns.

The use and installation of polythene pipelines for water supply should conform to BS 6730 (for above-ground use) and BS 6572 (for underground use). Both these standards cover pipes from 20 mm to 63 mm OD. A standard is in preparation for pipes of larger diameters.

Pipes to both these standards are suitable for pressures up to 12 bar. Polythene pipes for water services above ground are black in colour and for underground use blue. Polythene in its natural state is colourless but owing to the ultraviolet rays of the sun it degrades and hardens causing it to crack. To prevent this a black colouring agent is added.

The colour blue was chosen for plastic underground water services to fulfil a commitment made by all public utility companies to enable their services to be readily identified, as several fatalities have occurred as a result of the mistaken identity of electrical cables. It is worth noting the different colours used for underground polythene services. They are as follows:

| | |
|---|---|
| Water | Blue |
| Gas | Yellow |
| Electricity | Black and Red |
| Telecommunications | Grey |

In comparison with some other plastics, polythene has low mechanical strength, a tendency to 'creep', i.e. expand the tube bore under pressure, and is very light, having a density of 900 kg/m$^3$. Medium-density polythene for water supplies is usually marketed in coils of 24, 50 and 100 m in length.

*Identification of polythene and polypropylene*
Polythene and polypropylene are both members of the polyolefin group of plastics. They have a waxy appearance and touch, and when ignited, burn with a steady flame very similar to that of a wax candle. Because most polyolefins used in plumbing are less dense than water, they can also be identified very easily as they will float, unlike many other synthetic plastics.

On freezing, water expands approximately 1/10th of its volume and this pressure can cause water-filled vessels or pipes to burst, but the elastic property peculiar to polythene makes it superior in its resistance to damage by frost. Polythene is unsuitable for jointing using solvent welding or adhesives. The most common jointing methods used are fusion welding, or one of the many special fittings developed by manufacturers for this type of pipe. Non-manipulative compression fittings as used for copper tubes (but one size larger than the nominal bore of the polythene pipe) can be used with an appropriate liner. Some of the many types of joints used with pipes in the polyolefin family of plastics are shown in Chapter 5.

*Polypropylene* This material is in the same family group as polythene. It is a tough, lightweight plastic of high rigidity and surface hardness. Its main characteristic is its ability to withstand high temperatures, and in this respect it is superior to PVC, ABS or polythene. Articles made of polypropylene can be used with boiling water for short periods of time. For this reason most manufacturers of plastic waste systems produce traps made of this material.

Some also produce a range of pipes made of polypropylene for discharge systems which are jointed by means of 'O' rings. Having similar characteristics to high-density polythene, it is also used extensively for the manufacture of cold water storage and feed cisterns. When used for pressure pipework, fusion-welded and mechanical joints are the principal method of jointing.

*Polybutylene* Being another member of the polyolefin family of plastics, it has similar characteristics to those previously mentioned, and pipework systems made of this material can be used for both hot and cold water supplies. Its melting temperature is 125 °C, so that, with some reservations, it can be connected directly to hot water equipment such as boilers and hot store vessels. As with polythene, if the pipe is frozen it will expand without bursting and revert to its original shape at temperatures above freezing point. It is a poor conductor of heat, as are most plastics, and this does have the advantage that hot water in deadlegs will not cool as quickly as that in metal pipes. Hot water and heating pipes should, however, be insulated to the same standards as those of copper and steel pipelines. Based on simulated tests, this material is claimed to have a life expectancy of 50 years at pressures between 12 bar at 20 °C and 6 bar at 90 °C when installed in domestic systems.

Should failure of a thermostat produce temperatures for short-term periods of up to 100 °C, the material will not fail, but its long-term life may the shortened.

Because pipes made of materials in the polyolefin family of plastics permit the permeation of oxygen through their walls, polybutylene is produced both with or without a protective barrier. Barrier pipes should be used in heating systems to prevent

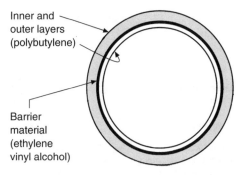

**Fig. 4.17** Section of laminated barrier pipe. This prevents the ingress of oxygen through the pipe walls. Its main use is for heating systems where no inhibiter is used.

corrosion where, for various reasons, the installer does not wish to use an inhibitor. Barrier pipe contains within its walls a material called ethylene vinyl alcohol (OVHO) which effectively prevents the ingress of gases shown in Fig. 4.17. As with other polyolefins, polybutylene can be identified in a similar way to that of polythene and polypropylene. The British Standard for this material is BS 7291 Parts 1 and 2. BS 5955 Part 8 relates to the installation of thermoplastic pipes for hot and cold water services.

*Cross-linked polythene* This material is another type of plastic that has been developed for both hot and cold water services. Its main characteristics are very similar to those of polybutylene. Tubes are obtainable in both coils of up to 100 m and straight lengths of 2 m or 3 m in nominal sizes of 15 mm and 22 mm. This material is designed to last for 50 years, but this is dependent on its working temperature. The fittings used with both this material and polybutylene may also be used on copper tubes. They are a mechanical joint employing the grab ring principle as shown in Fig. 4.11.

*Cross-linked polythene/aluminium pipe* This material is comparatively new in the UK but has been in Europe for several years. As with other materials in the polyolefin family of plastics, it is suitable for all water and heating installations. It is unique in that it is a laminated pipe (see Fig. 4.18(a)), the intermediate layer being

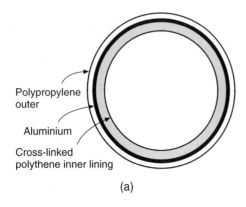

Polypropylene outer

Aluminium

Cross-linked polythene inner lining

(a)

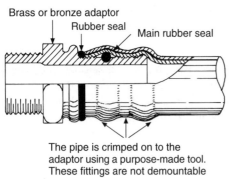

Brass or bronze adaptor

Rubber seal

Main rubber seal

The pipe is crimped on to the adaptor using a purpose-made tool. These fittings are not demountable

(b) Special joint used with some types of plastic/aluminium pipe

**Fig. 4.18**   Plastic/aluminium pipe.

aluminium giving it some degree of rigidity, unlike many other plastic pipes. The method of jointing is shown in Fig. 4.18(b). The material is marketed in straight lengths of 3 and 5 m or in coils up to 50 m long, and can be cut using similar tube cutters as those used with copper tube. The smaller sizes can be bent by hand, but for accurate work a bending machine, using the same principles as those used for bending LCS.

*Polyvinylidene fluoride (PVDF)*   This material has similar characteristics to those of high-density polythene where systems made of other materials would not survive. It is mainly used for industrial and manufacturing applications where its physical properties have advantages over other plastic materials. As with all other polyolefins it is unsuitable for solvent welding, the methods of jointing being the same as those used for polythene and polypropylene.

*Polyvinyl chloride*   Possibly the most common of the synthetic plastics used for sanitary pipework is PVC to BS 4514. This Standard requires that the material should not soften below 70 °C for fittings and 81 °C for pipes, and it should be capable of discharging water of higher temperatures for short periods of time. PVC, unlike many other plastic materials, will not burn easily, a fact which can be used for its identification. If a doubt exists, cut a thin strip from a piece of scrap pipe or gutter and hold over a lighted match. It will ignite initially and give off a heavy vapour, which excludes the oxygen to the flame causing it to be smothered. For this reason PVC is said to be self-extinguishing, and while it is not the only plastic having this property, it could reasonably be assumed on site that it is indeed PVC. It is obtainable either plasticised (softened) or unplasticised, the former being flexible and not used for building services. There are two types of unplasticised PVC, both of them being specified for water and chemicals above ground where there is a risk of impact damage. The requirements of standard PVC are covered by BS 3515 and BS 3506 which deal with 'high impact' strength pipe. The impact strength (ability to withstand a blow) of both materials is greatly reduced as temperatures approach 0 °C and great care must be taken when handling in cold weather or cracking will occur. For continuous operation the maximum temperature is around 60–65 °C and if this is exceeded the mechanical properties are lowered in a similar manner to polythene.

It is a light material (density 1,360 kg/m$^3$) that can be handled easily, a 6 m length of 75 mm diameter pipe weighing only 8.5 kg. Its major application in the plumbing industry is for sanitary pipework and rainwater goods. The maximum length of pipes for sanitation is usually listed as 4 m with diameters from 32 to 150 mm nominal bore. The usual method of jointing PVC pipe up to 50 mm is the use of solvent cement, the only exceptions normally being the connection between the discharge pipe and trap, using either a rubber olive and nut or an 'O' ring joint. Both of these allow for expansion. Connections to the main stack may be made using solvent welding, a rubber 'O' ring or olive, the latter two being used where provision for expansion is necessary.

*Acrylonitrile butadiene styrene (ABS)* This is a rigid material having a much higher impact strength than PVC at low temperatures. Its maximum operating temperature is 80 °C and it can be used for pressures up to 15 bar or 150 m head. Although a British Standard is not available for this material, it should comply with licence no. 7691 in respect of BS 5391 for pipes and licence no. 7962 in respect of BS 5392 for fittings.

Although it is not suitable for hot water services, it does have advantages over PVC in that it can withstand higher water temperatures for a longer period of time. It also retains its impact strength at subzero temperatures which gives it more resistance to mechanical damage. For these reasons many manufacturers produce a range of pipes and fittings for discharge pipework, although this material is more expensive than PVC. It is also marketed as a pressure pipe for cold water services but is not in general use for this purpose in the plumbing industry.

The usual method of jointing ABS is by solvent welding. It has a rather more matt exterior finish than PVC, which has a polished or shiny appearance. If it is ignited it burns with a bright white flame which gives off particles of carbon, very similar in fact to that of a welding blowpipe when it is burning only acetylene.

*Synthetic rubber* There are many types of synthetic rubber, some of which contain a proportion of natural rubber. The most common synthetic rubber used in plumbing is called Neoprene which is usually used for the manufacture of 'O' ring expansion joints. It is more stable against oxidation than natural rubber and is only affected slightly by oils.

Another of its main features is its resistance to sunlight, an important factor when pipes are situated externally. Unlike most types of rubber, it is flame resistant and self-extinguishing.

*Polystyrene* This material has many applications in industry for moulded components but its principal use in the mechanical services industry is for insulation where it is known as expanded polystyrene. It is obtainable in both sheet and sectional form for pipe insulation, is usually white

in colour and has the appearance of a coarse rigid foam. It is very light and brittle and must be handled with extreme care. During manufacture it is made in large blocks which are cut to size by an electrically heated wire. It is best cut on site with a very sharp knife. It has been superseded to some extent by alternative insulating materials owing to its high fire risk.

*Acrylic* This is one of the acrylic resin group of plastics which is obtainable in clear sheet form or mouldings where it is extensively used for sanitary appliances. It can be cut with a fine-tooth saw and drilled in the usual way, providing sharp drill bits are used. It is not so resistant to scratching as components made of metal, and extra care must be taken when it is handled during installation.

*Nylon* This is a very strong and tough thermoplastic material with very good heat-resistant qualities. It is one of the more expensive plastics to produce, which is probably why it is not produced in tubular form for plumbing services. It is extensively used, however, for small components in taps and valves, the seatings in float-operated valves being typical. It is also used for gearing and bearings in shower mixers and water meters, where due to its self-lubricating properties, little maintenance is necessary.

## Jointing materials

Joints made on plumbing components, e.g. threads, require some material to make good the small gaps between the mating surfaces. Even machined joints such as those on radiator unions are more reliable if a suitable jointing material is employed.

For many years the mechanical services industries used proprietary paste and hemp for this purpose, but because the oil used in their manufacture was generally vegetable based, it was found they supported the growth of harmful bacteria in pipelines. It was also found that joints on gas services using similar materials could be permeated by natural gas. For these reasons it is important that the correct jointing medium is used for the type of work to be carried out by checking the instructions on the container.

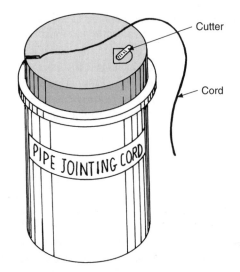

**Fig. 4.19** Nylon cord for threaded joints.

*Polytetra-fluoroethylene (PTFE)* This material is supplied in the form of white tape and is suitable for water services, including potable water and gas. It is used to seal threads on metal and plastic materials, is clean to use and is non-toxic. Unlike pastes it is not affected by oxidation, enabling any joint to be easily taken apart. The only reservation to its use is that the pipe threads must be in good condition, e.g. no stripped threads, and the male and female ends must be free from distortion. Another alternative to traditional methods is the use of a thin white multi-filament nylon cord with an inert coating. It is simply wrapped around the external thread of a joint in a similar way to a hemp and paste joint. It is also a very convenient material for repacking glands on taps and valves. It can be used with all metal threads and is suitable for gas, hot and cold water services and compressed air, see Fig. 4.19.

### Jointing pastes

These are best described as thick pastes and have traditionally been used with hemp to fill the gaps between threads. In fact if the threads are well formed, hemp should not be necessary. It was never good practice to use it on steam services, as at high temperatures it could burn out.

For potable water special pastes have been developed based on silicone compounds and filler that will not support bacteria. They can be used with a special plastic fibre which swells when wet and is used as a substitute for hemp. Most pastes that are suitable for potable water are also suitable for gas heating and low-pressure steam services. Traditional paste-jointing compounds are still available but they must not be used for potable water services. The inclusion of graphite to a compound increases its lubricative qualities and enables joints to be broken more easily. Manganese-based compounds are especially suitable for flange joints, as they set more quickly and the joints are harder.

Special jointing compounds are necessary for joints on oil pipelines; oil-based pastes are not suitable as fuel oils will soften and permeate through the joints. Suitable compounds include finely ground ferrous oxide (rust) which is bound with alcohol and phenolic resins. Owing to their alcoholic content they are flammable and give off a harmful vapour. For this reason they should be used with care and in a ventilated environment.

### Soldering fluxes

*Fluxes* The object of using a flux is to prevent oxidation of the joint during the soldering process. Oxygen in the air will oxidise and tarnish metals, and soldered joints cannot be made satisfactory under such conditions. One of the major prerequisites of soldering is the work must be thoroughly clean.

Fluxes traditionally used by the plumber are dealt with more fully in Chapter 5, where they are related more specifically to the requirements of sheet weatherings using soft-solder alloys of tin and lead. The use of tin, copper, and in some cases tin/silver solders, can cause problems due to their high melting temperatures in connection with some grease-based fluxes. What actually appears to happen is the grease burns at these high temperatures, which tends to blacken the copper pipes and fittings being soldered. It must be said that in many cases the fault is sometimes due to the flame being too large, and in the case of butane or propane blowlamps the pressure may be too high. If the equipment is provided with a regulator, try reducing the pressure, which will result in a softer flame giving a more even heat, and will, in many cases, solve the problem. Flux manufacturers are aware of this and many are now producing fluxes which are more compatible with the higher temperatures required for soft-solder alloys used for hot and cold water supplies.

Grease-based fluxes, not being soluble in water, are not recommended for pipe jointing whatever type of solder is used, as they have been found to be a common cause of corrosion. Paste fluxes which are water soluble are now readily available, and any residue may be washed out with water when the work is commissioned.

Fluxes in general may be divided into three main groups:

(a) Strong oxide-dissolving fluxes made with organic salts, i.e. zinc, ammonium and stannous chlorides which are sometimes combined with various acids. Most fluxes used for capillary joints are of this type and a close look at the ingredients should make the reader realise how important it is to remove any flux residues after a joint is soldered.
(b) Less active fluxes made of stearic and lactic acids and helagons, e.g. various hydrochlorides.
(c) Non-aggressive resin usually based on undissolved natural pine resin. These fluxes are more suitable for use with soldering irons, (copper bits) rather than a flame.

### Soft solders
Lead-free solders have been dealt with earlier in this chapter. Those described here are traditional plumbing solders, but because they contain lead they must not be used for water work supply. They are mainly used for odd soldering jobs on non-ferrous metals and repairs to sheet metal weatherings. Solders containing lead are manufactured to BS 219 and are an alloy of tin and lead, although for some solders small amounts of antimony are also included. Tin is very expensive and to reduce costs antimony is used as a substitute, but this can only be included in amounts of up to 6 per cent of the tin content or the properties of the alloy are affected. Antimony must never exceed 0.5 per cent of the tin content in solders to be used for zinc and alloys of zinc such as brass. A wide

**Table 4.9** Common types of solder used by the plumber.

| BS 219 grade | Melting point (°C) | | Composition (%)* | | |
|---|---|---|---|---|---|
| | Solid | Liquid | Sn | Pb | Sb |
| D (Wiping solder) | 185 | 248 | 30 | 68.3 | 1.7 |
| F (Bit solder) | 183 | 212 | 50 | 49.5 | 0.5 |

*Sn = tin; Pb = lead; Sb = antimony

range of solders are available but those detailed in Table 4.9 are the most common types used by the plumber, Grade D being used for the repair of lead gutters, flats and wiped solder dots and Grade F for general purpose copper-bit work.

When solder is in a liquid state both tin and lead are molten. On cooling, crystals of the excess metal begin to form in the molten solder and this provides a consistency between liquid and solid states that is referred to as 'pasty' or 'plastic'. This plastic condition is essential for the formation of wiped joints where the plumber has to have sufficient time to shape the solder. In other solders, such as those used for zinc, shaping is not necessary and the alloy is required to set as soon as it is deposited on the parent metal. The cooling periods can be seen in the equilibrium diagram shown in Fig. 4.20, the temperature range between liquid and solid being called the 'plastic range'. It can easily be seen that a wide plastic range is necessary for wiping solder but for copper-bit work a solder is needed which will set as the copper bit moves forward.

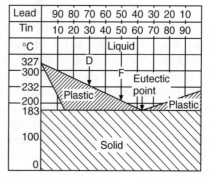

**Fig. 4.20** Equilibrium diagram for lead/tin alloys.

Equilibrium diagrams are used to record the changes that occur during the solidification and subsequent cooling of alloys. In Fig. 4.20 'D' indicates the proportion of lead and tin in Grade D wiping solder. It should be noted that it remains plastic over a wider temperature range than Grade F solder ('F' in the diagram). An increase in the proportion of tin in the alloy reduces its plastic range. The *eutectic point* in an alloy indicates two things:

(a) its lowest melting point
(b) the point at which it solidifies without going through a plastic range

It will be noticed from a study of the diagram that the *eutectic point* has two important characteristics.

(a) An alloy of 63 per cent tin and 37 per cent lead has no plastic range; it both melts and solidifies at the same temperature.
(b) The lowest temperature at which an alloy of these two metals will melt is 183 °C. (The word *eutectic* does in fact mean 'easily melted'.)

## Corrosion

Corrosion is a chemical or electrochemical action which causes the decay or 'eating-away' of metals. Chemical corrosion is mainly caused by oxidation, acidic or alkaline attack. Electrolytic corrosion, another form of corrosion, is not readily appreciated. In a practical situation both chemical and elecrolytic corrosion often take place simultaneously.

*Electrolytic corrosion*
A simple experiment demonstrating electrolytic corrosion is illustrated in Fig. 4.21 where the placing of the two metals in a water-filled beaker is shown to produce a simple electric cell or battery. Electrically charged particles called ions flow from cathode to anode through the electrolyte producing a current that will register on the ammeter. This action eventually causes the destruction of the anode.

The rate of electrolytic corrosion is increased if the electrolyte is acidic or the two metals are widely separated in the list of metallic elements shown in

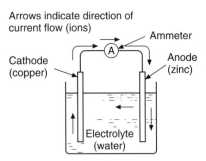

**Fig. 4.21** Electrolysis. The flow of electrically charged ions from the anode to the cathode eventually results in the complete destruction of the anode. A similar situation can exist if two dissimilar metals are used together in weathering or water supply work. Water acts as the electrolyte and the anode is destroyed.

**Table 4.10** Electrochemical series in metals used in plumbing.

| HIGH | Magnesium |
| | Aluminium |
| | Manganese |
| | Zinc |
| | Chromium |
| | Iron |
| | Nickel |
| | Tin |
| | Lead |
| LOW | Copper |

Table 4.10. The higher placed metal always forms the positive electrode (anode) and the lower the negative electrode (cathode). A practical application is the use of sacrificial anodes in hot water storage vessels.

*Oxidation*
Oxidation is a form of corrosion that involves the combination of the oxygen in the air with another substance and can be seen when freshly cut metal is exposed to the atmosphere. The cut sample quickly 'dulls' due to the action of oxygen producing a surface coating which is an oxide of the metal involved, e.g. copper forms copper oxide. The action of moisture and certain chemicals in the atmosphere such as carbon and sulphur dioxide on the metal leads to the formation of a 'skin' that is commonly referred to as a *patina*.

With non-ferrous metals, i.e. those that do not contain iron, the patina adheres to the metal forming a protective coating which is effective in preventing further attack. In the case of most ferrous metals, the skin (rust) formed by oxidation flakes off, leaving a new, unprotected, metallic surface beneath open to further corrosion.

### Dezincification

The action of certain water and soil conditions (usually in soft-water areas) can cause dezincification of duplex brass which is a zinc-rich alloy. This is a particular form of electrolytic corrosion owing to the fact that zinc and copper are widely separated in the electrochemical series. An aggressive water or soil provides an electrolyte which accelerates the destruction of the anodic metal (zinc) leaving the copper porous and brittle – the alloy has dezincified, i.e. has had its zinc content removed.

To overcome this problem there are several alternatives. Gunmetal, an alloy of approximately 85 per cent copper, 5 per cent tin, 5 per cent lead and 5 per cent zinc, may be used. This is an expensive alloy and is normally only used for heavy-duty fittings. Compression fittings made of brass but specially treated, and referred to as dezincification resistant, are a cheaper alternative. Such fittings are identified by the letters DR (dezincification resistant) marked on them. The most commonly used fittings are, however, made of copper, which is unaffected by dezincification.

From the foregoing it will be seen that caution must be exercised if pipework components and fittings of differing metals are used in the same installation. Special care must be taken to avoid using, for example, galvanised LCS with copper, especially in soft-water areas.

## Protection of materials against corrosion

Corrosion may be described as the gradual eating away of a material by chemical action.

The nature of the water to be conveyed, the chemical nature of the ground in which it is to be laid, or the environment above ground, will influence the choice of a suitable pipe material to minimise the effects of corrosion.

### Steel

Black steel (i.e. not galvanised) is not acceptable for hot and cold water supply as it is liable to corrosion both internally and externally. Water supply pipes must be galvanised but gas supplies can be conveyed in black steel pipes although a non-corrosive external covering is required for buried pipes. Most new gas service pipes, however, are now made of polythene, identifiable by its bright yellow colour, and for use in the building, copper tubes have superseded steel except for large industrial or commercial installations.

### Copper

Almost all waters are 'cuprosolvent', i.e. liable to dissolve copper, this property being due to the same conditions that cause plumbosolvency. Only rarely does cuprosolvency reach a level where it is harmful to health, but if the copper content of water is in excess of $1.5 \text{ mg/3 cm}^3$ an unsightly green strain is deposited on sanitary fitments. When used underground in certain soils, corrosion can occur and it may be necessary to wrap with a protective tape, such as a self-adhesive plastic type or that made of cotton, impregnated with petroleum jelly and an inert silicious filter (Denso tape is a well-known trade name for this type of material). An alternative is to use plastic-coated pipe where copper tubes are to be installed in plaster, brickwork or concrete.

### Stainless steel

This metal is unaffected by corrosion with the exception of water containing high chloride concentrations which affect the protective oxide film. This can be prevented by using metal having a higher chromium content such as austenitic 18/8 steel. It should be noted that good-quality stainless steel is non-magnetic.

### Plastic pipe

All plastic pipes are unaffected by the normal corrosive effects of water and soil conditions. Where brass compression fittings are used underground for jointing, it will be necessary to wrap them for protection against corrosion.

## Flow characteristics of pipe materials

An important consideration when designing pipework installations is the *frictional resistance* caused by the pipe walls. This resistance can be appreciated by comparing the speed of someone running in a race in the normal manner with the speed of the same person running a race while in contact with a wall. Obviously the result of rubbing on the wall will cause a considerable reduction in speed. Pipe walls have the same effect on water flow, i.e. the water is 'rubbing' on the pipe, and this is further influenced by the rough or smooth nature of the material. The effect of this is known as 'frictional resistance' or frictional loss. Where the influence of smooth and rough pipe surfaces is compared, the flow of water from a 25 mm diameter pipe under a head of 10 m would be approximately 22 litres/minute for copper and plastic pipes having smooth internal surfaces, and 18 litres/minute for galvanised LCS which has a rougher inner surface.

The effects of friction and subsequent loss of head or pressure are also influenced by the number and types of fittings in the pipe run. This resistance is measured in terms of equivalent pipe length. To give one example, the frictional loss caused by water passing through a 28 mm elbow will be the same as a length of the same diameter pipe 1 metre long. Where possible machine-made or large-radius bends should be used as their resistance to the flow of water is negligible.

## Pipework fixings

Care must be taken to ensure pipes are adequately supported according to their strength and rigidity. Details of the horizontal and vertical spacing of the fixings and a selection of the various types of fixings available are dealt with in Chapters 6 and 7 relating to cold and hot water supplies.

## Water storage vessels

Cisterns, tanks and cylinders are all containers used for the storage of water. Suitable definitions of these terms are as follows.

*Cistern*   An open top circular or rectangular cold water storage vessel with a loose lid or cover.

*Tank*   A closed rectangular vessel used for the circulation and storage of hot water or oil.

*Cylinder*   A closed cylindrical vessel with parallel sides and domed ends. Used for the circulation and storage of hot water.

All vessels used for water storage must be constructed of a material that will not impart taste or odour and be able to resist any corrosive action by the water.

### Steel water storage vessels

*Cisterns*   Cisterns of galvanised steel are manufactured to BS 417 and are available in capacities ranging from 18 litres up to 3,364 litres. Capacities of cisterns can be referred to as *actual* or *nominal*, the latter being based on calculations using extreme overall dimensions, while the actual capacity takes the depth to be that of the water below the water line. They are made in Grades A and B which relate to the thickness of the steel sheet used, and are of either riveted or welded construction. Loose covers are supplied in the same material but of a lighter gauge. Greater capacities are catered for either by coupling two or more of these vessels together or by using sectional steel cisterns, which can be built into any required size from 1 metre square sections. The sections have flanged edges which are bolted together and are erected on site. These cisterns, which are made to BS 1564, must be protected by galvanising, metal spraying or by bonding rubber or plastic coatings to the plates and fittings.

### Plastic storage vessels

Plastic is the most recent category of material to be utilised for the production of storage vessels. The type of plastic could be glass fibre, polythene or polypropylene. The British Standard dealing with polythene and polypropylene cisterns is BS 4213. Actual capacities of these cisterns range from 18 litres to 455 litres and they are comparatively light in weight. These vessels are not affected by electrolytic corrosion and are able to resist the effects of freezing because of their elastic properties. Some can also be flattened so that passage can be gained through a confined space and being black in colour are resistant to algae growth. Polythene cisterns are usually circular but rectangular patterns can be obtained made of polypropylene, this material being slightly more rigid.

*Cylinders*

Cylinders of galvanised steel are manufactured to BS 417 in a range of capacities from 73 litres up to 441 litres. The metal thickness is related to maximum working head pressure, Grade A (3.2 mm), Grade B (2.5 mm) and Grade C (2.0 mm) being suitable for heads of 30 m, 18 m and 9 m respectively. These cylinders are of the direct pattern but there is also an indirect type made to BS 1565. The nominal water content of the indirect pattern may be from 136 litres up to 455 litres. Further details of the two types of cylinder are given in Chapter 4.

The only vessels commercially manufactured from sheet copper are hot water storage cylinders. They must comply with BS 3198 to ensure the maximum treat transference from the heat exchanges. Cylinders are specified in accordance with the maximum working head of the system, three grades being available for heads of 25, 15 and 10 metres: Grades 1 to 3 respectively. Capacities of cylinders to BS 699 are from 74 litres to 450 litres, although *preferred sizes* having capacities of 116, 120, 144 and 166 are those normally stocked for installation in new dwellings. In addition to these four sizes, 98 and 200 litre capacities are also available.

Specially adapted cylindrical storage vessels for unvented hot water systems are available made of steel, stainless steel or copper. They are in effect a self-contained unit incorporating the pressure vessel and the associated controls. Those made of steel are protected internally by a plastic coating bonded on to the steel which ensures a long service life.

**Storage calculations**

Calculations in respect of water storage are undertaken by the plumbing designer who takes into consideration the legal requirements involved, the number of consumers, type of building, rate and regularity of supply, and the consequence of exhausting the storage supply. Some reference is made to these factors in Chapter 8, but the subject of design is outside the scope of this book.

The plumber is, however, required to undertake calculations which involve the following:

(a)  specifying the dimensions of a vessel to hold a given quantity of water
(b)  calculating the capacity of a storage vessel from given dimensions
(c)  deriving the mass of a storage vessel either empty or when filled

To solve these problems certain factors must be understood. One cubic metre (1 m$^3$) of water contains 1,000 litres. One litre of water has a mass of 1 kg, therefore 1 m$^3$ of water has a mass of 1,000 kg. Diameter (D) is the distance across the centre of a circle (or cylinder), and the radius (r) is half this distance, while the circumference is the distance around the extreme edge of a circle.

Pi ($\pi$) is equal to 3.142 and is the number of times the diameter of any circle can be divided into the circumference.

The following formulae are used in calculating the size and capacity of a vessel.

*Area of a rectangle*   Multiply length by breadth.
*Circumference of a circle*   Multiply $\pi$ by the diameter.
*Area of a circle*   Multiply $\pi$ by the radius squared ($\pi r^2$).
*Area of a cylinder side*   Multiply circumference by height.
*Volume of rectangular vessel* (i.e. a tank)   Multiply area of base by depth of vessel (usually in metres).
*Volume (m$^3$) of cylindrical vessel*   Multiply area of end by length.
*Mass of material in vessel*   Multiply total volume (m$^3$) by mass/unit volume (i.e. water has a mass of 1,000 kg/m$^3$).
*Capacity of vessel in litres*   Multiply volume (m$^3$) by 1,000.
*Mass (kg) of water in vessel*   Multiply capacity (litres) by 1 kg (i.e. 1 litre of water has a mass of 1 kg).

*Examples*

1. A galvanised steel storage cistern measures 1.5 m by 2.0 m and is 1.0 m deep. Calculate the capacity of the vessel in litres when filled to the top edge.

$$\begin{aligned} \text{Volume} &= \text{Area of base} \times \text{depth} \\ &= 1.5 \text{ m} \times 2.0 \text{ m} \times 1.0 \text{ m} \\ &= 3.0 \text{ m}^3 \\ \text{Capacity} &= \text{Volume} \times 1{,}000 \\ &= 3.0 \times 1{,}000 \\ &= \underline{3{,}000 \text{ litres}} \end{aligned}$$

(Note that this would have a mass of 3,000 kg.)

2. A copper hot water cylinder measures 1.8 m high and is 0.6 m in diameter. Calculate the mass of water in the cylinder.

$$\begin{aligned} \text{Volume} &= \text{Area of end} \times \text{height} \\ &= \pi r^2 \times \text{height (radius is half of} \\ &\quad \text{diameter, i.e. 0.300 m)} \\ &= 3.142 \times 0.3 \times 0.3 \times 1.8 \\ &= 0.51 \text{ m}^3 \\ \text{Mass} &= \text{Volume} \times 1{,}000 \text{ kg} \\ &= 0.51 \times 1{,}000 \\ &= \underline{510 \text{ kg}} \end{aligned}$$

(Note that the capacity of the cylinder is 510 litres since 1 litre of water has a mass of 1 kg.)

3. A galvanised LCS storage cistern has a nominal capacity of 3,000 litres and base measurement of 1.0 m by 2.0 m. What is the depth of the vessel?

$$\begin{aligned} \text{Capacity} &= \text{Volume} \times 1{,}000 \\ 3{,}000 &= \text{Volume} \times 1{,}000 \\ \therefore \text{Volume} &= 3{,}000 \div 1{,}000 \\ &= 3 \text{ m}^3 \\ \text{Volume} &= \text{Area of base} \times \text{depth} \\ 3 \text{ m}^3 &= (2 \times 1.0) \times \text{depth} \\ 3 \text{ m}^3 &= 2 \times \text{depth} \\ \text{Depth} &= 3 \div 2 \\ &= \underline{1.5 \text{ m}} \end{aligned}$$

Sizes and capacity can also be obtained by reference to the British Standards for each type of vessel.

## Further reading

Much useful information can be obtained from the following sources:

BS EN 877 Cast iron pipes and fittings, their joints and accessories for the evacuation of water from buildings. Requirements, testing, quality control.

BS EN 1329-1 Plastics piping systems for soil and waste discharge (low and high temperature) within the building structure. PVC-U Part 1 Requirements for pipes, fittings and the system.

BS EN 1451-1 Plastics piping systems for soil and waste discharge Polypropylene (PP) Part 1.

BS EN 12666-1 Plastics piping systems for soil and waste discharge Polyethylene (PE) Part 1.

*Copper fittings and tubes*

Mueller Industries Inc., Oxford Street, Bilston, West Midlands, WV14 7DS. Tel.: 01902 499700.

Peglers Ltd, PO Box 182, St Catherines Avenue, Doncaster, South Yorkshire, DN4 8DF. Tel.: 01302 329777.

Yorkshire Imperial Metals, PO Box 166, Leeds, LS1 1RD. Tel.: 0113 2701104.

*Steel pipes and fittings*

George Fischer Ltd, Paradise Way, Coventry, CV2 2ST. Tel.: 02476 535535.

*Plastic pipes for water services*

Durapipe S & LP, Horton Cranes, Cannock, Staffs, WS11 3HS. Tel.: 01543 279909.

Geberit Ltd, Aylesford, Kent, ME20 7PU. Tel.: 01622 717811.

George Fischer Ltd – see Steel pipes and fittings.

John Guest Ltd, Horton Road, West Drayton, Middlesex, UB7 8JL. Tel.: 01895 449233.

Hepworth Plumbing Products, Hazelhead, Crow Edge, Sheffield, S36 4HG. Tel.: 01226 763561.

*Plastic materials for sanitary pipework*

Caradon Terrian Ltd, Aylesford, Kent, ME20 7PJ. Tel.: 01622 778811.

Hepworth Plumbing Products – see Plastic pipes for water services.

Hunter Building Products, Nathan Way, London, SE28 0AE. Tel.: 0181 855985.

Marley Plumbing, Dickley Lane, Lenham, Maidstone, Kent, ME20 7PJ. Tel.: 01622 85888.

*Plumbing solders and fluxes*
Fry's Metals, Tandem House, Harlow Way,
 Beddington Farm Road, Croydon, Surrey,
 CR9 4XS. Tel.: 020 8665 6666.

*Jointing pastes*
Fry Technology UK Ltd – see Fry's metals.
Loctite UK Ltd, Watchmead, Welwyn Garden City,
 Herts, ALY 1JE. Tel.: 01707 358800.

*Cast iron drainage*
Glenwed Drainage Pipes and Fittings, Sinclair
 Works, PO Box 3, Ketley, Telford, Shropshire.
 Tel.: 01952 262500.

## Self-testing questions

1. Explain the term 'alloy' in relation to metals.
2. List three alloys used in plumbing and name their constituents.
3. State the colour used to identify LCS pipes made to BS 1387 and relate the colours to the wall thickness.
4. Explain the term 'galvanising' and how it is used to protect steel against corrosion.
5. Explain why stainless steel has a high resistance to corrosion.
6. Explain why the maximum working pressure is important when selecting a hot storage vessel.
7. A cold water storage cistern measures 2.0 m long × 1.5 m wide × 0.8 m deep. Calculate its actual capacity in litres when the water level is 0.075 m below the top edge.
8. Describe the difference between manipulative type B fittings and non-manipulative type A compression fittings used for copper tubes. State which of these can be used on (a) R220 (Table Y) and (b) R290 (Table Z) tubes.
9. Explain the term 'dezincification' of fittings used for water supply.
10. Describe two methods of producing cast iron pipes for discharge pipework and underground drainage.
11. The following are all methods of making joints on synthetic plastic pipes:
    (a) synthetic rubber 'O' rings
    (b) fusion welding
    (c) solvent welding
    (d) mechanical joints employing grab rings or rubber seals
    Which of these methods may be used for (i) polybutylene and cross-linked polythene water services and (ii) PVC discharge pipework? Which of these jointing methods permit pipework to expand and contract?
12. (a) State the purpose of a flux when making soldered joints.
    (b) Specify the type of flux recommended for copper pipework installations.

# 5 Working processes

After completing this chapter the reader should be able to:

1. Have a working knowledge of the Building Regulations.
2. Understand the basic principles of scales and construction drawings.
3. Describe the various methods of making soft-soldered joints and the processes employed.
4. Select suitable methods of jointing for all types of pipework materials.
5. Describe the basic principles of making bends in various pipework materials.
6. State the working principles of pipe-bending machines and their requirements for maintenance.
7. Describe how to adjust the working parts of pipe-bending machines to produce acceptable bends.
8. Understand the purpose of taps and their use in tapping and threading.

## The Building Regulations

Most construction operations must comply with these Regulations, some of which affect work carried out by plumbers and heating fitters. The following text is intended to introduce this important subject, but for detailed information the reader should refer to the relevant approved documents.

Until 1875 few localities had any formal control over building construction. However, owing to the appalling state of housing, a Public Health Act was passed which empowered all local authorities to make bylaws relating to construction. These bylaws were designed to protect the public in relation to safety, fire prevention, health and sanitation in all types of buildings. Unfortunately these bylaws varied from area to area and this presented many difficulties for architects and building contractors, especially those who operated in different parts of the country.

In order to rationalise the system the first National Building Regulations came into force in 1966 replacing all the local authority bylaws, except those in London which were exempt until circa 1972.

The actual Regulations are written in legal language not easy to understand and interpret by the layperson and from 1985 they were linked with a series of 'approved documents' which explain and illustrate very clearly how they should be interpreted to meet the legal requirements.
The revised edition documents are as follows:

A   Structure
B   Fire safety
C   Site preparation and resistance to moisture
D   Toxic substances
E   Resistance to the passage of sound
F   Ventilation
G   Hygiene
H   Drainage and waste disposal
J   Heat producing appliances
K   Stairs, ramps and guards
L   Conservation of fuel and power
M   Access and facilities for disabled people
N   Glazing

Approved documents G, H, J, L are those that mainly apply to the work of plumbers and heating fitters. Document B, Fire safety, is also important, relating to flues and hearths for heating boilers and fires.

### Materials and work

Regulation No. 7 of the Building Regulations requires any work to be carried out in an efficient manner using the appropriate materials. Its range applies to any work carried out in connection with all the approved documents. It can be assumed that materials complying with the standards listed in Chapter 4 will meet the requirements of this Regulation.

### Drawings

The recommendations for construction drawings are the subject of BS 1192 Part 1 which deals with general principles. Figures 5.1(a)–(d) show the usual way in which building drawings are presented. This type of drawing is called 'orthographic', and as it shows the true face of the plan and sides (called elevations), they can be drawn to scale thus avoiding cluttering up the drawing with a host of dimensions and dimensional lines. This figure is drawn in what is called the third angle as shown by the symbol. Orthographic drawings can be produced using the first angle method which is similar but differs slightly in how the elevations are shown. For more detailed information on construction drawing practice a study of BS 1192 Part 1 will be helpful.

### Using a scale rule

A scale represents the dimensions of an object or distance proportional to its actual size. To illustrate this in very simple terms Fig. 5.2 shows a line 100 mm long. If this line is reduced proportionally to a scale of 1:5, its length will be exactly one-fifth of its original length but would be shown on a scale rule as 100 mm. The use of scales enables a designer to show a component or structure to be shown in a convenient format, e.g. working drawings which enable a craftworker to take accurate measurements for transfer to a full-size job. Most building projects are drawn to a reduced scale but there are cases, especially in engineering, where a very small component may be shown to an enlarged scale. For example, a scale of 2:1 would mean the drawing is twice the size of the actual component.

Figure 5.3 shows the ground floor plan of a small dwelling house drawn to a scale of 1:100. To practise using a scale rule the reader should check the measurements shown.

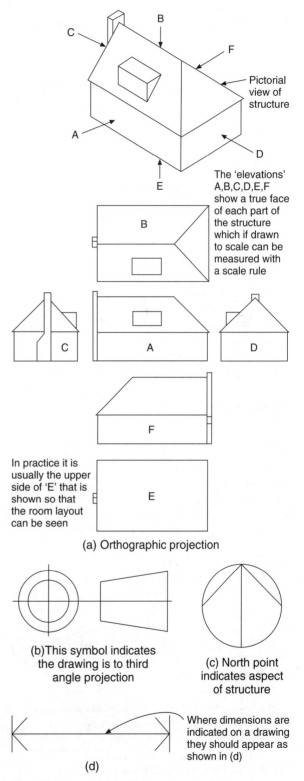

Pictorial view of structure

The 'elevations' A,B,C,D,E,F show a true face of each part of the structure which if drawn to scale can be measured with a scale rule

In practice it is usually the upper side of 'E' that is shown so that the room layout can be seen

(a) Orthographic projection

(b) This symbol indicates the drawing is to third angle projection

(c) North point indicates aspect of structure

(d)

Where dimensions are indicated on a drawing they should appear as shown in (d)

**Fig. 5.1** Construction drawings and symbols.

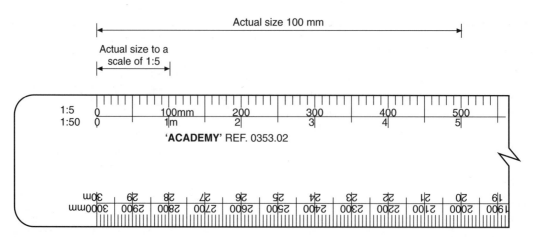

**Fig. 5.2**   Use of a scale rule.

This figure illustrates the ground floor of a small dwelling drawn to a scale of 1:100. Using a pair of dividers or good-quality compasses and the scale rule illustrated, check the measurements shown.

Measurements in metres

| | |
|---|---|
| Dining room | 4.0 × 3.5 |
| Kitchen | 3.5 × 2.8 |
| Lounge | 4.5 × 5.0 |
| Hall | 5.0 × 2.30 |

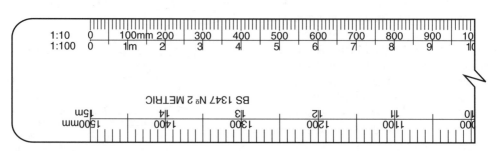

**Fig. 5.3**   Practise using a scale rule.

BS 1192 Construction Drawing Practice Part 1 quotes the approved scales for construction scales as follows:

| | | |
|---|---|---|
| Location drawings | 1:2500 | Usually copies from an Ordnance Survey map to |
| Block plans | 1:1250 | indicate the position of a proposed building in an existing area |
| Site plans | 1:500 1:200 | Show general layout of a site in relation to an existing area but may indicate more details such as drains, sewers and public utility services. |
| General location | 1:200 1:100 1:50 | These scales are generally used for working drawings and show dimensions of rooms and positions of, for example, sanitary appliances, rainwater pipes and drains. |
| Component drawings | 1:50 1:20 | Show individual specific parts of a drawing where more detail must be shown. |
| Detailed and assembly drawings | 1:20 1:10 1:5 1:1 | May show specific details, e.g. position of valves, water services, and layouts of sanitary pipework where extreme accuracy in detail and setting out is required. |

## Connections to existing lead pipe

The use of lead pipe for water services in post-war housing has been rare, mainly because of its high costs and the fact that other materials, especially copper tubes where they are exposed, are neater in appearance and easier to keep clean. Quite apart from the foregoing, because lead is a poisonous substance and soft acidic waters are capable of dissolving it, public health authorities have been averse to its use not only for water pipes but also in paint and leaded petrol. Since 1986, water authorities have forbidden the use of lead pipes not only for new work but for repairing existing services. To replace all lead water services would be too costly but as their service life comes to an

end they must be replaced with copper or a suitable plastic material. Irrespective of any regulations, all water authorities treat water to ensure that lead will not be taken into solution and it is extremely unlikely that consuming water supplied through lead pipes is likely to result in any fatalities.

### Lead to copper mechanical joints

Where repair work becomes necessary and the consumer cannot afford to replace all the lead pipes in the system, the most economic method is to use one of the mechanical joints shown in Figs 5.4(a) and (b) that are available for making connections to the new copper or plastic pipe used to replace the defective section. The ends of the fitting for copper or polythene pipes are made with compression, or push fit joints. Connections to existing lead pipes present greater difficulty as lead, being soft, is not a suitable material on which to make a compression joint; it would simply pull out when subjected to longitudinal stress and any lateral movement would almost certainly result in a leaking joint.

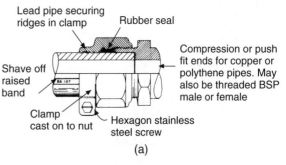

(a)

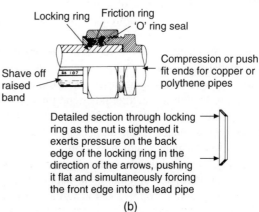

(b)

**Fig. 5.4**   Connections to existing lead water services.

These problems are prevented in the two fittings illustrated. Figure 5.4(a) shows a clamping arrangement which is tightened when the compression nut has been screwed home. The ridges inside the clamp bite into the lead when the hexagon screw is tightened, preferably with a suitable spanner rather than a screwdriver. This type of fitting is excellent for underground use.

Figure 5.4(b) shows a similar fitting having a different method of positively securing the lead pipe. As the nut on the lead end of the fitting is tightened, the conical-shaped locking ring is forced flat causing the inside edge to bite into the lead pipe thus preventing any lateral withdrawal of the pipe.

In all cases where lead pipe is jointed using these fittings, a few simple points must be observed:

(a)  The lead pipe must be as round and straight as possible, the seating area being free from deep dents or score marks.
(b)  The pipe must be cut square and any burrs removed. Entry into the fitting will be assisted if the outer edge is slightly chamfered.
(c)  The raised bar showing the BS code of the pipe should be shaved off to ensure it is a sliding fit into the socket; it must not be loose or slack.
(d)  Fittings having a locking ring have a positive stop when the nut is fully tightened. This eliminates overtightening and ensures the locking ring secures the lead.

**Soft-soldered joints – general principles**

Soft soldering is a process of joining two metals together by the use of lead/tin and copper or silver/tin alloys. Only the latter can be used for water supplies. Soft solder has a lower melting point than the metals being jointed. The process is different to welding where the edges of the metals to be jointed are brought to melting point and then fused together. In soldering the filler material flows between the unmelted metal surfaces to which it chemically adheres combining to form a compound 'intermetallic' layer as indicated in Fig. 5.5. Tin is a metal that is capable of combining with certain other metals such as copper, lead and zinc to form an intermetallic compound and this fact forms the basis of soft-soldering alloys.

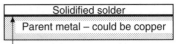
Intermetallic layer of solder and parent metal brought about by chemical action in the soft soldering process

**Fig. 5.5**  Detail showing the adhesion of soft solder to parent metals.

The three preliminary stages in forming a soldered joint are:

(a)  thoroughly clean the surfaces
(b)  apply a flux
(c)  heat the joint to the desired temperature

It is vital for each of these three stages to be carried out precisely so they are described in detail.

(a)  Cleaning is achieved by the use of steel wool, wire brush, file or, in the case of lead sheet and pipe, a shavehook. The aim is to mechanically remove all grease, dirt, traces of corrosion and oxide film, as solder will not adhere to non-metallic substances.
(b)  Applying the flux has four basic functions as follows.
    (i)   To assist in the 'wetting' ('tinning') process in which the solder is made to flow easily over the surfaces being joined, allowing bonding or alloying with the parent metal at every point. If 'wetting' does not take place the solder will remain on the surface as beads or globules, and even if the surface 'held together' the joint would not be sound or watertight.
    (ii)  To chemically clean the surface of the parent metal by removing or reducing any oxide film that remains after mechanical cleaning. (Reduction is the reverse chemical process to oxidation, i.e. it is the removal, rather than the addition, of oxygen.)
    (iii) To prevent metal surfaces from oxidising while they are heated to soldering temperature when they are more reactive towards oxygen.
    (iv)  To float away any foreign matter from the soldering operation.
(c)  Heating the joint to the required temperature is achieved on pipe joints by the use of oxy-fuel gas, propane or natural gas blowpipes.

## Jointing copper tube

Copper tubes have thinner pipe walls than most comparable materials. Copper is light, corrosion resistant and has a much smaller outside diameter than most other tube materials of a similar nominal bore. Except for those of fully annealed temper, copper tubes are quite rigid and do not require continuous support. This rigidity also permits them to be spaced off any surface to which they are fixed enabling the surface to be cleaned and painted.

The three main types of fittings used for making joints in copper tubes have been described in Chapter 4 but the methods of jointing, using these fittings, are described here.

### Capillary solder joints

Probably the capillary solder fittings are the most popular owing to their reliability, neatness of appearance and low cost in comparison with most fittings of the compression type.

The tolerance between the tube and fitting is vitally important, and if either the tube or fitting is distorted, capillary action will not take place and result in a leaking joint. Reference should be made to Chapter 11, Figs 11.2–3, which explain the basic principles of capillarity. When working with fully annealed tube, which is usually supplied in coils, the tube is slightly flattened and before a joint is made the ends of the tube should be restored to their correct bore using a mandrel which may be obtained from the tube manufacturers. Figure 5.6 illustrates the method used to make solder capillary joints. When the joint has been completed, any surplus solder should be wiped off with a dry cloth before it solidifies. Do not forget to remove any surplus flux when the joint is cooled.

Straight joints in copper tubes can be formed by making capillary sockets on plain ends of tube by using a socket-forming tool as described in Chapter 2. The soldered joint is made by using the same techniques as those used for end feed fittings.

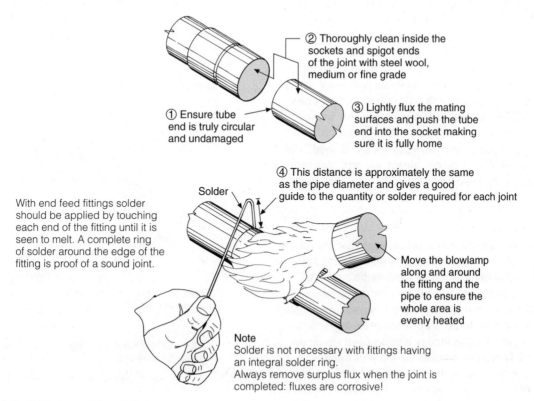

① Ensure tube end is truly circular and undamaged

② Thoroughly clean inside the sockets and spigot ends of the joint with steel wool, medium or fine grade

③ Lightly flux the mating surfaces and push the tube end into the socket making sure it is fully home

④ This distance is approximately the same as the pipe diameter and gives a good guide to the quantity or solder required for each joint

Solder

With end feed fittings solder should be applied by touching each end of the fitting until it is seen to melt. A complete ring of solder around the edge of the fitting is proof of a sound joint.

Move the blowlamp along and around the fitting and the pipe to ensure the whole area is evenly heated

Note
Solder is not necessary with fittings having an integral solder ring.
Always remove surplus flux when the joint is completed: fluxes are corrosive!

**Fig. 5.6**   Soldering capillary fittings.

Unlike lead/tin solder alloys, joints made with tin/copper solders are difficult to remake after the joint has been soldered. It is essential that the pipe and fitting are correctly located before the joint is made, as it will be almost impossible for alterations to be made later.

### Compression joints

Reference should be made to the illustrations shown in Chapter 4 and Table 4.7 which indicate the type of fittings which can be used on the three main types of copper tube used in the industry.

*Manipulative and non-manipulative joints* The manipulative type of compression joint will resist high internal pressures due to the swaged end formed on the tube. Under service conditions the tube will split before the fitting pulls off. When making these joints, do not forget to put the nut on the tube before the swaged end is formed.

Non-manipulative joints, as their name implies, require no swaging on the tube end. They rely instead on the compressive effect of the nut on the olive for their soundness. Figure 5.7 illustrates the method of making these joints.

Although most fitting manufacturers do not recommend any form of jointing paste to be used on these joints, this may be found necessary in practice, especially with fittings with a hard brass olive. PTFE tape is very suitable for this purpose.

Two very important points of preparation must always be remembered whatever joint is used. The tube must always be cut square and any burr formed internally by cutters or externally by a hacksaw must be removed by a file or reamer.

### Branch joints

When branches are inserted into existing copper pipework, it is sometimes difficult to spring the pipe apart sufficiently to admit the new fitting. 'Slip' straight joints and tees of the capillary type are available for this situation and have no internal stop, allowing them to be slipped along the pipe. When using one of these fittings it is necessary to make sure it is centralised over the cut pipes before making the joint in the usual way (see Fig. 5.8).

Where it is necessary to cut into existing supplies and make new joints it is essential to remove any

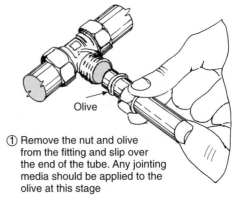

Olive

① Remove the nut and olive from the fitting and slip over the end of the tube. Any jointing media should be applied to the olive at this stage

② Push the tube into the fitting making sure it is pressed fully home against the stop

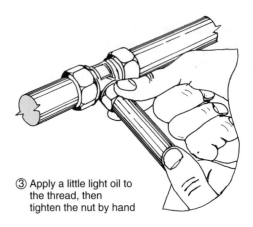

③ Apply a little light oil to the thread, then tighten the nut by hand

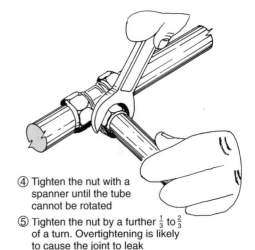

④ Tighten the nut with a spanner until the tube cannot be rotated

⑤ Tighten the nut by a further $\frac{1}{3}$ to $\frac{2}{3}$ of a turn. Overtightening is likely to cause the joint to leak

**Fig. 5.7** Making non-manipulative compression joints.

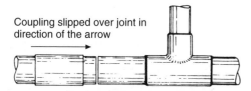

Coupling slipped over joint in
direction of the arrow

**Fig. 5.8** 'Slip' straight coupling for copper tubes.
Coupling has no stop which permits it to slide over
the pipe. Ensure centre of fitting is over meeting point
of pipes.

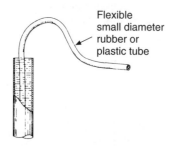

Flexible
small diameter
rubber or
plastic tube

**Fig. 5.9** Removing water by siphonage from a pipe prior
to soldering.

traces of water that may be laying in the pipes.
Water will prevent the solder from reaching melting
point and thus result in a defective joint. In such
circumstances a compression joint may be used but
if for some reason this is not desirable there are two
ways of overcoming the problem:

(a) Attaching a hose over the cut end of the pipe
    and blowing the water out through a tap
    downstream of the cut.
(b) Where there is no such convenient tap, water
    can be withdrawn from the pipe as shown in
    Fig. 5.9.

### Jointing steel pipe

The smaller sizes of steel tubes used in plumbing
for hot and cold water services are almost invariably
joined by cutting and screwing using BSP (British
Standard Pipe) threads to BS 21. Pipes are usually
threaded on site by hand dies or power machines,
and when using the latter the appropriate safety
regulations must be strictly observed. A thread cut
on the outside of a pipe is called a male thread
while the internal thread in a fitting is referred to
as female.

*Threaded joints*

The cutting head on threading machines and some
types of hand dies is adjustable. Each set of dies is
designed to cut two or more thread sizes, and being
adjustable they allow for a little wear to take place.
Care must be taken to ensure that the marks on the
face of the die are correctly set so they will remove
the correct amount of metal from the pipe. Failure
to do this will result in the thread being too loose or
too tight in the fitting. When the size of the dies are
changed, the numbers 1 to 4 stamped on them must
correspond to the same number on the die stock to
avoid damaged or stripped threads. Some stocks
employ block dies, one die being required for every
tube diameter. These are not adjustable and when
they become worn must be replaced.

As a comparatively large amount of metal is
removed when a thread is cut, to avoid the dies
becoming overheated or threads stripped, cutting oil
or paste must be freely applied during the threading
operation.

The normal type of dies are designed to cut a
tapering thread which when screwed into the socket
of the fitting causes it to expand slightly. This
ensures that both thread surfaces are in close
contact, making a watertight joint. A well-formed
thread cut with correctly adjusted dies should enable
a parallel threaded fitting to be screwed on the pipe
by hand for three to four threads. When it has been
fully tightened, only two or three threads should be
exposed.

Care should be taken to avoid damage to the
fittings with grips and wrenches during the
tightening operation. With elbows or tees a piece
of pipe may be screwed into the socket at 90° to
the joint being made, the fitting being tightened
home using this short length of pipe. In this
way the excessive use of grips can be avoided
(see Fig. 5.10).

Prior to making these joints the male thread is
smeared with a proprietary jointing paste before
being tightened home. Some plumbers use hemp
with the jointing paste but this should not be
necessary if good threads are cut and new fittings
are used. If hemp is used any 'whiskers' remaining
visible on the joint after tightening should be
removed with an old hacksaw blade. Hemp should
not be used for joints on potable water pipes.

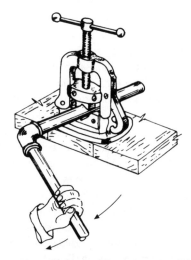

**Fig. 5.10** Tightening steel fitting where possible without the use of pipe grips. Use a short length of pipe as a lever.

A cleaner method of making joints is the use of PTFE tape, or PTFE/nylon-based string, which although suitable for normal joints on gas, water and steam pipes, should not be used for making joints where the thread is slightly distorted. A typical instance where this occurs is the welded connection fitted into sheet steel radiators. The welding operation may distort the thread slightly and often the only effective method of making a watertight joint in such a situation is by using jointing paste and hemp. PTFE has a high resistance to chemicals and a working temperature range of −80 °C to 250 °C. It has a very low coefficient of friction which enables the joints to be tightened up easily.

LCS pipes are cut with a hacksaw or wheel cutters. When a hacksaw is used the work should be rigidly clamped in a pipe vice to avoid breaking the saw blade. The application of a little cutting oil or paste will lengthen the blade life and reduce the manual effort required to cut the pipe. Wheel cutters, due to their compressive action, leave a burr in the bore of the pipe and unless this is removed it will obstruct the flow of water or gas. All cut ends should be deburred with a tapered reamer, most threading machines having one fitted as standard. If hand methods are used a suitable reamer may be obtained for fitting into the chuck of a carpenter's brace (see Chapter 2, Fig. 2.5(e)).

*Flange joints*

Joints on large steel pipes and in industrial installations where pipework may have to be periodically renewed are often joined with flanges which are more easily removed than screwed joints (see Fig. 5.11(a)). The flanges are fitted to the pipe with BSP threads, or may be fusion welded provided the pipe is not galvanised. The material used as a gasket between the flanges will depend upon the fluid the pipe is carrying. Insertion rubber is suitable for many applications and consists of a sandwich of one or more layers of loosely woven

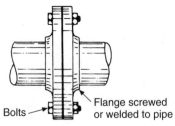

(a) Flange joint can be used with large size steel, copper, and plastic pipe installations

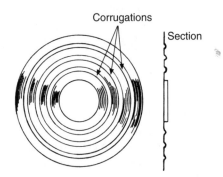

(b) Corrugated rings for flange joints

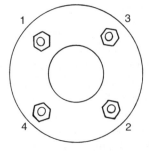

(c) Order of tightening nuts to avoid stressing the flange

**Fig. 5.11** Flange joints.

linen between heat-resistant rubber. It is also possible to obtain synthetic rubber sheets from which suitable gaskets may be cut, this material having no need for reinforcement linen. Purpose-made corrugated brass rings are another alternative (Fig. 5.11(b)). The corrugations give slightly on tightening and make a sound joint.

When flanges are bolted up, the correct sequence of tightening the nuts is important to ensure an even pressure on mating faces. Figure 5.11(c) shows the order of tightening a small flange with four bolts. Assuming the nuts are all hand tight and the flange faces and the pipes are correctly aligned, the nuts should be turned approximately one revolution at a time through the sequence shown. This will avoid undue stress at any one point in the joint.

## Jointing synthetic plastics

The most common of these materials in use for underground water services is medium-density polythene, methods of jointing being fusion welding or compression/mechanical joints made of gunmetal, high-density polythene or polypropylene. Fusion welding, requiring specialist equipment, is mainly used on gas and water main work and, in some cases, chemical discharge pipework.

British Standards list two grades of medium-density polythene tubes coloured blue for underground use and black for installation above ground. Liners are inserted into the end of the tube, their purpose being to support the tube against the crushing effect of tightening the cone. Typical compression fittings for adapting polythene to a male iron thread are shown in Figs. 5.12(a) and (b).

To make a joint on (a) the nut is placed over the end of the pipe and the liner inserted, a flange on the liner preventing it being pushed too far into the bore. The olive or cone is then pushed over the end of the pipe before it is inserted into the socket of the fitting. The nut should be tightened as far as possible by hand and completed by one and a half turns with a spanner. Do not overtighten as this will cause the edges of the cone to cut into the polythene. When the joint is tight it should not be possible to twist the pipe within the fitting.

Figure 5.11(b) shows a push fit connection. Unlike simple 'O' ring joints of the type used for discharge pipes, which would not be capable of withstanding high internal pressures, this joint is designed to tighten as the water pressure increases. Reference to the illustration shows that owing to the tapering shape of the body any increase in pressure tends to force the 'grip' and 'O' ring more tightly into the body. To make the joint, the polythene tube should be bevelled (a special tool is available for this purpose) and a mark made on the pipe to indicate the length of pipe to be inserted into the fitting. Do not make the mark with a sharp instrument. The liner is pressed into the bore of the pipe, the pipe end then being inserted into the socket with a slight twist. When the pipe is fully home the original mark on the pipe should correspond to the end of the fitting. As further proof of a sound joint, two points of resistance will be felt as the pipe is pushed through the grip and 'O' rings.

Should it be necessary to dismantle the joint, special extractor tools must be used. They are inexpensive and available from the fitting supplier. The fitting illustrated has a bronze body but those for connecting plastic to plastic are made of high-density polythene, or polypropylene.

The connection shown in Fig. 5.12(a) is a copper compression joint which is adapted for connection to polythene pipes. An alternative is the use of fittings made entirely of plastic, usually high-density polythene or polypropylene. The body can be used with a variety of inserts to accommodate connections to different pipe materials. The type shown in Fig. 5.12(c) shows the fitting with inserts for copper and medium-density blue underground water service pipe.

### Plastic pipes for hot and cold services

The two main materials that have been developed for this purpose are polybutylene and cross-linked polythene, the latter being used extensively by specialist firms for underfloor heating. Both types of pipe have the same external diameter as copper tube and can be adapted for direct connection to this material. The above-mentioned plastics have similar characteristics relating to pressure resistance and maximum working temperatures. Until recently the maximum diameter available was 22 mm but one company is currently producing fittings and tube of 28 mm OD.

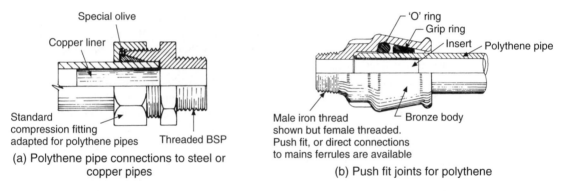

(a) Polythene pipe connections to steel or copper pipes

(b) Push fit joints for polythene

(c) Polygrip coupling

A standard body can be used for most connections between a wide range of differing pipes, e.g. lead, copper and various grades of polythene. Adaptors for this material are colour-coded to ensure the correct insert is selected. The illustration shows the adaptors necessary to make a joint beween R250 (Table X) copper tube and blue medium-density polythene.

**Fig. 5.12** Pipe connections.

The basic principle of joints used for these materials is best described as mechanical. They are push fit and employ what is called a grab ring which is designed to embed into the pipe when the joint is made. On no account should the fingers be pushed into the joint – the grab ring will ensure it is difficult and painful to remove them!

The water seal in all joints of this type is an 'O' ring made of heat-resistant synthetic rubber. When cutting pipes on any plastic in the polyolefin group, a tool of the type shown in Fig. 2.6(b) should be used as it produces a clean square cut.

Prior to making a joint using these materials a support sleeve must be inserted into the end of the pipe to give a greater degree of rigidity at the joint.

Figure 5.13 shows a joint marketed by Hepworth Building Products using the Hep 'O' system. It is made of polybutylene and is marked throughout its length as shown, which enables the installer to check visually that a satisfactory joint has been made.

The rubber seal makes the joint watertight while the grab ring is employed to prevent the withdrawal of the pipe under pressure. These fittings are designed for use with cross-linked polythene but they may also be used as a substitute for traditional fittings on copper tube.

It is important that the pipe is pushed fully home into the fitting as failure to ensure this will result in a serious leak when the system is filled with water under pressure.

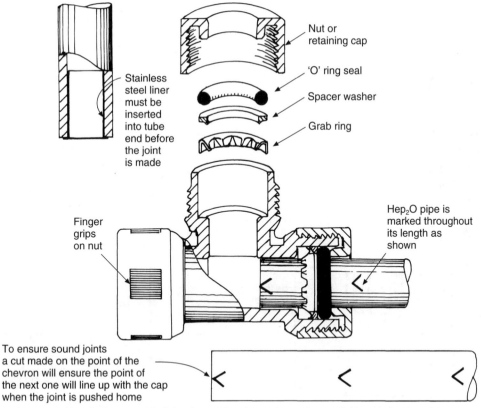

Stainless steel liner must be inserted into tube end before the joint is made

Nut or retaining cap

'O' ring seal

Spacer washer

Grab ring

Finger grips on nut

Hep₂O pipe is marked throughout its length as shown

To ensure sound joints a cut made on the point of the chevron will ensure the point of the next one will line up with the cap when the joint is pushed home

Hepworth Hep₂O demountable fitting for polybutylene pipes; also available with fixed retaining caps.

**Fig. 5.13**

*Connection to other pipework material*
In cases where polybutylene or cross-linked polythene tube has to be joined to steel pipes, a range of adaptors, having male or female BSP threaded ends, are marketed. Where it is necessary to connect them to medium-density polythene, e.g. the entry of a water service into a building, stop valves are available having the inlet end adapted for medium-density polythene pipe, the outlet end being suitable for direct connection to polybutylene or cross-linked polythene. Because these materials have a much better resistance to high temperatures than other plastics they can, in some cases, be connected directly to boilers. Figure 5.14 illustrates the recommendations of Hepworth Building Products.

Although both these materials are relatively rigid, any bends made will tend to spring back to their original shape because, like polythene, they have a plastic memory. Any bends made will be of a large radius and they must be well secured with suitable

pipe clips. Hepworth Ltd produce a special former illustrated in Fig. 5.15 which not only provides a good fixing for bends but enables a smaller radius to be used.

Both types of pipe have sufficient rigidity for surface mounting for very short runs although more fixings must be employed than for metal pipes. However, because plastic pipework of any material lacks the mechanical strength of metal, it should not be used in circumstances where it could be subjected to mechanical damage. Long exposed surface-mounted runs of pipe are not recommended, as while they may be fitted in neat straight runs, an increase in temperature will cause them to expand and undulate (look wavy) which is unsightly.

Figures 5.16(a) and (b) show manufacturers' recommendations which allow for expansion when installing crossed-link polythene and polybutylene pipes for water services in ducts and intermediate floors. Provision for expansion must be at the

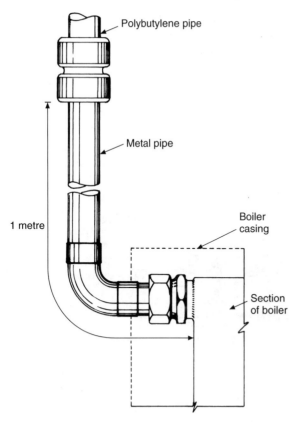

**Fig. 5.14**   Polybutylene connections to boilers. Metal pipework must be used if the connections are inside the boiler casings, if the heat source is within 350 mm of the polybutylene pipe or if the boiler has no positive thermostatic control, e.g. solid fuel boilers.

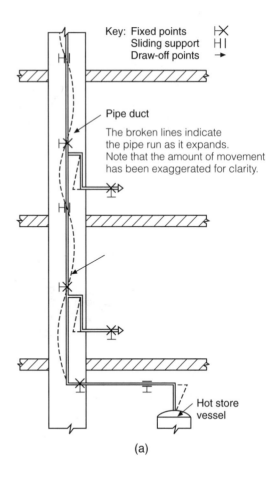

(a)

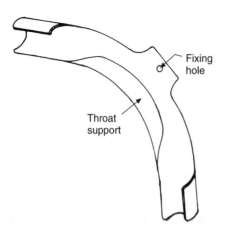

**Fig. 5.15**   Steel former for supporting bends made with Hep 'O' polybutylene pipe.

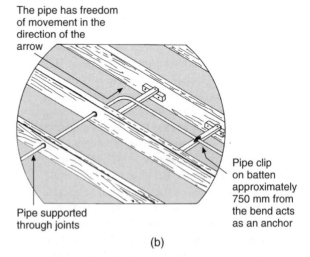

(b)

**Fig. 5.16**   (a) This shows how provision is made for expansion in hot water services using cross-linked and polybutylene pipes. (b) Support for plastic pipe in intermediate wooden floors.

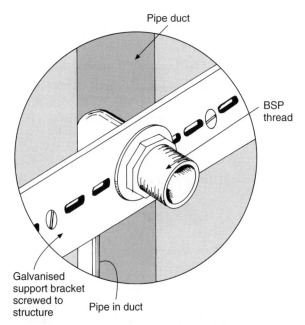

Pipe duct

BSP thread

Galvanised support bracket screwed to structure

Pipe in duct

**Fig. 5.17**   Support for plastic pipe connections to appliances.

forefront of the installer's mind when fitting these materials in order to avoid serious problems at a later date. When used in solid floors they should be run in a duct to enable them to be withdrawn if necessary. The flexible polythene ducting shown in Chapter 7, Fig. 7.31(a,b) is an ideal material to use for this purpose. Because plastic pipework used for water services lacks the rigidity and strength of metals, special brackets will be required to support taps and valves; the general principles of such a support is shown in Fig. 5.17.

*Fusion-welded joints in plastic pipe*
One of the most satisfactory ways of jointing polythene for water and gas mains is by means of fusion welding. Special tools containing an electrical heating element are used for this purpose and a pictorial illustration of a hand-held tool for smaller pipe sizes is shown in Fig. 5.18(a). By changing the bushes and spigot heads a variety of pipe sizes can be jointed. All these tools are designed for use with an electrical intake control box. This enables accurate assessment of the temperature and the period of joint heating time to be made automatically; an indicator light shows when the correct joint temperature is achieved.

Obviously a larger joint will require a longer period of time to achieve fusion temperature and this can be varied by the timer on the control box.

The voltage on which these tools operate varies, but for on-site work it should not exceed 110 volts. In the event of there being no mains supply a portable generator can be used. To make a satisfactory joint the tube end should be cut square with a suitable cutter and any burrs or woolliness adjacent to the cut removed with a sharp knife. The surface film caused by the ultraviolet rays of light must then be removed as it has a detrimental effect on fusion-jointing processes. It may be scraped off with a sharp knife but a more accurate method is to use a cutting tool as shown in Fig. 5.18(b).

After first ensuring the spigot and bush are of the correct size for the pipe being jointed, a check should be made to ensure they are quite clean and dry. The socket of the fitting and pipe end must then be positioned on the spigot and bush and the machine switched on. The temperature and heating recommendations must be followed exactly. When the indicator light shows, the fitting and pipe are removed from the tool and the pipe pushed firmly into the socket of the fitting. All fusion joints must be allowed to cool completely before a pressure test is conducted, and a general rule is to wait at least an hour after the final joint has been made. A diagrammatic illustration showing how these joints are made can be seen in Fig. 5.18(c).

Another method of making fusion-welded joints is shown in Fig. 5.18(d). A special fitting is used which has an electrical heating element moulded integrally in each socket. The joint is made by connecting a low-voltage electrical supply to the heating element made integrally in the socket. This heats and fuses the mating surfaces of the socket and the pipe simultaneously. These joints are equally successful on both high- and medium-density polythene and are also used for polypropylene, a plastic used in industry where a more rigid material than polythene is required.

A method of fusion welding using a hot inert gas such as nitrogen or hot air is sometimes employed to make joints on thermoplastics. It has some industrial uses but is not generally used in the plumbing industry because of the special equipment required and the readily available alternatives which are more suitable for site work.

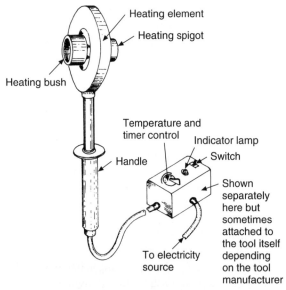

(a) Pictorial illustration of a fusion-jointing tool for polythene and polypropylene

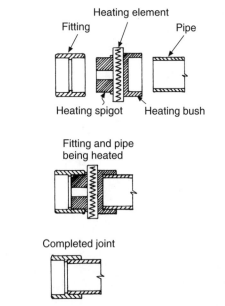

(c) Fusion-jointing polythene and polypropylene

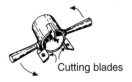

(b) Pipe-peeling tool for plastic pipe

To remove the surface film this tool is rotated round the end of the pipe.

**Fig. 5.18**   Fusion-jointing equipment.

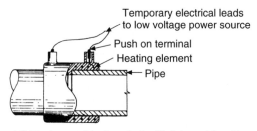

(d) Fusion-welded socket with integral heating element for polythene or polypropylene

### Jointing plastics used for discharge pipework systems

Solvent welding is the usual method of jointing PVC and ABS discharge pipes. A typical example of a solvent-welded joint is shown in Fig. 5.19. When making a joint both the internal surface of the socket and the pipe end should be degreased with cleaning fluid. The mating ends are then lightly brushed with solvent welding cement and pushed home. This cement is not an adhesive; its action is to permeate the surfaces to be joined and soften them, thus permitting fusion of the surfaces. On exposure to the atmosphere the liquid content of the cement evaporates causing the joint to harden. Any surplus cement must be quickly removed as soon

as the joint has been made or it will solidify and be almost impossible to remove without damage to the joint.

This method of jointing is best described as a cold welding process. Although the cement sets in a very short period of time, 12 to 24 hours should elapse before the joint is subjected to full working

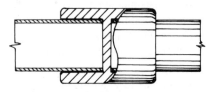

**Fig. 5.19**   Solvent welded joint.

pressures. When joints of this type are used for pipes having a large diameter a slower setting cement is used to enable the larger surfaces to be covered before the cement solidifies.

These joints obviously make no provision for expansion within the sockets, but as such short lengths of pipe are normally used on a well-designed system, any expansion is usually taken up by bends and the compression joints on traps connections. Most systems, however, include a special expansion coupling where excessively long pipe runs cannot be avoided.

*'O' ring joints*    The predominant type of joint used for long runs such as the main discharge pipe is the 'O' ring which, by leaving a gap of approximately 12 mm between the spigot and the base of the socket (see Fig. 5.20), provides for any expansion when hot water is being discharged, or for any variation in air temperature which may occur. The socket of each length is secured with a bracket and acts as an anchor, so that the expansion is accommodated within the joint. The pipe must be capable of moving freely through any intermediate brackets.

The method of making 'O' ring joints is similar whatever type of plastic material is used. When it is necessary, pipes should be cut with a fine-tooth saw, a hacksaw being most suitable for this purpose. The spigot of the pipe to be inserted should be chamfered and marked with a pencil to determine the length of insert into the socket allowing for the expansion gap. When cut it should then be lubricated with a silicone based lubricant prior to being pushed into the depth of insert indicated by the predetermined mark. Always make sure that the 'O' ring is not distorted and is secure in its groove before making the joint. PVC and polypropylene are suitable materials for this method of jointing.

*Adjustable bends*    To provide greater flexibility with their material many manufacturers produce adjustable bends (see Fig. 5.21(a)). These enable a standard bend to be cut and fitted with a special adaptor, thus allowing bends to be fabricated to any angle.

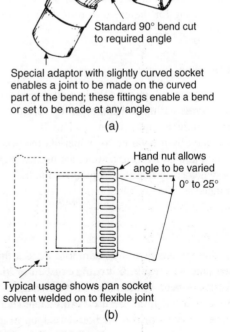

Standard 90° bend cut to required angle

Special adaptor with slightly curved socket enables a joint to be made on the curved part of the bend; these fittings enable a bend or set to be made at any angle

(a)

Hand nut allows angle to be varied

0° to 25°

Typical usage shows pan socket solvent welded on to flexible joint

(b)

**Fig. 5.21**    Adjustable bends.

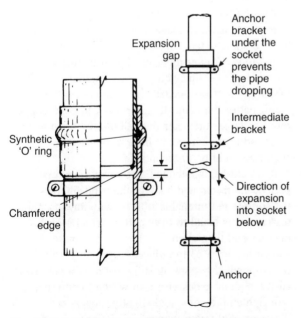

Expansion gap

Anchor bracket under the socket prevents the pipe dropping

Intermediate bracket

Synthetic 'O' ring

Chamfered edge

Direction of expansion into socket below

Anchor

**Fig. 5.20**    'O' ring joint. This method of joining PVC pipes provides for freedom of movement to allow for the effects of expansion.

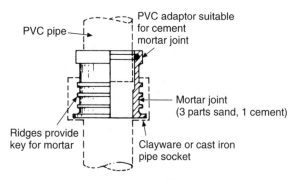

**Fig. 5.22** Adaptor for PVC discharge pipes to pipes of other materials.

Figure 5.21(b) shows a flexible joint for 100 mm nominal bore discharge pipes. It is made of PVC and is designed to allow flexibility between the angle of a WC outgo and the branch discharge pipe into which it is fitted. They are especially useful when connecting to a range of 'p' trap WCs where, owing to the fall on the 'float' or main branch, the angle of each WC branch pipe will vary slightly.

*Jointing to cast iron and clayware* Plastic discharge pipes are joined to cast iron or clayware drains by means of PVC connecting pieces, which provide for a cement or mastic joint to be made while allowing for the expansion of the plastic pipe (see Fig. 5.22).

## Pipe bending

One of the most important skills a plumber should possess is the ability to bend pipes of various materials quickly and accurately. This is complicated by the variety of materials the plumber handles and the differing techniques used to make bends whether by hand or in purpose-made machines.

## Bending by hand

It will be obvious that any attempt to bend a thin-walled pipe will result in it being crushed or kinked before it has been pulled through many degrees unless certain methods are used. To enable such pipes to be bent, various methods of supporting or filling the inside of the pipe are employed to ensure

against collapse during bending. Of all the materials a plumber uses, only the heavier grades of polythene have pipe walls of sufficient thickness to enable them to be bent without internal support. One other possible exception is the method of hot bending steel pipes, but even these pipes tend to flatten slightly when the bend is made, and on the occasions when this method is used, the sides are squeezed inward by pressure from a vice while still at red heat, causing the section of the bend to assume a truly circular shape throughout its length. With the increased use of hydraulic bending machines this method of bending is little used except in the cases where a radius, different from that produced by the machine former, is required and a screwed or weldable fitting is unsuitable.

### Bending springs for copper tube

Spring bending is a useful method of bending small-diameter pipes, especially if the radius of a machine made bend is unsuitable for any reason. The spring, which should be a running fit inside the pipe bore, is located centrally at the point where the bend is to be made. The most common method of pulling the bend is across the knee, taking care to shift position slightly to avoid too sharp a bend, which may result in causing the throat to wrinkle. The physical strength required to make a bend using this method does to some extent limit its use to the smaller sizes of the tube. Copper tubes of 22 mm diameter or more should be annealed prior to bending, to reduce the physical effort needed and to avoid rippling on the throat of the bend. It is not usual to bend copper pipes of larger diameter than 28 mm by hand as the large radius which would have to be used to prevent rippling would be unacceptable.

It should be appreciated that when a bend is pulled, the throat and back of the bend tightens on the spring, and to enable it to be withdrawn without damage, it is usual to over pull the bend by a few degrees and then open it to the desired angle as shown in Fig. 5.23 to reduce the gripping effect of the bend. If the spring still cannot be withdrawn easily, a slight clockwise turn with a tommy bar will tighten the coils, and slightly reduce its diameter facilitating its removal. Prior to bending, the spring should be oiled to facilitate its withdrawal. Springs can be used to make bends

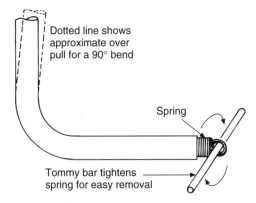

Dotted line shows approximate over pull for a 90° bend

Spring

Tommy bar tightens spring for easy removal

**Fig. 5.23**   Withdrawing a spring.

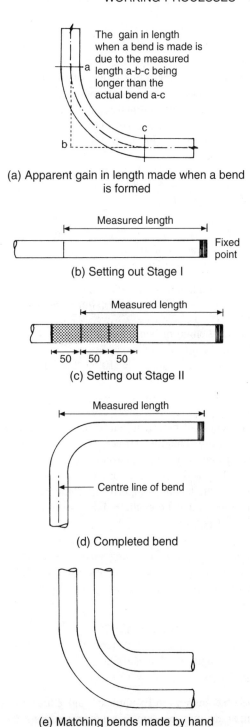

The gain in length when a bend is made is due to the measured length a-b-c being longer than the actual bend a-c

(a) Apparent gain in length made when a bend is formed

Measured length

Fixed point

(b) Setting out Stage I

Measured length

50    50    50

(c) Setting out Stage II

Measured length

Centre line of bend

(d) Completed bend

(e) Matching bends made by hand

The radius of each bend is different, so the bends are spaced evenly apart. If one bend is made by machine the other must be made by hand so that it follows the curve.

**Fig. 5.24**   Making bends.

in the middle of long lengths of pipe by fastening a stout wire extension to the loop at the end of the spring, fencing wire being very suitable for this purpose. String should not be used as it will not permit the coils of the spring to be tightened should this be necessary to enable it to be withdrawn.

Bends having a radius of less than 6 × diameter are not recommended on copper tubes because of the possible difficulty of removing the spring.

Other materials such as sand or a low-melting-point alloy containing lead, tin and bismuth were at one time used in the plumbing industry for bending copper pipes of large diameter to a relatively small radius. The thinner wall thickness now used has made this very difficult, and where such bends are required they are usually made by companies which have the special bending machines necessary for the work.

*Setting out*

Hand-made bends of all materials should be 'set out' to give the same accuracy of position and radius as those made by a bending machine. The main point to understand about all bends is the apparent 'gain of material' when the bend is formed. A study of Fig. 5.24(a) will show the reasons for this. The distance from 'a' to 'c' through 'b' along the broken line is in effect the measured length of the bend, but when it is actually pulled, its path follows the arc a–c which is a shorter distance than the measured length.

To make allowances for this – and to ensure the bend is pulled in the right position in relation to a fixed point – the length of pipe actually occupied by the bend in relation to the measured distance can be calculated in the following way. The first step is to decide on the centre-line radius of the bend. In the absence of any other specification it is assumed to be four times the nominal diameter of the pipe, this being written as 4D. To find the length of pipe occupied by a 90° bend the following formula may be used:

$$\frac{\text{Radius} \times 2 \times 3.142}{4}$$

*Example 1*   Assuming a 25 mm diameter pipe is to be bent to a radius of 4D, find the length of pipe which will be occupied by the bend.

$$\text{Radius of bend} = 4 \times 25$$
$$= 100 \text{ mm}$$

$$\therefore \text{ Length of bend } = \frac{\text{Radius} \times 2 \times 3.142}{4}$$

$$= \frac{100 \times 2 \times 3.142}{4}$$

$$= 157.1 \text{ mm}$$

The length of pipe required to form this bend will be 157.1 mm. This method is very accurate but is rather a long way round to ascertain a simple length. In practice an approximation is made by assuming the length of pipe forming the bend to be one and a half times the radius, assuming a 4D radius, of course.

*Example 2*   Assume the same size of pipe bent to the same radius as Example 1, but use the following formula: length of bend equals one and a half times the radius:

$$\text{Radius} = 4 \times 25$$
$$= 100 \text{ mm}$$
$$\text{Length of bend} = 100 \times 1.5$$
$$= 150 \text{ mm}$$

When the answers to Examples 1 and 2 are compared it will be seen that the approximate method is incorrect by only 7.1 mm and is sufficiently accurate for normal site calculations.

To make the bend, mark off the required length from a fixed measuring point (probably the end of the pipe) to the centre line of the bend as shown in Fig. 5.24(b). Divide the calculated length of pipe which will form the bend by 3, which in this example gives three equal parts of 50 mm. From the original centre line, mark 50 mm forward and 100 mm back as shown in Fig. 5.24(c). This shows the relationship of these measurements to the centre line and the fixed point or end of the pipe. When the bend is pulled it must be confined to the areas indicated by the shading, which will ensure its centre will be at the correct distance from the fixed point as shown in Fig. 5.24(d). This method can be used for making bends of less than 90°, but it is necessary to understand that a bend of 135° will occupy only half the length of pipe required for a right-angled bend. By using this technique, bends can be made as accurately by hand as with a bending machine, with the added advantage that, unlike machine bends, which have a fixed radius, the radius can be altered to make following or matching bends if necessary (see Fig. 5.24(e)).

**Bending by machine**

*Light-gauge copper tubes*
Copper tube is probably the most widely used material for both hot and cold water supply and domestic heating and it is essential that a plumber can bend it quickly and accurately. While a bending spring is useful for small repair jobs or perhaps making a bend on a pipe that is already fixed, mechanical means of bending are essential on large installations if they are to be economic in time and materials.

There are many differing types of machines available, from the hand-operated machines used for bending small-diameter tubes to those of large powered type which can bend pipes of up to 150 mm in diameter. A selection of the smaller benders used for site work is illustrated in Fig. 5.25.

Figure 5.25(a) shows a simple tool for bending soft copper tubes of 8 and 10 mm diameter, this being simply a former around which the tube is bent by hand. Owing to the comparatively large-radius bends that are made on these small-diameter pipes no support is required at the back of the bend.

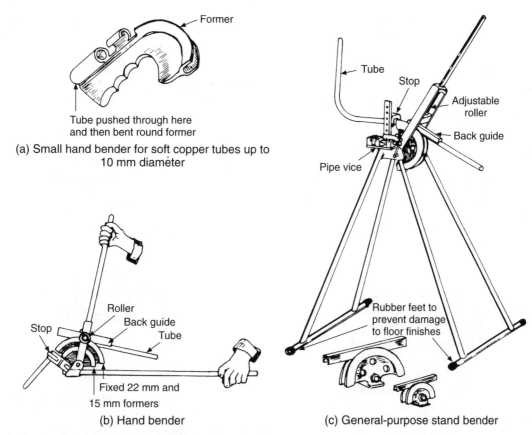

**Fig. 5.25**  Small pipe-bending machines for general site work.

The small hand bender shown in Fig. 5.25(b) is a very useful tool both to the jobbing plumber and the heating fitter for bending 15 mm and 22 mm tubes. The machines are very light and can easily be carried in a tool bag. They are most useful when working in occupied premises as they take up little space and can easily be carried from room to room.

Figure 5.25(c) shows a larger machine capable of bending up to 35 mm pipes and is quite suitable for general site work. Similar types are made for bench fitting.

For bending pipes of 42 mm to 54 mm special machines are made which are in effect geared down to reduce the manual effort required to make bends of this size. The type shown in Fig. 5.26 employs a screw thread arrangement for rotating the tube former, giving it a similar effort–weight effect to that of the simple lever. This machine can also be equipped for bending mild-steel tubes to BS 1387 of up to 25 mm nominal diameter.

This is a rotary bender where the former is made to rotate against a fixed roller. By operating the handle, the former is turned in the direction of the arrow causing the tube to be bent

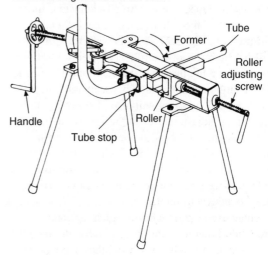

**Fig. 5.26**  Horizontal bender.

Bends on tubes of larger diameters than 54 mm are made on special equipment called mandrel machines which support not only the outside of the tube but also the inside by means of a travelling mandrel. Bends of small radius can be produced on these machines, a typical example being the fabrication of copper tube traps.

*Bending light-gauge stainless steel pipes*
Light-gauge stainless steel pipes may be bent in all the machines mentioned with the exception of the hand bender which has a roller supported on one side only which would not sustain the stresses imposed by the tougher and more rigid stainless steel pipe. It is generally recommended that the capacity of benders should be reduced by one size when bending this tube, i.e. a bender with a capacity of up to 28 mm copper tube should not be used for bending stainless steel in excess of 22 mm diameter. Steel back guides, not those of aluminium alloy, should always be used with stainless steel.

*Components of bending machines*
Figure 5.27(a) shows the two principal components of a typical rotary bending machine of the type used in the plumbing industry. The former supports the throat of the pipe while the back guide rotates around the bend as pressure is applied by the roller. Figure 5.27(b) shows a sectional view of how these components support the tube during the bending process and prevent distortion taking place, provided the machine is correctly adjusted. Incorrect adjustment of the roller can have two effects, as illustrated in Fig. 5.27(c). Excessive throating will be the result of the roller 'A' being too tight, compressing the throat of the bend into the former. Slight throating of the tube is acceptable but too much is unsightly and will also obstruct the flow. The wrong adjustment of the roller 'B' will produce ripples on the throat of the bend due to the lack of support at this point. Some machines have a guide fixed to the handle which, when lined up to the tube, indicates the correct roller position. The roller, is, of course, adjustable.

There are other machines which have no provision for roller adjustment, and while they produce good bends when new, they have no provision for wear of the working parts. To obtain

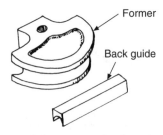

(a) The essential components of a rotary bender

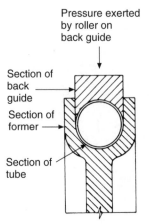

(b) How the tube is supported during bending

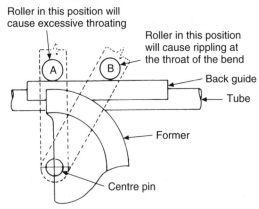

(c) Defects to bends caused by incorrect roller position

**Fig. 5.27**   Action of a pipe-bending machine.

good bends with these older machines, especially when the larger sizes of tube are used, may necessitate the insertion of a thin piece of steel between the roller and back guide to avoid rippling the bend.

*Maintenance*

Very little maintenance is required to keep tube-bending machines in good working order. They should be cleaned regularly and lightly oiled with thin machine oil. The parts most prone to damage are the knife edges of the back guides. Any deviation from a sharp straight edge will transmit dents and cuts to the tube during bending. Light distortion can be erased with a fine file but in cases of severe damage, they are best replaced. Care should be taken to avoid dropping the guide when the handle of the machine is raised after a bend has been made. When not in use protect the guide by wrapping in a clean soft cloth.

## Machine-bending processes

There are two basic operations to be fully understood and mastered in order to use a bending machine effectively, one being the ability to form 90° or square bends, the other being to bend offsets. Both of these operations must be carried out accurately to given measurements.

*90° or square bends*

The measurements of square bends given on a drawing are usually taken from the centre line of the pipe, as illustrated in Fig. 5.28(a). It is often more convenient on site to express the measurement back to back, as shown in Fig. 5.28(c). Those taken from the inside to back as at Fig. 5.28(b) will be the same as (a) and are only given in this way to enable the tube to be squared across from the inside or the outside of the former.

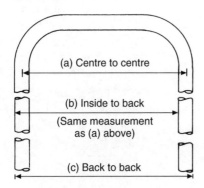

Fig. 5.28   Methods of taking measurements for tube bending.

The example in Fig. 5.29(a) shows a bend made from a fixed point to the back of the bend. A pencil mark is made from the end of the tube, in this case 300 mm, the end of the tube being referred to as the fixed point (see Fig. 5.29(b)). The tube is set up in the machine with the mark squared off from the outside of the former. With the back guide in position and the roller correctly adjusted the bend is ready to be pulled (see Fig. 5.29(c)). If the measurement has been quoted to the centre of the bend, then half a tube diameter must be added to the original measurement. To give an example of this, assume the tube diameter to be 22 mm, and the given measurement 300 mm from the fixed point to

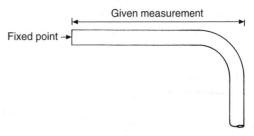

(a) The given measurement here is from a fixed point to the back of the bend

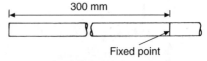

(b) Marking tube for bending

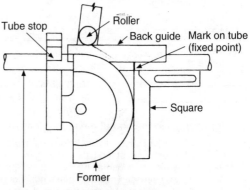

(c) Squaring across the mark to the outside of the former

Fig. 5.29   Bend made from a fixed point.

the centre of the bend; then this measurement would be 311 mm from the fixed point to the back of the bend.

To make a return bend or to bend the pipe again in another place, the technique is the same, the only difference being that the first bend has now become the fixed point. Figure 5.30 shows the set up for making a return bend from the centre-line measurements but actually measuring from the inside of one bend to the back of the other bend (this is equivalent to measuring centre to centre – see Fig. 5.28(a)).

### Offsets

There are various methods for measuring offsets. The method shown here will give accurate results for small offsets and are most suitable for use with

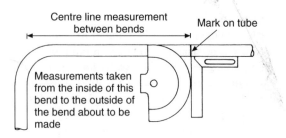

**Fig. 5.30**   Setting out return bends.

bench benders. The first set is made to the desired angle as shown in Fig. 5.31(a). In many cases the angle is not critical and can be judged by eye, but if a fixed angle is essential, it should be taken from the actual job using a bevel. When making very small offsets take care not to make the first set with too large an angle, or when the second one is pulled

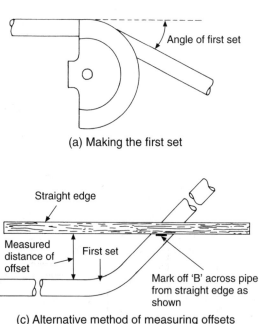

(a) Making the first set

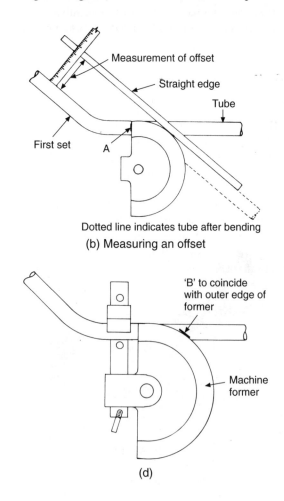

Dotted line indicates tube after bending

(b) Measuring an offset

(c) Alternative method of measuring offsets

This method of bending offsets to centre-line measurements may be used when using stand type bending machines where difficulty may be found using the method shown in (b). It must be stressed however that it may not be as accurate. It is important to note that the channel of the former must be as deep as the outside diameter of the pipe and is not suitable for some types of hand benders. It is assumed that the first set has been formed to the required angle.

(d)

**Fig. 5.31**   Making an offset.

it will be found that they run together, often kinking the tube. Having made the first set the tube is reversed and pushed through the machine, a straight edge being set against the tube former, parallel to the tube, as illustrated in Fig. 5.31(b). The required offset measurement is taken from the inside of the tube to the inside edge of the straight edge. Note the relationship of the second set shown as a broken line parallel to the straight edge. It will be seen that the measurements are actually taken from the inside to the back of the tube, this being the equivalent of a centre-line measurement. When the tube has been adjusted against the stop of the machine, it is advisable to mark the tube at the edge of the former, as shown at 'A' in Fig. 5.31(b). This permits the tube to be correctly repositioned in the machine if it is accidentally moved after the measurements have been taken.

While the foregoing is not impossible with stand-type machines, it is not quite so easy to use a straight edge. The method of measuring offsets shown in Figs 5.31(c) and (d) will produce offsets to a reasonable degree of accuracy. It must be stressed, however, that if an accuracy of ±2 mm is required then the method using a straight edge must be employed.

### Passover bends

It is often necessary to make bends to clear obstructions across a run of pipe. Bends of this type are called 'passovers' and fall into two categories. The type shown in Fig. 5.32(a) occurs when a branch is made in a service pipe and has to pass over another service pipe above. The measurements in this case are taken in the same way as for an ordinary offset, the only difference being that the second bend is overpulled until the desired angle is obtained to fit the tee.

The other type of passover is sometimes called a 'crank set' because of its similarity to the crank on an engine, this being shown in Fig. 5.32(b). A 2 or 3 millimetre gap should be left between the throat of the passover bend and the obstruction, which avoids any possibility of action between dissimilar metals or chafing between the two pipes. The required clearance is determined after the centre set has been made, the measurement being taken as shown in Fig. 5.33(a). Marks 'A' and 'B' are then

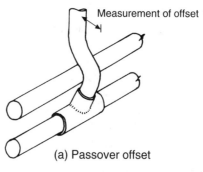

Measurement of offset

(a) Passover offset

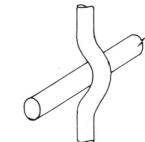

(b) 'Crank' passover bend
Used to clear an obstacle such as another pipe.

**Fig. 5.32** Types of passover.

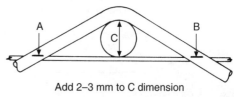

A    C    B

Add 2–3 mm to C dimension
(a) Measurement of crank passover

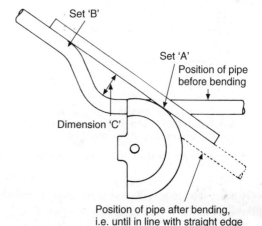

Set 'B'

Set 'A'
Position of pipe before bending

Dimension 'C'

Position of pipe after bending, i.e. until in line with straight edge
(b) Making the final bend in a crank set

**Fig. 5.33** Making a crank passover bend.

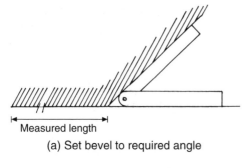

Measured length

(a) Set bevel to required angle

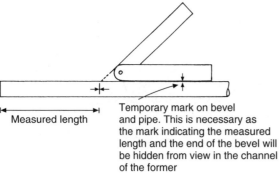

Measured length

Temporary mark on bevel and pipe. This is necessary as the mark indicating the measured length and the end of the bevel will be hidden from view in the channel of the former

(b) Line up to top edge of bevel on marked pipe

When these temporary marks line up the end of the bevel will be correctly located in relation to the centre of the set. The pipe can now be bent to the required angle.

Measured length

(c)

This method can be adapted as an alternative to that shown in Fig. 4.24(a) and (b) for measuring offsets on tripod bending machines.

**Fig. 5.34** Accurate location of sets in copper tubes.

made across the tube with a straight edge as shown in the illustration before replacing the pipe in the machine. When mark 'A' is at a tangent to the former the pipe is in the correct position and should be pulled to the angle indicated by the broken line.

The pipe is then reversed in the machine and mark B lined up to the former. A straight edge should be used at this stage to check both the clearance of the passover and the alignment of marks A and B. The last set can then be pulled to complete the bend.

*Location of sets*
It is sometimes necessary to locate the exact position of a single set in relation to that of another, a typical example of that being when a pipe has to be bent around the angles of a bay window.

Some plumbers make a mark on the former to locate the measured mark shown in the illustration. This does not always give accurate results as marks on the former will vary as to the degree of the angle to which the tube is to be bent. For absolute accuracy when using this method a simple drawing is required indicating the degree of the angle. This method is illustrated in Figs 5.34(a)–(c). In the absence of a bevel a 600 mm steel folding rule can be used to determine the angle.

## Steel pipe bending

Steel pipes to BS 1387, because of their thicker walls, do not need to be fully supported during the bending process, and a method known as press bending is used. This involves the use of a hydraulic bending machine of the type shown in Fig. 5.35.

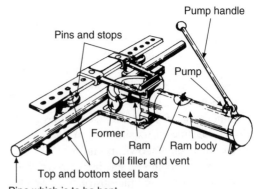

Pump handle

Pins and stops

Pump

Former

Ram / Ram body

Oil filler and vent

Top and bottom steel bars

Pipe which is to be bent

**Fig. 5.35** Hydraulic press bender for steel pipes. The action of moving the pump handle backwards and forwards is to pressurise the oil which, acting against the piston, moves the ram and former against the pipe which is to be bent.

## Hydraulic pipe-bending machines

Some idea of hydraulic power can be seen on most building sites where bulldozers and trenching machines are commonplace. The power source of all hydraulic machinery is the ram which in many ways is similar to a pump. Liquids, for all practical purposes, are incompressible and as such are capable of exerting a similar force to that of a steel bar when they are subjected to pressure. Figure 5.36(a) shows the basic principle of a hydraulic ram which can be explained simply as follows. The fluid is pumped into the cylinder of the ram and exerts pressure against the piston which is forced outward to work many differing types of equipment. Figure 5.36(b) illustrates the pump details of the ram.

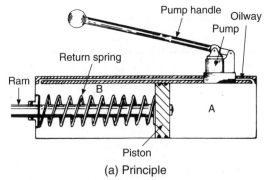

(a) Principle

When the pump is operated oil is pumped from B and A, forcing the ram outward and tightening the spring. By opening a bypass valve on the pump (see (b)) the pressure exerted by the spring pushes the piston back, simultaneously allowing the oil to be displaced from A to B.

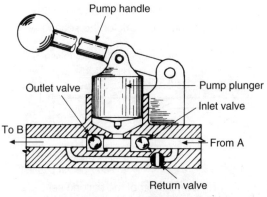

(b) Detail of ram pump used for hydraulic bending machines

**Fig. 5.36**   Basic principles of a hydraulic pump.

In the case of a hydraulic pipe-bending machine, a former of the appropriate size for the pipe to be bent is selected and fitted on the end of the ram. When the pump is operated the ram moves forward pressing the tube against the stops, which are secured by pins through two steel bars. These bars have a series of holes throughout their length which permit the stops to be moved to accommodate various pipe sizes. The appropriate holes are clearly marked indicating for which pipe size they are suitable. The position of the stops must always be checked before a bend is made to ensure that the pins are in the correct holes. Failure to do this could cause very serious damage to the machine such as broken formers, a bent ram or damaged seals, all of which are expensive to repair.

Some machines require the bleed screw (which is usually the filler cap) to be in the open position during bending to allow the machine to 'breathe' during operation. When it is necessary to top up with fluid, always use the correct grade of hydraulic oil as recommended by the maker as the wrong oil could degrade the synthetic rubber seals rendering them useless. This information is usually marked on the machine, near the filler as a rule. Do not overfill or the excess fluid will be pumped out through the bleed screw when the pump is operated.

### 90° bends

The method or system of using press benders to make accurate bends is very simple. To make a 90° bend from a fixed point, e.g. the threaded end of a tube, mark off the required measurement of the centre line which in Fig. 5.37(a) is 600 mm. Bearing in mind the gain in length when bends are made, one pipe diameter (in this example 25 mm) is deducted from the measurement, which leaves 575 mm. This should be marked off from the end of the tube. The pipe should be placed in the bender with this mark corresponding to the centre line of the former as shown. Having checked the stops are in the correct position, the bend can now be made. Owing to the fact that even LCS is slightly springy, the bend should be overpulled by about 5° to allow for springback. When the bend is complete it should be checked for measurement as shown in Fig. 5.37(b).

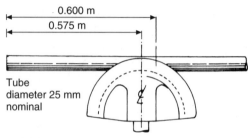

(a) Location of pipe in former for 90° bend

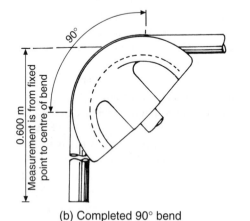

(b) Completed 90° bend

A 90° set square (workshop size) can be laid over the bend to check the accuracy of bending.

**Fig. 5.37** Using a hydraulic press bender.

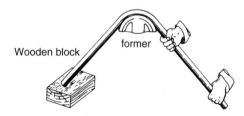

**Fig. 5.38** Removing a former from the pipe. Former must not be allowed to fall on hard ground or it may be damaged.

It will sometimes be found difficult to remove the former from the pipe when the bend is completed, especially when the former is new, but this need not be a problem if the pipe is held as shown in Fig. 5.38 and the pipe end is tapped on the wooden block until the former becomes disengaged. Small formers may be held by hand to prevent them dropping on to the ground, larger ones should be protected by allowing them to fall on soft material.

(Care must be taken to avoid the former dropping on the operative's feet when it falls, this being a typical example of why shoes with steel toe caps are necessary.) Some machine manufacturers recommend greasing the tube where the bend is made. While this is messy, it does aid the bending operation and enables the former to be removed more easily.

*Offsets*

To make offsets in steel pipes, the marking out method is very similar to that used for copper tubes. The required measurement for the first set is marked off on the pipe and placed in the machine with no deduction, as shown in Fig. 5.39(a). The measurement indicated is 0.450 m from the

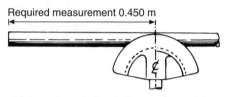

(a) Location of pipe in former for making the first bend

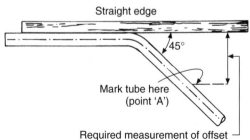

(b) Setting out the mark for the second set

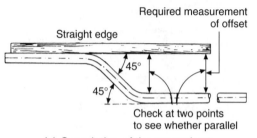

(c) Completion of the second set

The angle of 45° which is shown is only approximate, but both angles must be exactly the same.

**Fig. 5.39** Making offsets in steel pipe.

end of the pipe to the centre of the set. After the set has been pulled and removed from the machine, a straight edge is placed against the back of the tube (Fig. 5.39(b)), and the measurement of the offset marked from it to point 'A' on the tube. This mark should coincide with the centre of the former when the pipe is replaced in the machine. After the second set has been made it is important to ensure that both ends of the pipe are parallel, which is done by checking the measurements from a straight edge laid across the back of one of the bends (see Fig. 5.39(c)). The remarks concerning the setting out of bends, sets and passovers in light-gauge copper tube also apply to steel tube. Once the methods governing the accurate bending of 90° bends and offsets have been mastered, this knowledge and skill can be applied to most other applications.

## Tapping and threading

### Use of taps for making female threads

A complete set of taps for a given size of thread are shown in Chapter 2, Fig. 2.36. Taper taps are used for making threads in through holes, the second and plug taps are used when forming threads on blind holes, see Fig. 5.40. Before tapping, a hole must be drilled not less than the minor diameter of the thread, see Chapter 2, Fig. 2.37. This is important as taps are made of very hard brittle steel and are easily broken unless the correct drill is selected and great care is taken, especially for those of less than 6 mm in diameter.

In all cases when drilling blind holes, if possible use a slightly oversized drill, drilling the holes deeper than is necessary for the thread. This allows room for the swarf to collect at the bottom of the hole. It is recommended that the tap is removed occasionally during the threading process to allow the swarf to be removed.

During a tapping operation it is important to get the 'feel' of the tap. This means that when it is felt to be tighter in the hole, turn it back anticlockwise to clear the swarf; *never force a tap* or it will break. Always use the correct cutting oil, especially when tapping copper or aluminium – failure to do this is responsible for many broken taps as these materials, being relatively soft, are not easy to thread.

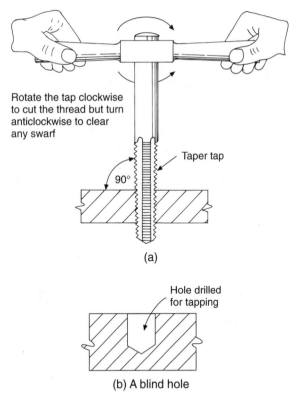

Rotate the tap clockwise to cut the thread but turn anticlockwise to clear any swarf

Taper tap

90°

(a)

Hole drilled for tapping

(b) A blind hole

**Fig. 5.40** Use of taps for forming female threads.

### Removing sheared-off studs

Studs are used to secure components such as access covers to a boiler or clearing eyes on cast iron drainage, etc. It is sometimes necessary to remove these covers to replace a defective seal or gain access to the inside of a boiler, e.g. to descale the inner surfaces. The nuts securing the cover can become very corroded and when attempting to undo them the stud itself shears off. Prevention is better than cure and if the nut appears to be very tight, first apply some penetrating oil and tap the flats of the nut with a hammer taking care to support the other side, see Fig. 5.41 Heating the nut may also help as the expansion tends to loosen the corrosion products. These procedures are also useful for disconnecting all types of nuts, e.g. connections to cylinders and boilers. If the worst happens and the stud shears off the following procedures should be followed.

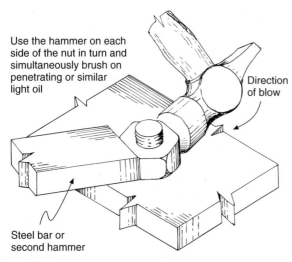

Use the hammer on each side of the nut in turn and simultaneously brush on penetrating or similar light oil

Direction of blow

Steel bar or second hammer

**Fig. 5.41** Loosening corroded nuts.

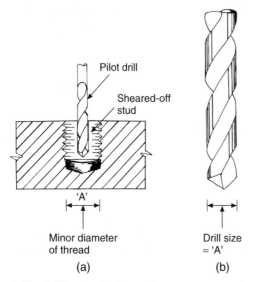

Pilot drill

Sheared-off stud

'A'

Minor diameter of thread

Drill size = 'A'

(a)                    (b)

**Fig. 5.42** Drilling out broken studs.

(a) Carefully centre punch the stud and drill a pilot hole in its centre ensuring the drill is at 90° to the surface supporting the stud. See Fig. 5.42(a).

(b) Select a drill having the same diameter as the minor diameter of the thread and drill through the stud again keeping the drill at 90° to the surface supporting the stud. See Fig. 5.42(b).

(c) It may be possible to remove the remainder of the stud using an extractor, see Fig. 5.43, but if the stud is obviously corroded and unlikely to move, use the procedures described for tapping a blind hole.

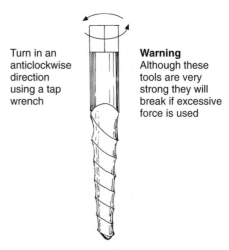

Turn in an anticlockwise direction using a tap wrench

**Warning**
Although these tools are very strong they will break if excessive force is used

**Fig. 5.43** Screw extractor. These tools are made in a variety of diameters to suit various thread sizes. To use, drill a hole in the centre of the broken stud large enough to permit entry of the extractor, then turn as shown.

## Soft soldering

### Wiped soldered joints

The use of lead/tin solders is not permitted for water supplies, but this method of jointing has some applications in lead sheet work and may still be used for repair work; for example, in lead gutters or flats where a wiped patch can be used used to repair a fatigued crack. Unless the very old method of using a pot and ladle is employed a blowlamp will be necessary and it is important to stress the risks of using a flame on or near very dry timber. A safer method is to solder on a patch using a large soldering iron (copper bit). The use of lead welding techniques is preferable if the equipment is available, as solder, having a different coefficient of expansion, will eventually crack round the edges of the patch. It must always be made clear to a client that any form of patching will have a limited life span as it does not remedy the cause of the crack. This will almost certainly be due to insufficient provision for expansion and contraction, thickness of the lead sheet used and poor fixing. Figure 5.44 illustrates the preparation and soldering method used for patching sheet lead weathering.

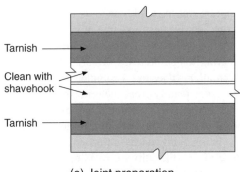

### (a) Joint preparation

Plumber's black is used to ensure a clean straight edge between the wiped solder and lead sheet. Note that if plumber's black is difficult to obtain, the section of a cut potato rubbed on the lead prior to shaving can be used as an alternative.

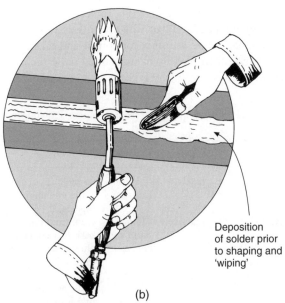

(b)

① Flux the joint using tallow
② Heat the lead while rubbing the solder stick on it to 'tin' it
③ Add the solder to the tinned surface depositing sufficient to achieve the profile shown in (c)
④ Reheat the solder until it is in a pasty condition and carefully shape using a small wiping cloth

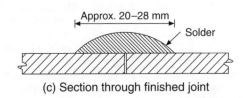

### (c) Section through finished joint

**Fig. 5.44**  Joining or patching lead sheet using wiping solder (Grade D).

## Copper-bit work

Copper bits are usually specified by mass (weight) and the heavier they are the greater will be their heat capacity, which is necessary when soldering on large areas of metal having a high rate of conductivity. Reference should be made to the section on solders in Chapter 4.

Prior to commencing the soldering operation it is necessary to tin the soldering 'iron', and this is done by heating it to the required temperature (when it will melt the solder), cleaning the faces by filing, then rubbing it on sal ammoniac crystals (or some other flux) and finally applying it to a stick of solder. This should result in the end of the bit being completely covered with a layer of solder. It must not be overheated as this will burn off its tinned face and necessitate cleaning and retinning. A good guide to the correct temperature to which it should be heated is to watch the flame. Immediately a green tinge appears, its temperature will be about right for the majority of soldering operations.

A diagrammatic close-up of the soft-soldering operation using a copper bit is shown in Fig. 5.45. The action of the boiling flux is shown removing the oxides from the metal to expose an oxide-free surface enabling the solder to 'tin' – or 'wet' – the parent or base metal and form a thin surface alloy with it.

Figure 5.46 illustrates a method of soldering using a soldering iron.

It should be noted that gas-heated soldering bits are available from most manufacturers of portable gas equipment. The combined burner and copper bit

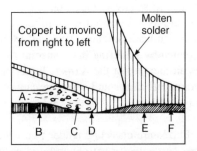

**Fig. 5.45**  Soft soldering with a copper bit. A: Flux on the oxidised metal. B: Oxidised metal. C: Boiling flux removes oxides. D: Bare metal exposed. E: Surface alloy of tin and base metal (intermetallic layer). F: Solidified solder.

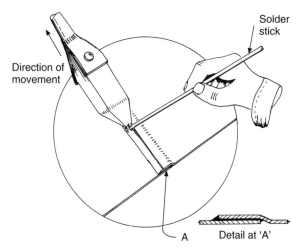

**Fig. 5.46** Soldering sheet metal using a copper bit.

are readily adaptable to the hand-held shank, and where extensive soldering operations are carried out they usually save both time and fuel.

*Design of soldered joints*

Joints made of soft solder of Grade F (refer to Chapter 4, p. 91) rely for their strength on capillary attraction in exactly the same way as when making soldered joints on copper tube. When soldering sheet metal an overlap is necessary to allow capillarity to take place. The design for brazed joints shown in Book 2 of this series is suitable for most of those employing soft solder, as the techniques are, with few exceptions, identical. Brazing is sometimes called hard soldering and if a percentage of silver is used in the filler metal, it is often referred to as silver soldering.

Copper and its alloys used in plumbing can be soldered as can lead. Stainless steel and LCS can also be jointed by soldering but a special flux is necessary in the case of the former. Joints or repairs to aluminium, its alloys and cast iron cannot be effectively soldered.

*Fluxes* Fluxes are classified tender two general categories, corrosive (or active) and non-corrosive (inactive). All fluxes are to some extent acidic as this is necessary to enable them to dissolve and remove metallic oxides on the surface to be soldered. Active fluxes will attack metals at any

temperature but those classified as inactive, such as tallow, are not very reactive at normal air temperatures. They are, however, sufficiently active at soldering temperatures. Corrosive fluxes must be removed after soldering is completed as the residues absorb water readily from the atmosphere and form acids that will continue to attack the metal. Removal is affected by washing in hot, diluted, soda solution followed by a further hot water wash. Because of the difficulty of avoiding corrosion with zinc chloride it must never be used for electrical work or any other operation which cannot be washed effectively. Non-corrosive fluxes do not require to be removed owing to their low reactivity at normal temperatures.

Details of the fluxes in general use in the plumbing industry are given in Table 5.1.

It should be noted that while resin was traditionally used as a flux for copper and brass it has been superseded by manufactured paste fluxes of which there are many types now produced.

**Table 5.1** Fluxes in general use for wiping and copper bit work.

| Flux | Type | Function | Origin |
|------|------|----------|--------|
| Tallow | Inactive | Soldering lead | Organic fat of cattle containing oleic acid |
| Resin* | Inactive | Tinning brass and copper | Gum from pine tree bark that contains abiatic acid. Used in powdered form or mixed with spirit or formed into a paste. |
| Manufactured paste fluxes | May be active or inactive | Tinning brass or copper | Various deoxidants bound in Vaseline or similar pastes that remain fluid at soldering temperatures |
| Zinc chloride† | Active | Zincwork | Zinc dissolved in hydrochloric acid until acid is 'dead'. Commonly called 'killed spirits'. |

*Resin-based fluxes are used in flux-cored solder for soldering electrical components
†Zinc chloride, although still used, has been superseded by liquid fluxes specially formulated for zinc work

## Further reading

Much useful information can be obtained from the following sources:

*Copper tube bending*
Record Tools Ltd, Parkway Works, Sheffield,
    S9 3BL. Tel.: 0114 2449066.
Hillmore Ltd, Caxton Way, Stevenage, Herts,
    SG1 2DQ. Tel.: 01438 312466.

*Steel pipe bending*
Tubela Engineering Company Ltd, 2–6 Fowler
    Road, Hainault, Essex, 1GB 3UP.
    Tel.: 020 8500 1253.

*Pipes and fittings (see Chapter 4)*
Philmac Pty Ltd, Diplocks Way, Hailsham, East
    Sussex, BN27 3JF. Tel.: 01323 847323.
    (Plastic–metal; couplings and adaptors.)

## Self-testing questions

1. State the main purpose of the Building Regulations.
2. Using the scale rule shown on page 100, and dividers or compasses, measure the length of the water service pipe shown in Fig. 5.47 from the main to its entry into the building.
3. State the reason for 'rounding up' the ends of coiled soft copper tube prior to making capillary joints.
4. Explain why the correct sequence of tightening the nuts should be employed when flange joints are made.
5. State the reason for slightly overbending steel pipes when a hydraulic bending machine is used.
6. Describe why it is necessary to deburr steel and copper pipes after they have been cut with a wheel cutter.

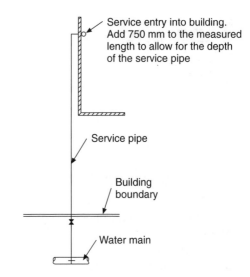

Service entry into building. Add 750 mm to the measured length to allow for the depth of the service pipe

Service pipe

Building boundary

Water main

**Fig. 5.47**   Question 2; scale 1:100.

7. Which type of compression fitting must be used with R220 (Table Y) copper pipe?
8. State the cause of rippling the throat of a copper tube bend made with a bending machine.
9. Explain why it is important to position the pins in the correct holes of the bars on a press bender.
10. State the effect of using the wrong type of oil in hydraulic-bending equipment.
11. Using the approximate method, calculate the length of 40 mm pipe occupied by a 90° bend made to a radius of four times its diameter.
12. Name two methods of jointing PVC pipes.
13. Explain why a lubricant is necessary when cutting threads on steel pipes.
14. Name the tap used to finish tapping a thread in a blind hole.
15. List the procedures used to remove a sheared-off stud securing an access door on a boiler.

# 6 Taps and valves

After completing this chapter the reader should be able to:

1. Describe the working principles of taps and valves for domestic hot and cold water supplies.
2. Describe the methods used to repair and maintain taps and valves in good working order.
3. Recognise different types of float-operated valves and understand their working principles.
4. Specify suitable types of valves for specific purposes.

## Taps and valves – general requirements

The Water Regulations require that all fittings must be suitable for their purpose, made of corrosion-resistant materials having sufficient strength to resist normal and surge pressures. They should also be capable of working at appropriate temperatures. Seals and washers must be easily accessible for renewal when necessary. Taps and fittings used for both hot and cold water supplies should conform to BS 1010 Parts 1 and 2 and BS 5412. High-resistance taps should comply with BS EN 200 and BS 6920, the minimum inlet pressure being 0 5 bar. Float-operated valves should comply with BS 1212 Parts 1, 2, 3 and 4. All taps and valves are usually made of brass pressings or castings and some taps are chromium plated to improve the appearance and to facilitate easy cleaning. More recently a thermosetting plastic called acetal has become increasingly popular for the manufacture of taps. Marketed under various trade names, it is an engineering plastic which is hard wearing and capable of being machined in a similar way to metal. It has a lower impact strength than metal, is more easily damaged and not so resistant to abrasion as brass. It is mainly these factors which have at present prevented it becoming a more serious competitor to metals. Certain components of modern taps and shower mixer valves are made of synthetic plastic, especially nylon which has good wearing qualities. It is also resistant to heat and limescale.

Polypropylene is another plastic material that has been used successfully for the manufacture of float-operated valves, especially the diaphragm type which, because of its design, lends itself to the limitations of this particular material.

### Screw-down taps

Prior to 1986 only screw-down taps were permitted on water supplies taken directly from the mains, as these are designed in such a way that they cannot be closed quickly, thus avoiding water hammer. This is a term used to describe noise, usually a banging sound or noisy vibrations throughout the system, which occurs when the flow of high-pressure water is suddenly arrested. Not only is this noise disconcerting but water hammer can be damaging to pipework, especially where it has been weakened owing to exposure to frost.

Since the 1986 Water Bylaws, however, quick closing, usually quarter-turn taps, are permitted, but the reader should be aware that by installing them they may cause water hammer. To avoid this, especially in areas where water is supplied by the water authorities at very high pressure, it may be

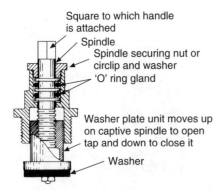

**Fig. 6.1** Headwork of tap having non-rising spindle. This type of headwork is used by many manufacturers of mixer taps for modern bathroom equipment. To enable the 'O' rings to be replaced the spindle securing nut or circlip is removed. The spindle is rotated thus unscrewing it from the washer plate unit when it can be lifted out of the headwork.

necessary to provide pressure-reducing valves or water hammer arresters. These are dealt with in Book 2 of this series.

Screw-down taps are made in a variety of patterns depending upon the fitting they supply. However, the headwork of all screw-down taps and valves is basically the same, incorporating a threaded spindle which has to be turned through several revolutions to effect its closure. Many modern taps are made having non-rising spindles. This is an advantage as it reduces wear on the spindle as it passes through the gland. It also enables manufacturers more freedom in the aesthetic design of taps (see Fig. 6.1).

*Bib taps*

The bib tap is a typical example of a screw-down tap, being designed to be fitted into bosses screwed to the wall over such appliances as butlers' or cleaners' sinks. They are also used on standpipes when the bib is threaded, which enables a hose union to be fitted. When bib taps are provided over a sink, always be sure that they are fitted high enough to enable a bucket or bowl to be comfortably placed beneath. A sectional elevation of a bib tap is shown in Fig. 6.2(a). The shroud which covers the headwork is called an easy-clean shield and it serves to prevent the build up of dirt and corrosion in the head. The spindle, gland and jumper shown in the illustration

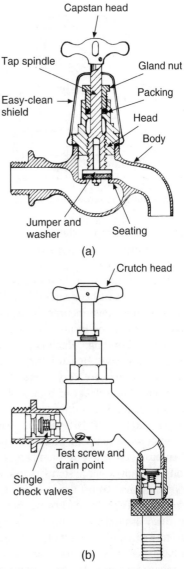

**Fig. 6.2** (a) Bib tap. The action of all screw-down taps or valves is the same. When the spindle is turned it engages on the threads in the head allowing the gradual opening or closing of the jumper. (b) Hose union bib tap with integral double-check valves. These taps should only be used in existing buildings to replace those having no back flow protection.

of the bib tap are common to all types of screw-down valves. When the capstan head is turned, the threads on the spindle engage the threads inside the head, causing the spindle to move upwards or downwards, thus opening or closing the water supply.

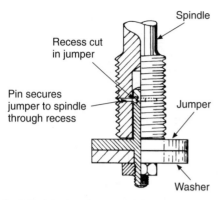

**Fig. 6.3** Fixed jumper in screw-down tap.

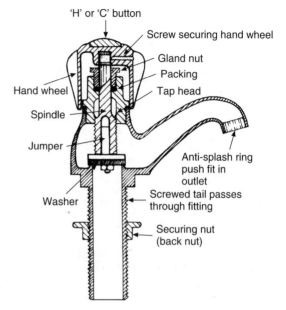

**Fig. 6.4** Pillar tap.

Another type of bib tap is shown in Fig. 6.2(b). Usually fitted externally these taps are fitted with a union on the spout to which a hose can be connected. Because the use of hoses constitutes one of the major risks of back siphonage, all water supplies to these taps must incorporate a double-check valve. They may be installed either in the pipeline inside the building, which is preferable, or in the tap itself as shown. Because they cannot be effectively drained, they are very prone to damage in frosty weather. The subject of back siphonage is dealt with fully in Book 2 of this series.

Until the revision of CP 1010 all cold water taps were fitted with loose jumpers. Hot water taps have always been provided with fixed jumpers as the pressure on most hot water systems is low and sometimes insufficient to lift the jumper. By pinning it to the spindle, this problem is overcome. As so many taps fitted in modern plumbing systems are also fed from a low-pressure supply, i.e. the storage cistern, it has been found necessary to provide fixed jumpers in those taps used for cold supplies for the same reason. Figure 6.3 shows a detail of the usual method of fixing jumpers.

### Pillar taps

These are also of the screw-down pattern, but unlike bib taps, they are designed for fitting directly into the sanitary appliance and secured to it by a back nut. Chapter 7 deals with the methods used for fixing these taps to ensure a watertight joint. The pillar tap illustrated at Fig. 6.4 is of modern design, the easy-clean shield being made in such a way that it also serves to turn the tap on or off. Many of

these modern taps employ acrylic plastic for the manufacture of this type of head. Pillar taps of a similar design but with extended bodies are used with sink units. The upward sweep of the outlet enables sufficient clearance under the tap for filling kettles and culinary equipment.

### Stop valves

Yet another member of the family of screw-down valves is the stop valve (see Fig. 6.5(a)). These are fitted in a straight run of pipe, their purpose being to control the supply of water into a building, or control one fitting or a group of fittings independently of others in the same building. They are available with union ends as shown suitable for copper, or they can be obtained with threaded ends conforming to BS 21 which enables connection to LCS pipes. Stop valves are also made with capillary and compression ends for copper tubes made to BS EN 1057. Most copper-fitting manufacturers also provide adaptors so the compression-ended stop valves can be used with various types of plastic pipework.

The stop valve illustrated at Fig. 6.5(b) incorporates a drain-off cock, and is made in such a way that the drain-off cock can be fitted and be accessible irrespective of the position of the stop valve. Such valves are intended for use as the main

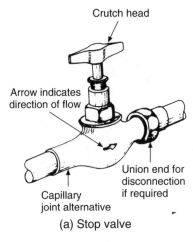

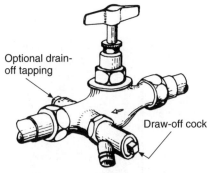

(a) Stop valve

(b) Stop valve with drain-off cock

When the stop valve is turned off, the pipe line it serves can be drained as a precaution against frost.

**Fig. 6.5**  Stop valves.

control of the water supply on entry into a building, the drain-off enabling all the water in the pipework to be emptied as a precaution when the building is left without heat in periods of cold weather.

Stop valves for use below ground must comply with BS 5433. They are made of corrosion-resistant material such as gunmetal or bronze, have a loose jumper and are of heavier construction than those complying to BS 1010. An alternative to valves complying with BS 5433, which are now permissible, is a quarter-turn valve having working principles similar to those of the ball valve shown on p. 135 in Fig. 6.9. As they are used for control purposes only, they will not give rise to persistent water hammer. An additional advantage is there is little or no pressure drop through ball service valves as they are classified as 'full way', unlike those of the screw-down pattern.

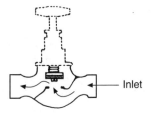

**Fig. 6.6**  Fitting stop valves. Care must be taken to fit the stop valve in the correct direction of flow.

Screw-down stop valves are stamped with an arrow indicating the direction of the water flow. Figure 6.6 shows the correct installation of a stop valve and the flow of water through it.

*Lock shields*

These are fitted to taps not intended for public use to prevent interference from unauthorised persons. They are provided on stop valves and bib taps (used by cleaners, etc.) in schools, hospitals and public toilets. Figure 6.7(a) shows the head of a tap fitted with a lock shield. The spindle terminates as a

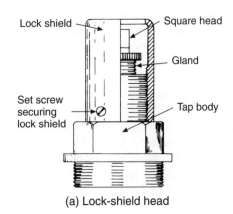

(a) Lock-shield head

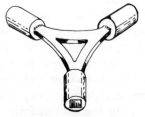

(b) Isle of Man key

Ends are of different sizes – for use with lock-shield valves and taps.

**Fig. 6.7**  Lock shields.

square below the top edge of the lock shield which is screwed over the head of the tap and secured with a set screw. These taps are provided with a loose key which should be handed over to a responsible person for safe keeping and use when the tap is installed.

A three-headed key (Fig. 6.7(b)), sometimes called an Isle of Man key, is one of the tools that should be part of a plumber's tool kit. Each of the three ends are recessed with squares of differing sizes for operating lock-shield valves if the original is not available when required.

### Ceramic disc taps

Taps of this type are quite different in operation to the screw-down pattern which were designed in Victorian times to replace the plug cock commonly in use at that period. Figure 6.8 illustrates in diagrammatic form the working principle of a simple disc valve. Unlike screw-down valves they have no seating or washer. These are replaced by two ceramic discs, one static, being fixed to the valve body so it cannot rotate; the other can be turned through 90°, usually by a lever.

The discs are provided with waterways which line up when the tap is turned on. Like most modern taps they have non-rising spindles which eliminate leakage from glands and it is claimed that discs, unlike rubber washers, will last the lifetime of the tap.

The principle of the ceramic disc is also used for mixing taps on wash basins. They are more complex than the simple illustration shown and should it be necessary to dismantle the tap, it is advisable to refer to the maker's instructions.

### Non-concussive taps

These are special taps which are designed to save water in public buildings and operate only when downward pressure is exerted on the head. The original types were quite simple, the taps being opened by compressing a spring, closure being effected automatically by the spring when pressure was released on the head. It was found, however, that they were very noisy in operation, the sudden closure causing water hammer (see Book 2, Chapter 3), especially in areas of high pressure. Modern non-concussive taps are rather more

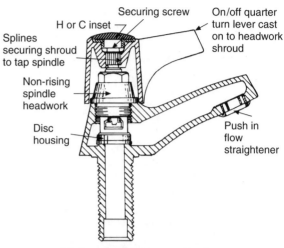

(a) Ceramic disc tap – pillar type

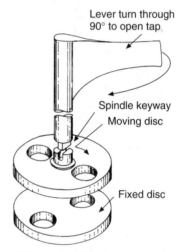

(b) Detail illustrating the operating principles of a simple disc tap

Note that although the discs are shown apart here, in practice the faces of each are close together in the housing. When the tap is in the off position no water can pass. By rotating the lever through 90° the holes in the discs line up allowing water to flow to the tap outlet.

**Fig. 6.8** Ceramic disc tap.

complicated, closure being effected gradually by means of a hydraulic valve inside the tap. In all cases where repair and maintenance of these valves becomes necessary, it is important to refer to the manufacturer's instructions, as wrongly adjusted, they are not non-concussive and will give rise to water hammer.

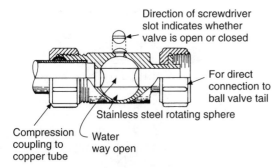

**Fig. 6.9** Ball-type servicing valve.

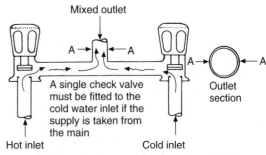

(a) Mixer tap for use when both hot and cold supplies are at equal pressures

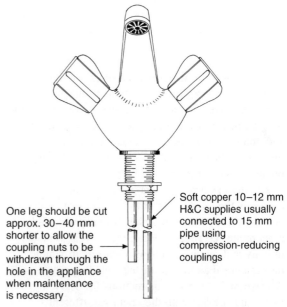

(b) Monobloc type mixer for single-hole wash basins

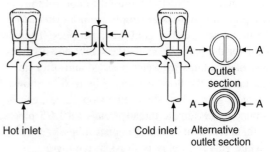

(c) Mixer tap suitable for unequal pressures

**Fig. 6.10** Mixer taps.

*Service valves*

These are fitted to plumbing fixtures such as flushing cisterns and storage cisterns so that maintenance work can be carried out without shutting down the entire system. Stop valves to BS 1010 and gate valves to BS 5154 can be used, but a much neater and cheaper alternative is the use of a ball-type servicing valve (not to be confused with float-operated valves) which is similar to a plug cock but has a circular seating instead of one that tapers. They are made with a variety of connections and may be straight or angle pattern. The ball valve shown in Fig. 6.9 has a compression joint for copper tubes on one end, the other having a union end suitable for direct connection to a float-operated valve.

They may also be used to isolate individual draw-off fittings such as central heating pumps, washing and dishwashing machines, water heaters and softeners. They should be fitted as near as possible to the inlet of the appliance they serve.

*Mixer taps*

It is often desirable to use a mixed or blended supply of hot and cold water. Typical applications are where the water is blended in a special mixer valve, as in the case of a shower fitting or a bath or sink when hot and cold water is delivered through a common spout. (Shower mixers are generally more complicated than those used on baths and sinks and are dealt with more fully in Book 2 of this series.)

Figure 6.10(a) shows in diagrammatic form a mixer tap where the water mixes in the tap body. It will be seen that if this mixer is used with unequal pressures, the water having the higher pressure (usually the cold water) would flow back into the hot water service subjecting the hot store vessel to excessive strain and possibly causing it to overflow. Despite the foregoing, the Water Regulations permit their use with unequal pressures, and providing a single check valve is fitted on the cold inlet and the mixer has an inbuilt type 'A' air gap, they may be connected directly to the mains supply. Such mixers may now be used for sinks, wash basins and baths having both equal and unequal supplies providing appropriate precautions are taken against contamination. They can be used with advantage where the hot water is provided from an unvented system. Figure 6.10(b) shows a 'monobloc' mixer tap of the type used in sanitary appliances having a centre tap hole.

To meet the demand for taps capable of providing a mixed supply at differing pressures without the disadvantages previously mentioned, the 'Biflow' mixer has been developed (see Fig. 6.10(c)). It will be seen that by using this type of mixer, both hot and cold supplies are quite separate until they are discharged from the outlet. The outlets are usually made to swivel through 180° so they can be used with a double sink. The one weakness in their design is the joint between the swivel of the outlet and the body of the tap which is made watertight by a neoprene 'O' ring. These seem to wear very quickly and require frequent replacement which necessitates the removal of the swivel spout. Always make sure the instructions for servicing these taps are left with the client as different manufacturers recommend various methods for removal of the swivel.

### Gate valves

Gate valves to BS 5154 are used to control the low-pressure cold water supply from a cistern to the hot store vessel or cold down services. They are ideal for the purpose, as unlike screw-down valves, they have a full-way water passage and offer no resistance to the flow of water. Figure 6.11 shows a typical gate valve which, like stop valves, can be obtained with female threaded ends for LCS pipes or for direct connection to copper tubes.

Gate valves employ no washer, having a metal-to-metal seating. A circular wedge-shaped gate closes into a matching wedge-shaped seating

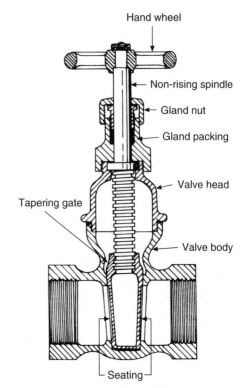

**Fig. 6.11** Gate valve suitable for capillary or compression connections for copper tubes or ends may be screwed for steel pipes with BSP threads.

when the valve is closed. They are always fitted with a wheel top and non-rising spindle. Some plumbers and fitters avoid installing these valves in an upright position, preferring a horizontal or even upside down position. This prevents any sludge or solids in the water from settling in the base of the seating and preventing the gate from closing.

### Plug taps

Sometimes referred to as 'quarter-turn' taps these are probably the oldest type of tap in use (see Fig. 6.12). Until the 1986 Water Bylaws they were not permitted on mains cold water services, but since this date they have found applications as service valves in some circumstances and are used almost exclusively by the gas industry for domestic appliances, main cocks on gas services and control taps on gas boilers.

The plug is operated by turning the tapered valve which is tightened into a corresponding circular

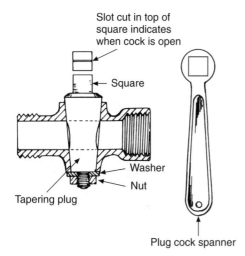

**Fig. 6.12** Plug cock. Turns on or off by a quarter turn. Used mainly on gas supplies and appliances.

tapering seating, a lock nut securing the valve into the body of the tap. Plug cocks do not require a great deal of maintenance; if they become difficult to turn, the plug can be removed by undoing the lock nut and applying a light smear of grease. Plug taps may be provided with a square head for which a suitable spanner is provided, and so that the position of the valve can be determined, i.e. whether it is in the open or closed position, a slot is cut into the square. When the slot is in line with the pipe run the valve is open; when it is at 90° to the pipe run, it will be in the closed position. Some of the smaller valves are provided with an elliptical-shaped finger plate instead of a square head.

### Drain-off cock

As the name implies, the chief use of these cocks (see Fig. 6.13) is to drain down hot or cold water services for maintenance purposes or to empty a system to prevent it becoming frozen. As these cocks are not in constant use it is often found that the washer has deteriorated, especially those used on hot water systems. It is good practice to replace the washer when the system is drained down as nothing is more annoying and time consuming than to find the cock leaks when the system has been refilled, thus involving another drain-down to replace the washer.

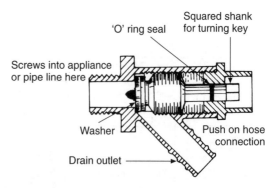

**Fig. 6.13** Drain-off cock.

### Automatic taps

Infrared sensing devices have been used for several years to control the volume of water needed to flush urinals, and this technology has now been extended to the use of taps and shower controls. Although these fittings can be installed in domestic properties, their main area of use is in public and commercial buildings where they are likely to reduce the volume of water used for hand washing and showers. They also have an advantage regarding hygiene in that it is unnecessary actually to touch the tap. This has advantages in hospitals, nursing homes and health clinics where cleanliness is absolutely necessary.

Figure 6.14 illustrates the typical installation of these taps, the basic working principles being that the sensor picks up a signal when the tap is about to be used. This completes an electrical circuit which energises a solenoid valve opening a supply of water to the tap. The electrical supply may be battery operated or mains with a suitable transformer to reduce the current to low voltage. It is recommended that a service valve and strainer are fitted to the inlet side of the solenoid valve.

Before installation it is important that the following points are considered:

(a) A mixed supply of water is necessary.
(b) Thermostatic control is essential to avoid scalding.
(c) Any mains electrical supply, if fitted, must comply with IEE Recommendations with no possibility of ingress of moisture.
(d) Maintenance checks must be made on a regular basis to ensure the taps are working properly. This is essential, especially in public buildings.

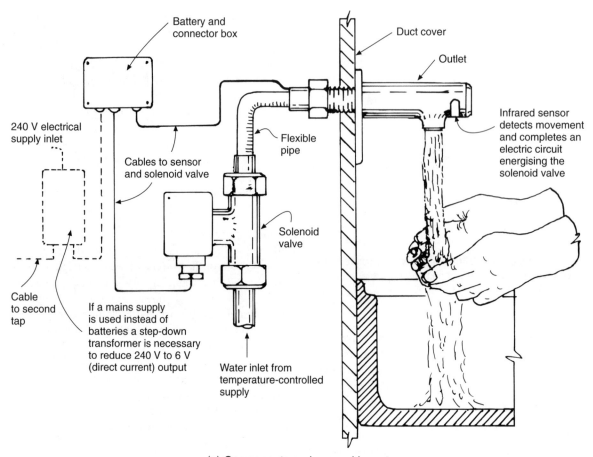

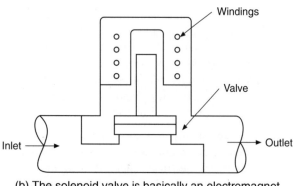

(a) Components and general layout

All electrical components must be accessible in a moisture-free environment. Special arrangements must be made to ensure safety with mains electricity!

(b) The solenoid valve is basically an electromagnet which, when energised, lifts the valve off its seating

(c) Wash basin outlet alternative

**Fig. 6.14**   Automatic taps.

## Rewashering and maintenance of taps

Generally speaking, the most common maintenance required to screw-down taps and valves is renewal of the washer. Soft rubber washers should not be used as these can cause water hammer, a loud knocking noise in the pipes.

The gland may also occasionally require repacking and PTFE tape will be found, in most cases, suitable for this purpose, but if on inspection the spindle itself is worn, the tap should be replaced. To gain access to the gland on those taps fitted with an easy-clean shield, the crutch or capstan head must be removed after first removing the set screw which secures it to the spindle. (Put the plug in the sink outlet first or the set screw is sure to go down it!)

If the head is still tight it is best removed from the spindle by using light taps on a copper drift with a hammer. A short piece of scrap copper tube is ideal for this purpose and, being soft, will not mark the chromium plating. When the gland has been repacked, lightly file the square on the spindle to allow the head to be replaced without resorting to force and secure by replacing the set screw.

The seatings of taps sometimes become pitted and, although the washer may have been renewed, a constant drip is still evident. It is possible to buy reseating tools which enable new seatings to be cut in old taps, but a defective seating in a tap is generally a sign of age and it will almost invariably be found that the spindles on such taps are badly worn. The most economic procedure in such circumstances is to renew the tap.

The gland seals on taps having non-rising spindles are usually of the 'O' ring type. To replace these the top of the operating handle or wheel must be removed by unscrewing the screw which secures it to the spindle. This will expose the tap headwork which must be unscrewed from the tap body. Great care is required here to avoid both damage to the sanitary appliance and the tap. Having removed the head the spindle may be taken out by pressing out the circlip housed in a groove on the spindle. This will expose the 'O' ring seals which should be replaced carefully by new ones. A little silicone grease will assist their replacement. During this operation the washer should be inspected for wear

and replaced if necessary, and the hexagonal guides, to which the washer plate is attached, should be checked to ensure there is no corrosion product present. It is a common failure with these taps that a build-up of corrosion will prevent the tap from functioning correctly.

One of the problems met by the plumber is the fact that taps not complying to BS 1010 have non-standard seal washers and ceramic discs, and unless the manufacturer is known, it can be difficult finding the right replacements.

Before taps are fitted to sanitary fittings they should be 'broken', that is to say the head should be removed from the body and a suitable oil or grease applied to the thread. This ensures that when rewashering is necessary at a future date, the head can be removed without resorting to excessive force. Many cracked basins are the result of attempts to remove the tap head to rewasher the taps. One method of supporting a pillar tap against the force applied by a spanner is to wrap a piece of cloth round the spout of the tap and carefully place a piece of pipe of suitable size over it to act as a lever against the pressure applied by the spanner (see Fig. 6.15). To be on the safe side, however, especially when considering wash basins, taps are best completely removed and held in a vice, care being taken to avoid damaging the chromium plating.

## Float-operated valves

These valves are best described as taps which close as the water level raises the float and automatically shut off the water at a predetermined level. They were commonly called ball valves until the ball service type of valve illustrated in Fig. 6.9 became popular. To avoid confusion they should be described as float-operated valves. There are two main types: Portsmouth and diaphragm. The British Standard for float-operated valves is BS 1212 Part 1 piston operated; Part 2 diaphram type, brass or gunmetal; Part 3 diaphragm type, plastic; Part 4 Specification for compact valves for WC flushing cisterns, e.g. the Torbeck pattern.

The smaller sizes of the Portsmouth valve have been superseded by the diaphragm type mainly for two reasons. Adjustment of the water level was made by bending the float arm, which is not

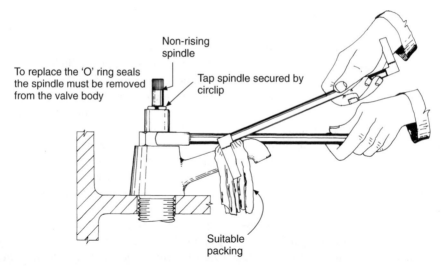

Non-rising
spindle

To replace the 'O' ring seals
the spindle must be removed
from the valve body

Tap spindle secured by
circlip

Suitable
packing

**Fig. 6.15**  Removing a tap head for maintenance. The spout is held with a basin (shetack type) spanner to prevent the body twisting in the sanitary fitting. Note the packing across the spout to avoid damage by the spanner to the surface plating. An alternative to the spanner shown is a suitable length of copper pipe held over the spout. Packing will still be necessary.

permitted on half-inch valves. The main reason, however, is due to the fact that, having a bottom outlet, the air gap requirements specified by the bylaws are difficult to achieve. In larger installations requiring float-operated valves supplied by pipes of 25 mm nominal bore or more, Portsmouth valves, usually of the equilibrium pattern, are still in use but special arrangements are made to accommodate them to ensure their use meets the requirements of the Water Regulations. Reference to air gaps and larger cold water installations is made in Book 2 of this series.

*Portsmouth valves*  The Portsmouth float-operated valve has a valve which moves in a horizontal plane, one of the best types of this pattern being that which complies with BS 1212 Part 1 (see Fig. 6.16). It has a removable seating, usually made of a tough nylon which has two advantages, one being it is easily interchangeable so that a high-pressure valve can be converted for medium or low pressure simply by changing the seating. The other advantage is that nylon is unaffected by cavitation, which is a form of erosion which causes the pitting of brass seatings. Cavitation is best described as the effect of air bubbles in high-pressure supplies bursting as they pass through the

restriction of the orifice. When this occurs with brass seatings a small quantity of metal is removed which, over a period of time, causes pits or cavities on the surface of the seating making it impossible for the rubber washer to seat properly and shut off the water. Portsmouth valves employ rubber washers which require occasional replacement by dismantling the piston. It is often found difficult to remove the cap which holds the washer in position on the valve body owing to the seepage of water corroding the thread. A light tap round the circumference of the cap will enable it to be removed more easily. When the washer has been replaced, all the many parts should be cleaned to remove any corrosion or fur that could cause malfunctioning of the valve when it is replaced.

*Diaphragm float-operated valves*  Diaphragm float-operated valves (see Fig. 6.17) should comply with BS 1212 Part 2 (brass body) and Part 3 (plastic body). They were used almost exclusively in flushing cisterns and due to the requirements of the Water Bylaws of 1986 they are becoming increasingly used in cold water and feed cisterns.

Unlike Portsmouth valves the diaphragm type will not permit back siphonage or back flow to take place even when completely submerged. They have

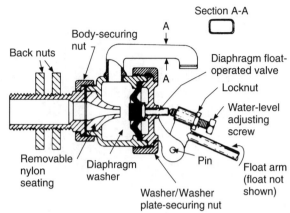

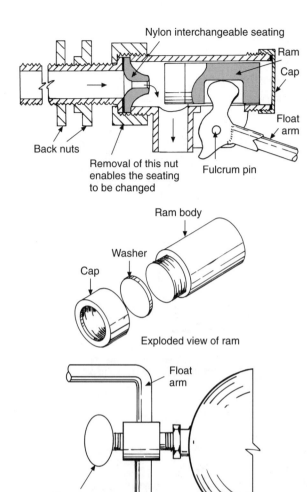

Water-level adjustment using Portsmouth float-operated valve. This applies to $\frac{1}{2}$ in valves only, as because of possible breakage of the float arm, the Water Regulations forbid its bending for adjustment.

**Fig. 6.16**   Portsmouth float-operated valve to BS 1212 Part 1.

**Fig. 6.17**   Diaphragm valve to BS 1212 Part 2 or 3.

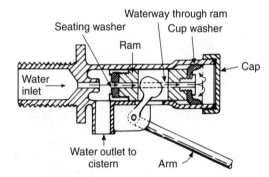

**Fig. 6.18**   Portsmouth-type equilibrium valve.

the advantage of fewer working parts than those previously described, and like the BS 1212 Portsmouth type, the seatings are interchangeable. These valves are made of brass or high-density polythene or polypropylene, both of which are unaffected by corrosion such as dezincification.

The water level can be finely adjusted by a screw, a further advantage over the Portsmouth type where adjustment was made by bending the arm – a difficult procedure, especially when the valve was

fitted in a narrow flushing cistern. Another advantage is the position of the outlet which can be situated on top of the valve thus reducing the risk of water pollution. The outlet can be moved through 180° to facilitate right- or left-hand entry into the cistern.

*Equilibrium valves*   Unlike those previously described which rely solely on the force exerted by the float to overcome the pressure exerted by the water, these valves utilise the pressure of the water to help close the valve. It will be seen in Fig. 6.18 that there is a waterway through the centre of the valve, one end of which carries a washer in the normal way, the other a cup washer, very similar in appearance to the type used in cycle pumps. Water passing through the hole in the valve exerts the same pressure on the cup washer (which tends to push the valve towards the seating) as the pressure of the incoming water tends to push it off. As these

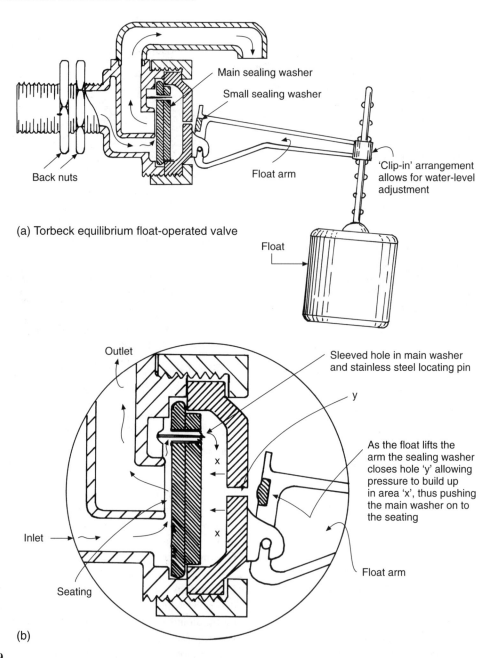

Main sealing washer

Small sealing washer

Back nuts

'Clip-in' arrangement allows for water-level adjustment

Float arm

(a) Torbeck equilibrium float-operated valve

Float

Outlet

Sleeved hole in main washer and stainless steel locating pin

y

As the float lifts the arm the sealing washer closes hole 'y' allowing pressure to build up in area 'x', thus pushing the main washer on to the seating

x

x

Inlet

Float arm

Seating

(b)

**Fig. 6.19**

pressures are equal they cancel each other out, hence the name equilibrium valve. The float on this type of valve has only to lift the arm whereas in the case of non-equilibrium types the effort provided by the float has to overcome not only the weight of the arm but also the pressure of the incoming water.

*Torbeck float-operated valves*  This type of valve is shown in Fig. 6.19 and is of the equilibrium type which enables a very small float to be used. These valves, being very small, are more easily fitted into cisterns of 6 litre capacity. The valve operates in the following way: as the cistern fills a limited amount

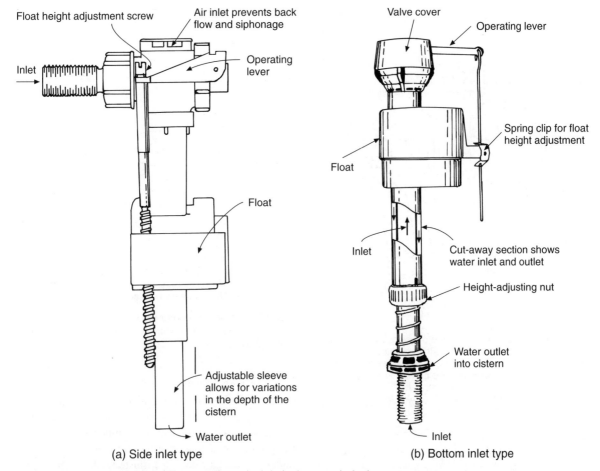

Fig. 6.20   Fluidmaster valves. The operating principle is the same in both types.

of water is permitted to pass into chamber x through a very small hole in the diaphragm which must be directly located on a stainless steel pin. The small hole y ensures that the pressure in x is lower than that on the inlet side of the washer until the float rises sealing off hole y by means of a small washer located in the float arm. Pressure now builds up in chamber x, and because of the larger surface area on the back of the diaphragm washer than that on the inlet side, it is pushed on to the seating closing the valve. The valve requires little maintenance but if it fails to close: (a) check the main washer for wear and ensure the pin is not binding on its sleeve; and (b) check orifice y.

*Fluidmaster float-operated valves*   To some degree the working principles of these valves are similar to the Torbeck valve, and like the Torbeck valve they occupy little space in a flushing cistern. Two patterns are made, one for side inlet and one for bottom inlet cisterns. They can be used with a traditional siphon or other types of flushing arrangements approved by the Water Regulations. Both types are shown in Fig. 6.20, and although they look different their working principles are the same. Discharge water from these valves enters the cistern at low level and little noise is heard as it fills. Both these valves are effectively protected against back flow and back siphonage. Figures 6.21(a)–(c) show their basic working principles which at first glance may look complicated. If, however, the position of the 'splines' in the stainless steel pin is noted when the valve is open and closed it will be seen how water is admitted and discharged from the pressure area.

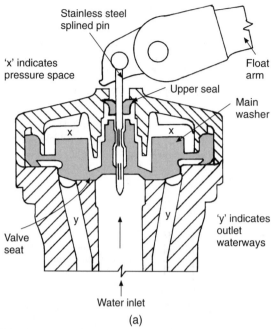

**'x' indicates pressure space**

Stainless steel splined pin

Float arm

Upper seal

Main washer

x   x

**Valve seat**

y   y

'y' indicates outlet waterways

Water inlet

(a)

This shows the valve in the closed position. Water, however, can pass through the splined pin to gain access to the pressure space 'x' where, owing to the larger surface area than the inlet, the pressure keeping the main diaphragm washer closed is greater than that forcing it up, thus keeping the outlet waterways closed.

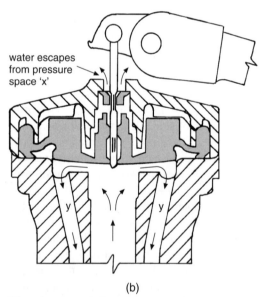

water escapes from pressure space 'x'

y   y

(b)

When the cistern is flushed the float arm falls lifting the splined pin. Water in the pressure space can now escape via the spline through the upper seal. The inlet pressure is now greater than that in the pressure space and lifts the main washer opening the outlet waterways.

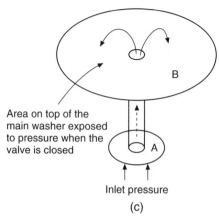

B

Area on top of the main washer exposed to pressure when the valve is closed

A

Inlet pressure

(c)

This illustration is not to scale but assume the area on the top of the main washer 'B' to be 0.00008 m² and that exposed to the inlet orifice 'A' to be 0.0007 m². Under a pressure of 1 bar the total pressure on 'B' will be 0.07 bar and that on 'A' will be only 0.008 bar. This illustrates how the valve is closed until pressure on 'B' is released through the upper seal when the float drops.

**Fig. 6.21** (a)–(c) Working principles of the fluidmaster valve. Note: The valve covers or shrouds are not shown in this illustration.

Like the Torbeck valve the float simply operates the arm, unlike the traditional valve where the buoyancy of the float has to overcome the pressure of water on the valve seat. It is important that the principles of total pressure over an area is understood. Figure 6.21(c) illustrates how the total pressure exerted on a small area, in this case the inlet, is overcome by that exerted on a larger surface, when both are subjected to the same pressure of water. Reference to the section on total pressure in Chapter 7 will enable the reader to understand this more fully. As with most float-operated valves, little maintenance is normally necessary apart from the occasional renewal of the main and upper seals. Limescale build-up in some areas may necessitate careful cleaning of the splined control stem, although the movement through the seals should prevent this.

Equilibrium valves are used with advantage in areas where very high mains pressures exist and where persistent water hammer may be encountered. All valves of over 50 mm nominal diameter are of the equilibrium type to reduce the size of the float that would otherwise be required to close the ram in an ordinary valve of this size.

*Floats*

These were at one time made almost exclusively of spun copper in two halves and usually soldered together. Acidic waters acting on the solder often caused it to disintegrate, resulting in the float becoming waterlogged and useless. While copper is still usual for the construction of large floats, polypropylene is almost exclusively used for those of small size. They are cheaper than those made of copper and generally give less trouble.

The main objections to float-operated valves which discharge at high level are (a) the noise caused by the water falling into the cistern and (b) high-velocity water from the main supply flowing through the orifice. Both causes of noise can be minimised by supplying flushing cisterns from the cold water storage cistern, especially in domestic premises. Prior to 1986 silencing pipes were allowed on any float-operated valve, but this is no longer permitted because of the possibility of back siphonage. The component shown in Fig. 6.22(a) is designed to reduce the velocity of the incoming water by breaking it up into finely divided jets. Figure 6.22(b) illustrates a flexible polythene silencing pipe which may be used on valves supplied by a cold water storage cistern.

Float-operated valves are designated as high pressure, medium pressure or low pressure and are marked HP, MP, LP. These terms relate to the size of the orifice inside the valve. Owing to the considerable force imposed by a high-pressure water supply on the cross-sectional area of the orifice there is a limit to the pressure against which a valve will close in relation to the orifice diameter. An increase in the size of the float would provide a greater opposing force to close the valve or a longer arm would provide more leverage, but neither of these two alternatives is a practical proposition, owing to the limited space into which they are often fitted, i.e. flushing cisterns. The pressures against which half-inch valves should close are given in Table 6.1.

**Table 6.1** Float-operated valve closing pressures.

| Valve type | Diameter of orifice | Max. pressure (Bar) |
| --- | --- | --- |
| High pressure | 3 mm | 13.80 |
| Medium pressure | 4–5 mm | 7.00 |
| Low pressure | 6 mm | 2.75 |

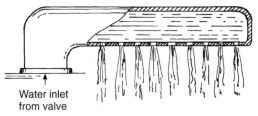

BS 1212 Parts 2 and 3 float-operated

### (a) Method of reducing noise in float-operated valves

This attachment has the effect of breaking up the water into a number of finely divided jets which reduce the velocity of the incoming water and the associated noise.

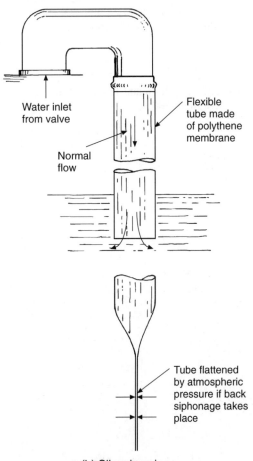

### (b) Silencing pipe

This arrangement permits the flow of water to discharge below the surface of the water in a manner similar to the original ridged silencing pipes. In the event of back flow causing a negative pressure, atmospheric pressure will flatten the tube as shown in the inset, thus preventing back flow or back siphonage. It should not be used in industrial premises or where the supply is taken from a service pipe.

**Fig. 6.22**

## Further reading

*British Standards (mentioned in text)*
BSI, 389 Chiswick High Road, London, W4 4ER.
   Tel.: 020 8996 7001
Copies of British Standards are available at public
   lending libraries for a nominal fee.

*Taps and valves*
Peglers (address as shown in Chapter 4).
Dart Valley Systems Ltd, Water Management
   House, Units 1 & 2, Alders Way, Yalberton
   Tor Industrial Estate, Paignton, Devon,
   TQ4 7QN. Tel.: 01803 529021.

## Self-testing questions

1. Describe the dominant feature of all
   screw-down taps.
2. Explain why it is necessary to fit lock-shield
   taps in public buildings.
3. State the purpose of 'breaking' a tap prior to
   fitting in a sanitary appliance.
4. State the main advantage of electrically
   controlled taps.
5. Explain why a gate valve is recommended for
   low-pressure supplies.
6. Describe the materials and methods used to repair
   the gland on a pillar tap having a rising spindle.
7. Describe two essential checks that should be
   made on a stop valve during a service visit.
8. Explain the working principle of float-operated
   equilibrium valves.
9. In relation to float-operated valves state the
   advantages of the diaphragm-type valve over
   that of the Portsmouth pattern.
10. Describe three possible causes of a float-
    operated valve failing to close.

# 7 The cold water supply

After completing this chapter the reader should be able to:

1. Identify the types of natural waters least likely to present problems in their subsequent purification.
2. State the physical properties of water.
3. Identify the causes of hardness in the water supply and its effect on pipework systems and components.
4. Understand the requirements of undergound service supply pipes and systems of supply in small buildings.
5. Describe the methods used to protect plumbing installations from frost damage.
6. State the need for correct positioning of tappings in storage cisterns.
7. List the main points requiring consideration in relation to the positioning of water storage cisterns in a roof space.
8. Describe the method of securing pipes of various materials to the building structure.
9. Describe the methods of making fixings to various types of building components.

## The supply of water

An adequate supply of pure water on tap is one of the prerequisites of modern living, a fact sometimes not fully appreciated by the average person. Water companies are required by law to provide a pure and wholesome supply of water, often referred to as 'potable' water, a term implying that it is fit for drinking and culinary purposes.

Before water is distributed throughout the water mains systems to individual premises, it must be collected and treated to rid it of any harmful water-borne bacteria and suspended organic and non-organic solids. Water companies constantly guard against pollution of main water supplies and every effort is made to ensure the supply of pure water to their consumers.

Prior to July 1999, water companies published bylaws which were enforceable with the backing of parliament and public funding; they were in effect owned by the state. When the water companies were privatised, Regulations were necessary to make the private companies directly responsible to the Secretary of State. The underlying purpose of the Water Regulations 1999, like the bylaws which preceded them, are as follows:

(a) prevention of contamination
(b) prevention of waste
(c) prevention of misuse
(d) prevention of undue consumption

It has been suggested that the previous legislation had a restraining effect on innovation, e.g. the introduction of new products and ideas. The 1999 Regulations do permit components and systems which would not have been permitted previously. The UK, being a member of the European Union, should have national plumbing standards compatible with those of Europe. These Regulations also reflect national responsibility to the environment which is especially important now in view of the effects of global warming and the need for conservation of water in an era where demand is increasing.

## Approved plumber scheme

The water companies are reluctant to police the Regulations, mainly because of the expense involved, but to ensure compliance they have set up a register of approved plumbers. Any suitably qualified person can apply to join the scheme, but they must meet certain criteria including a good knowledge of the Water Regulations. This should result in higher standards of work in the building services industry and people employing approved plumbers will have some redress in the case of poor work and non-compliance with the Regulations.

## Sources of water supply

### Rainwater

All water is derived from rainfall, which after use in various ways is returned to lakes, rivers or the seas, from where it evaporates to form water vapour and eventually falls again as rain. The term applied to this continuous succession of events is the 'Water Cycle', see Fig. 7.1.

Falling rainwater is quite pure, but as water is a natural solvent it is unlikely to be pure when it reaches the water storage areas such as the reservoirs and lakes. Rainwater falling through air which has been polluted by industrial gases, notably sulphurous compounds or heavy carbon dioxide concentrations, will absorb some of these gases and become a very weak acid, a quality which will increase its natural powers of dissolving other substances. This acidic rainwater as it runs through the earth's crust or as it flows over various rocks will dissolve and absorb further substances and, instead of being the pure water that it was in its original state, its nature is changed and it becomes soft or hard (i.e. acidic or alkaline) water, terms which will be explained later in this chapter.

The sources of supply from which water can be taken may be broadly classified under four main headings:

(a) wells
(b) springs
(c) upland surface water
(d) rivers

### Wells

Before the introduction of the mains water supplies provided by local water authorities, the main form of supply to each individual dwelling or small community was a well or pit dug into the ground, the water being extracted by a pump of some

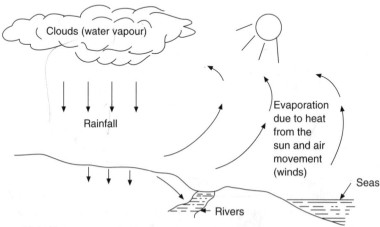

When the temperature of the atmosphere falls water vapour in the clouds condenses and falls as rain. Much of this is reconverted to vapour by warmth from the sun and air movement when it forms new clouds. Some, however, percolates through the earth and flows into rivers to produce the sources of water supply shown in Fig. 7.2.

**Fig. 7.1** The water cycle.

description. These wells were, and indeed still are, classified as being *deep* or *shallow*. Some water authorities still maintain deep wells and their associated pumping equipment to supplement their supplies taken from other sources. Others retain them for standby use in cases of drought.

The terms deep and shallow do not necessarily refer to the physical depth of the well, but whether or not its supply is derived from above or below the first impervious stratum of the earth. The earth is made up of differing layers of material, some of which permit the percolation of water and are called pervious, and some of which are more dense, usually a form of rock, through which water cannot pass and are referred to as impervious. Some water does find its way under this impervious strata by means of what the geologist calls 'faults' or 'slips' and outcrops.

*Shallow wells* are those which are dug into the earth above the first impervious stratum until a water-bearing vein is tapped. The water yielded by shallow wells is usually derived from the immediate locality and in bygone days was often in close contact with cesspits and refuse heaps, there being no waterborne systems of drainage. As a result the water was often heavily contaminated and where such wells are still in existence, water taken from them should be regarded with the utmost suspicion.

A *deep well* is one which penetrates the first impervious stratum. Figures 7.2(a) and (b) indicate the difference in the source of water provided by deep and shallow wells. Water found under impervious strata will usually have travelled long distances underground and as such will have been subjected to natural purification, making it pure and wholesome. Such a well, if sunk into the ground in an area where the first impervious stratum is relatively near the surface, may not be as deep as a well classified as shallow.

*Artesian wells* (see Fig. 7.2(c)) are formed as a result of the character of the surrounding strata and the water table. The term water table relates to the natural level of water under the earth. In the case of most wells this is below the outlet, which necessitates mechanical means of raising it to the surface. In the case of artesian wells, however, the water table is above the mouth of the well which forces water to the surface and may even form a natural reservoir or lake. If the water from such a well can be piped it can prove to be an effective source of supply for small communities. It is unlikely to dry up during prolonged periods of low rainfall because of the wide area from which it is supplied.

### Springs
Water obtained from this source, like well water, depends for its purity largely upon the distance it has travelled underground and whether or not it may be classified as a deep-seated spring. This term, as with well water, relates to its origin and whether it is derived above or below the first impervious stratum. While spring water is seldom capable of supplying more than a small number of isolated properties, a combination of several in the same area often form the source of rivers. For the different types of springs see Figs 7.2(d) and (e).

### Upland surface water
Several large water undertakings in the UK depend almost entirely upon this source of supply. This source is mostly found in the north and north western parts of the country where the isolated ranges of hills and mountains are favourable to the damming of streams to form impounding reservoirs. Lakes and natural reservoirs are also classified under this heading. Water from such a source is usually of good quality being comparatively free from human and animal contamination and suspended solids. As the majority of the water collected and stored in these reservoirs is derived from the 'run-off' of mountainous surfaces and has not percolated through the earth in great quantities, it tends to be soft, and if it has had any contact with moss or peat, it is likely to be acidic to the detriment of metals used in plumbing systems – as will be explained later.

*Artificial reservoirs* Owing to the increase of housing and industry in some parts of the country and the inability of natural sources of supply to meet the demand, artificial reservoirs have been created by flooding low-lying areas of land. The water stored in these reservoirs is derived mainly from rain and surface water, the quality being similar to lakes and natural reservoirs.

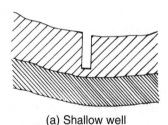

## (a) Shallow well

Water collected from immediate locality is often polluted by animal grazing and human habitation. A shallow well does not penetrate an impervious stratum.

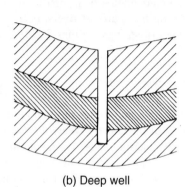

## (b) Deep well

Generally pure wholesome water is yielded which has been purified by natural means. A deep well has penetrated an impervious stratum.

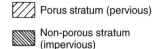
Porus stratum (pervious)

Non-porous stratum (impervious)

**Fig. 7.2**  Sources of water.

The arrows indicate entry of water under the first impervious stratum. These wells are often the source of lakes

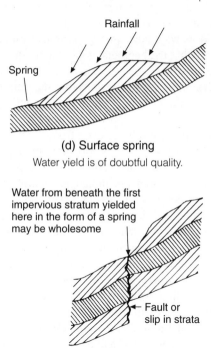

## (c) Artesian well

Water will be held underneath the impervious stratum to the level shown by the dotted line thus there is a head of water available to force the water out of the well under pressure.

Rainfall

Spring

## (d) Surface spring

Water yield is of doubtful quality.

Water from beneath the first impervious stratum yielded here in the form of a spring may be wholesome

← Fault or slip in strata

## (e) Deep-seated spring

### River water

Water taken from rivers provides the main source of supply where other forms of catchment are not available. The purity of river water is generally very suspect, especially if taken from the lower reaches. Rivers are liable to pollution by drainage from manured fields, effluents from sewage works, waste water from industrial processes and solid waste from factories and waste tips. The classification of water abstracted from rivers will, therefore, depend largely upon the source of the river and the characteristics of the tributaries which flow into it. Despite its disadvantages, many parts of the country rely solely on this source of supply.

### Characteristics and physical properties of water

Pure water is clear, tasteless and colourless. It is a compound of two gases, two parts hydrogen to one

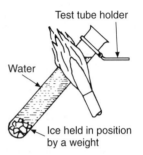

part oxygen. The chemical formula for water is given as $H_2O$.

The maximum density of water occurs at a temperature of 4 °C. This is to say the molecules in a given quantity of water are as small as possible and occupy the smallest volume possible at this temperature. Expansion of the molecules occurs if the water is heated or cooled above or below this temperature. Expansion takes place rapidly when the temperature drops to 0 °C and the water solidifies to become ice. An interesting fact emerges here: during the period of time that a change of state, i.e. liquid to solid, is taking place, heat is given off by the water while no visible drop of temperature is recorded by a thermometer. This heat is called 'latent' or hidden heat. The thermometer will not record any change of temperature until all the water has changed to ice. A total of 335 kJ must be given off by 1 kg of water at 0 °C before it changes to 1 kg of ice at 0 °C.

A similar occurrence takes place when water is heated to boiling point, i.e. 100 °C. At this temperature a change of state takes place in that the water begins to change from a liquid to steam which is a gas. Although heat is continually applied during this change, a thermometer will record only 100 °C until all the water has been converted to steam, and then the temperature will begin to rise, providing the steam is under pressure. It requires 2,258 kJ to convert 1 kg of water to 1 kg of steam. It will be seen that 1 kg of steam at 100 °C has a greater amount of energy stored in it than water at 100 °C. This being so, when the steam is converted back to water, it will release its heat energy. This fact is useful when steam heating of water is considered. A good example of this can be seen in cafés and restaurants where steam is injected into a cup of hot water so it will boil within seconds for tea-making purposes.

Yet another fact should be noted when the temperature of water is raised or lowered above or below 4 °C. In both cases an increase in volume takes place. When water turns to ice it expands by approximately a tenth, which is quite considerable. This causes pipes to burst when the water inside freezes and expands. Figure 7.3 shows how two ice plugs form in a pipe which is exposed to draughts causing the water between them to be compressed,

**Fig. 7.3** Frost bursts. Pressure by plugs of ice cause expansion thus weakening the pipe wall, finally causing it to split or burst. Note: water will not run from the burst pipe until it thaws.

**Fig. 7.4** Experiment to show water is a poor conductor of heat. Water at top end of test tube is boiling while ice remains unmelted at bottom of test tube.

first stretching the pipe walls and finally causing them to split. The damage caused by leaks when the temperature rises and the pipe thaws is often very serious and every effort should be made to position water pipes where they are least likely to become frozen, or when this is impossible, to provide the best possible insulation. Where pipes are fitted in very exposed positions, i.e. standpipes, farm buildings, etc., pipes in the polyolefin group of plastics can be used with advantage as these materials, owing to their inherent elasticity, will stretch as ice forms in the pipe and return to its original size on thawing.

Water, like most other fluids, is a poor conductor of heat and a simple experiment is often used to illustrate this point (see Fig. 7.4). The experiment shows how the water in the top of the test tube can be made to boil while the ice at the bottom remains unmelted.

### Relative density

The density of pure water at 4 °C is taken as the standard with which to compare the density of other materials. The term density relates to the mass per unit volume of a substance. For example, a thermolite building block, having a sponge-like

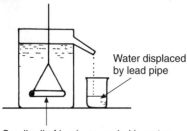

Small roll of lead suspended in water

**Fig. 7.5** Calculating the volume of an irregular shape.

structure which contains large quantities of air, will have less mass and a lower density than an engineering brick.

When comparisons are made between the densities of materials, it is important to ensure that the unit volume of each substance is the same. As an example, if the relative density of a piece of lead is to be found, its volume must first be ascertained. The simplest way to do this is to immerse it in a container of water carefully filled to the invert of the overflow as shown in Fig. 7.5. The amount of water displaced will be exactly the same volume as the piece of lead. If the water and lead are weighed, it will be found that the lead weighs 11.4 times the weight of the water, and its relative density is, therefore, said to be 11.4. Materials less dense than water, e.g. polythene, which has a relative density of approximately 0.9, will float.

Some knowledge of relative density is useful when comparing, for example, insulation products, as those having a lower density are likely to provide more effective insulation because of the volume of still air they contain.

### Volume

Volume is the word used to describe the space taken up by an object and two basic units of volume in the SI (Système International) metric system are the cubic metre and the cubic centimetre. The latter is very small and for purposes of calculations a rather unmanageable unit when dealing with the relatively large quantities of water met in plumbing calculations. If one can imagine a small cube, roughly the size of a small cube of sugar, measuring 10 mm on all sides, it will be realised just how small it is. Although the cubic metre is rather a large unit and many of the tanks and cisterns used in plumbing would not hold such a quantity of water, it is the most convenient unit to use to express volume. Volumes of less than 1 are expressed as a fraction of 1 m³.

An alternative, although technically not an approved SI unit, is the litre, this being the unit commonly used for plumbing and heating calculations. A litre of water at 4 °C will occupy a cube with sides of 100 mm square, and its mass will be equal to 1 kilogram. It is useful to remember that 1 m³ contains 1,000 litres. This is very convenient, because if the contents of a cistern are calculated in litres, not only is the volume obtained but also the weight of the water it contains. This has already been dealt with but is worth repeating here for clarity. Assume that the contents and the mass of water contained in a rectangular cistern are to be calculated. The cistern measures 0.450 m wide, 0.500 m high and 0.600 m long. Using the formula for finding volume:

$$\text{Volume in cubic metres (m}^3)$$
$$= \text{Width} \times \text{Height} \times \text{Length}$$
$$= 0.600 \times 0.450 \times 0.500$$
$$= 0.135 \text{ m}^3$$

As there are 1,000 litres in a cubic metre and its mass is 1,000 kg, by multiplying the volume in m³ × 1,000, i.e. 0.135 × 1,000 = 135, it will be found that the capacity of the cistern is 135 litres and the mass of water 135 kg. Furthermore, note how easy it is to multiply a number by 10, 100 or 1,000: the decimal point is simply moved one, two or three places to the right.

### Soft water

Water dissolves many of the metals and minerals with which it comes into contact, especially soft acidic waters which, if used with lead pipes for domestic supplies, can cause lead poisoning. Such water is said to be plumbosolvent.

Poisoning owing to the use of lead pipes is very uncommon today as lead and its alloys are no longer permitted for use in water supplies. Lead pipes may still be found in use in older properties but the water authorities are aware of the danger and in districts where water is plumbosolvent steps are taken to minimise the problem. Unfortunately as the cost of replacing all existing lead service pipes would be extremely high, they are simply replaced by alternative materials at the end of their service life.

Soft water also dissolves and takes into solution small quantities of copper and iron. As only small amounts are dissolved there is very little danger to health but copper-bearing water does tend to leave a bluish stain on sanitary fittings, while dissolved iron often discolours the water, especially in hot water systems. This can readily be seen if the sediment in the base of the hot water storage vessel is disturbed by a brisk circulation.

### Hard water

Much of the rain which falls percolates into the earth where it dissolves many of the minerals with which it comes into contact. Water that falls on and percolates through chalk and limestone is said to be 'hard'.

Unlike soft water which lathers easily, hard water makes the formation of lather very difficult. The reaction of the stearates in the soap and the lime content of the water produces an objectionable scum on its surface, and on the sides of sanitary fittings.

It should be made clear that there are two types of hardness called 'temporary' and 'permanent'. Both forms are undesirable as they present difficulties in laundering and in industries where steam plant is necessary for heating or manufacturing processes. Another disadvantage of hard water is the extra cost of soap and detergents that its use entails.

Permanent hardness occurs owing to the natural solvency of pure water which enables it to dissolve the sulphates of limestone. This type of water, however, is not such a problem in plumbing schemes as temporary hard water which is responsible for the formation of a hard scale which accumulates on the inside of boilers and circulatory pipes. This scale has the effect of insulating the water contained in the boiler from the source of heat, causing waste of fuel and, in the case of LCS boilers, being responsible for overheating of the boiler plates with the resultant possibility of buckling and distortion.

Temporary hard water occurs when the water, containing carbon dioxide gas absorbed as it fell as rain, dissolves the chalk and calcium carbonates which are very common in the earth's strata in many parts of the country. These carbonates are insoluble in pure water but, by the action of the carbon dioxide in the water collected from the atmosphere, the carbonates are converted into soluble bicarbonates. When such water is heated to approximately 65–70 °C the gases it contains, including the carbon dioxide, are given off. The water is then no longer able to retain the calcium in solution and it is precipitated in the form of scale or fur. The examination of the inside of a kettle used in a hard-water area will illustrate very clearly how serious this scale build-up can be. For this reason it is not advisable that water known to have a high temporary hardness content should be heated to high temperatures, 65 °C being the maximum recommended.

Although it would be too expensive to remove the hardness content of water completely, most water authorities in the hard-water districts do their best to reduce it to manageable proportions. Various methods are used for softening the water depending on the nature of the water concerned. The usual method used by water undertakings to remove some of the hardness in the water is the soda lime process. Extra lime is added to the water which amalgamates with that already held in solution. This causes the total lime content to reach saturation point and when this happens, both the lime which was added and some of that which was already in the water are precipitated. This has the effect of reducing the temporary hardness content. The addition of soda reduces the permanent hardness content of the water by acting on the calcium sulphates to convert them into insoluble carbonates which fall out of the solution as solid particles. Both these processes are carried out in sedimentation tanks, the water being kept as still as possible during the process.

The treatment of the water by water companies for removing hardness is not sufficient for all purposes and in some cases water is softened to a greater degree by private individuals using water softeners. This process is described fully in Book 2 of this series.

Water is classified as being hard or soft by the number of parts of calcium it contains per million

**Table 7.1**  Classification of water hardness.

| Classification | Parts per million |
|---|---|
| Soft | 0–50 |
| Moderately soft | 50–100 |
| Slightly hard | 100–150 |
| Moderately hard | 150–200 |
| Hard | 200–300 |
| Very hard | 300+ |

parts of water. As water may be both temporarily and permanently hard, both types are grouped together for this purpose to give the 'total' hardness content, as shown in Table 7.1.

## Water treatment

It is not intended to deal with this subject in great depth, but the plumber should have a general knowledge of the main processes of water purification.

### Sedimentation
When water is stored and allowed to remain undisturbed in a lake or reservoir, solid impurities suspended in it such as grit, mud and decaying vegetable matter sink to the bottom, this being known as *primary sedimentation*. Storage also has the effect of reducing the bacteriological content of the water, as lack of suitable food, competition with harmless organisms and low temperatures all have the effect of preventing the multiplication of harmful or *pathogenic* bacteria. Water companies, therefore, take advantage of this fact and storage is often used as a form of primary purification. When it is required, water is pumped from the storage reservoir, through a coarse metal strainer and allowed to flow into sedimentation tanks, where further controlled sedimentation takes place before the water is filtered.

### Filtration
*Slow sand filtration* is an old established method of water purification and is still used by many water companies. It consists basically of allowing the water to flow over a graded sand bed on top of which large colonies of minute vegetable growth

called algae have formed naturally. It is largely the action of the algae and filtration that purifies the water.

Owing to the slow action and the necessity constantly to replace the top 50–75 mm of the sand bed, *rapid sand* and *pressure filters* are gradually replacing those of the slow sand type. Rapid sand filters are very similar to slow sand types but operate under a greater head of water. There is also a difference in the depth of the sand and the grades used.

These pressure filters are usually housed in a steel container and are capable of dealing with 4,500 litres of water per hour. Filters worked at this rate very quickly become clogged, which necessitates frequent back-washing. This process is, however, entirely mechanical and does not involve the manual labour necessary to clean slow sand filters.

### Sterilisation
The final treatment of water to ensure its purity before it enters the main is its dosing with chlorine or a mixture of chlorine and ammonia. These are both sterilising agents which rapidly kill off any harmful bacteria which may have escaped filtration.

### Special treatment
This is carried out by water undertakings in areas where the water is either very hard or very soft. In many cases, especially in the case of the former, it would be too costly to produce a perfect water, but most water companies undertake to produce water of an acceptable quality. Figure 7.6 shows a

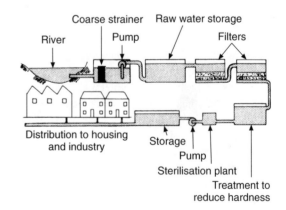

**Fig. 7.6**  Water treatment flow chart.

diagrammatic illustration of the main processes of water treatment.

### Pressure in pipes and fittings

It is necessary for a plumber to have a basic knowledge of the effects of pressure or force on the pipes and fittings being installed. The internal pressure to which a pipe or vessel is subjected will affect the strength of the material specified, i.e. a higher pressure will require the use of a thicker or stronger material. The earth exerts a positive 'pull' on everything on its surface which we call the force of gravity. When an object is said to be heavy two things are really being considered:

(a)  its weight in comparison with other objects
(b)  its mass in relation to the pull of gravity

It is the latter that is considered in relation to water pressure and may be defined as a force expressed in pascals (Pa) distributed over an area in metres squared ($m^2$). Thus the unit of pressure can be expressed as $Pa/m^2$. The pressure exerted by a column of water may be expressed in metres head and it is well known the higher a cistern is fitted the greater will be the pressure of water at the outlet. However, this does not indicate the force or pressure to which a pipe or vessel is subjected.

The downward force exerted by 1 kilogram of water (1 litre) may be expressed as 9.81 Pa and is shown in Fig. 7.7. If 10 of these cubes are placed on top of each other as shown in Fig. 7.8, giving a height of 1 metre, the intensity of pressure becomes

$$9.81 \times 10 = 98.1 \text{ Pa}$$

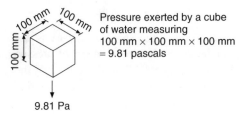

9.81 Pa

**Fig. 7.7**   Pressure exerted by 1 litre cubed (1 kg force).

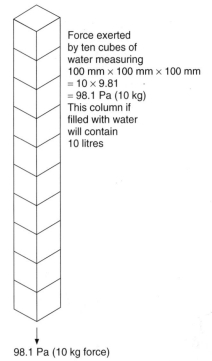

Force exerted by ten cubes of water measuring 100 mm × 100 mm × 100 mm
= 10 × 9.81
= 98.1 Pa (10 kg)
This column if filled with water will contain 10 litres

98.1 Pa (10 kg force)

**Fig. 7.8**   Pressure exerted by a column of cubes containing 1 litre of water.

If we now consider a cistern exactly 1 metre square (see Fig. 7.9) it will have a capacity of 1 cubic metre of water having a mass of 1,000 kg, the pressure on its base also being 1,000 kg, which when converted to pascals becomes

$$1,000 \times 9.81 = 9,810 \text{ Pa/m}^2$$

It will be seen that the pascal is a very small unit and its use leads to excessively large numbers – this makes them difficult to manipulate which can result in errors. For this reason it is usual to divide the number of pascals by 1,000 and call them kilopascals (kPa). Thus

$$\frac{9,810}{1,000} = 9.81 \text{ kPa/m}^2$$

This is a more convenient number to use for water pressure and is normally adopted for plumbing calulations. It will also be noticed that 9.81 is often approximated to 10 for many practical purposes, as it simplifies any calculation and has little effect on the final answer.

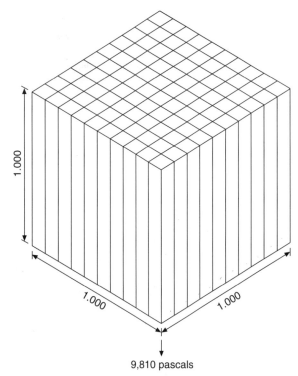

9,810 pascals

**Fig. 7.9**  Pressure exerted by 1 m³. There are 100 columns of water like that shown in Fig. 7.8 here, each exerting 98.1 Pa, so the pressure on the base of 1 m³ will by 98.1 × 100 = 9,810 Pa or 98.10 kPa/m².

The foregoing relates to the 'intensity' of pressure over 1 square metre owing to the height – in this instance a column of water. Where the area under consideration is more or less than 1 m, the terin m² is sometimes omitted simply and Pa or kPa only used to indicate a given pressure.

Figure 7.10 shows a common example in plumbing, namely that of the pressure on a tap fed by a high-level cistern. The pressure here is shown both as metres head and kPa/m². This example also introduces the bar as a method of expressing pressure.

*The bar*
In some cases involving very high pressures, e.g. compressed oxygen or mains water supply, a unit of pressure called the 'bar' is often used. Permissible working pressures on pipes are also given in bars. While the bar is not a unit officially approved by the scientists responsible for the evolution of the SI metric system, it is a very convenient unit

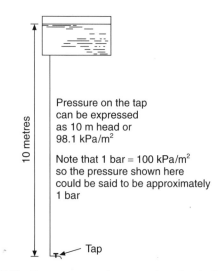

Pressure on the tap can be expressed as 10 m head or 98.1 kPa/m²

Note that 1 bar = 100 kPa/m² so the pressure shown here could be said to be approximately 1 bar

Tap

**Fig. 7.10**  Pressure exerted on a tap by a head of water.

commercially, 1 bar being equal to 100 kPa/m². To give a practical example of its use, the maximum working pressure to which 15 mm Table X copper tube should be subjected is 58 bars.

*The millibar*
Yet another unit of measuring pressure is the millibar (mb). Its actual value is a thousandth of a bar. One mb is a very small unit and exerts a pressure equivalent to 1 centimetre head of water. It is a very convenient unit for measuring very low pressures.

*Total pressure*
This relates to the pressure exerted over a given area subjected to a column of water of a given height. We know from the foregoing text that 1 cubic metre of water exerts a pressure of 9.81 kPa/m². Assume a vessel has a base measuring 3 m × 2 m and is 1 m in height; it will hold 6 m³ of water. The total pressure on the whole area of the base would be

$$6 \times 9.81 \text{ kPa} = 58.85 \text{ kPa}$$

The 'total' force or pressure on the base of the vessel is 58.85 kPa but the *intensity* of pressure would still be only 9.81 kPa. If the base of the vessel is the same but its height is increased to 3 m then the intensity of pressure would be

$$3 \times 9.81 = 29.43 \text{ kPa}$$

but the total pressure or force on the base area would be

$$29.43 \times 6 = 176.58 \text{ kPa}$$

Apart from the foregoing the reader should always remember:

(a) **a cube of water of 100 × 100 × 100 mm has a mass of 1 kg and a volume of 1 litre**
(b) **a cubic metre of water has a mass of 1,000 kg and a volume of 1,000 litres**

### Water supplies to consumers

When a new supply of water is required, an application must be made to the local water company who will provide the necessary forms to be completed by the building owner. When approval has been obtained and the fees paid, the supply will be laid to a point approximately 150 mm outside the consumer's boundary, terminating with a stop valve. All the work is generally done by the water company's own employees but in some cases it can be done by plumbing contractors, and for this reason it is necessary for the plumber to have a working knowledge of the materials and methods used for this type of work.

High-pressure water mains are made of cast iron, PVC (coloured grey) and polyethylene (coloured blue). These two latter materials are not strong enough to accept a BSP thread and special components are used to permit tapping into these mains. To make connections to existing asbestos mains a gunmetal clamp or saddle is used (see Fig. 7.11(a)). These may be bolted as shown or fixed by drive on dovetail wedges. Tappings to PVC and polyethylene mains are achieved in a similar manner, saddles of PVC being jointed to the main by solvent cement, and those of polyethylene by fusion welding using electrically heated elements. The ferrule itself with these two materials is moulded on to the saddle as a complete unit. The use of plastic materials makes the actual operation of tapping much easier as the ferrule body is made as a cutter. By screwing it down to open a supply of water to the service pipe, it simultaneously cuts a hole in the plastic pipe thus avoiding the necessity of the special machine required for tapping cast iron (see Fig. 7.11(b)).

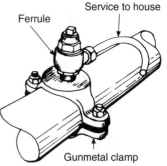

(a) Mechanical clamp for existing PVC and asbestos cement mains

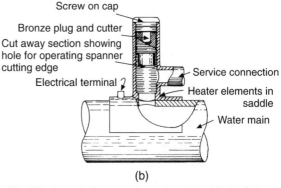

(b)

The joint is made between the main and saddle by fusion welding while the saddle is clamped on to the pipe. To cut a hole in the main the bronze plug is screwed down to its full extent with a hexagon spanner, forcing the cutting edge of the plug through the main; the disc cut from the main remaining in the plug. Having pierced the main the plug is screwed back as shown in the fully open position.

**Fig. 7.11** Ferrule connection to mains of various materials.

When tapping into cast iron mains a special machine (see Fig. 7.12(a)) is used which enables a new connection to be made without turning off the water supply and inconveniencing other consumers. The machine is constructed to be used in conjunction with the ferrule body shown in Fig. 7.12(b) which must be fitted into spindle 'A' of the machine after first screwing down the plug valve in the ferrule on to its seating. The other spindle 'B' carries a combined drill and tap (see Fig. 7.12(c)). The machine is then clamped on to the main over a leather washer and secured by an adjustable chain. The leather washer ensures a watertight joint between the main and the machine during the tapping operation. Pressure is imposed on the spindle 'B' by the screw which forces the combined drill and tap into the main, its

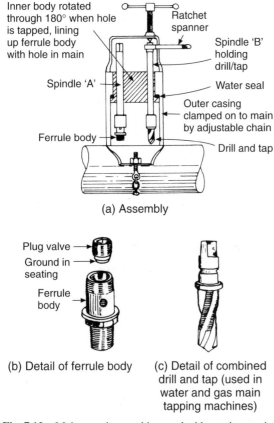

Inner body rotated through 180° when hole is tapped, lining up ferrule body with hole in main

Ratchet spanner

Spindle 'B' holding drill/tap

Spindle 'A'

Water seal

Outer casing clamped on to main by adjustable chain

Ferrule body

Drill and tap

(a) Assembly

Plug valve
Ground in seating
Ferrule body

(b) Detail of ferrule body

(c) Detail of combined drill and tap (used in water and gas main tapping machines)

**Fig. 7.12**  Mains tapping machine used with cast iron mains.

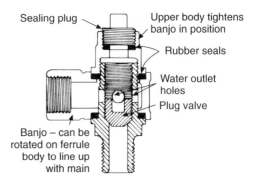

Sealing plug

Upper body tightens banjo in position

Rubber seals

Water outlet holes

Plug valve

Banjo – can be rotated on ferrule body to line up with main

**Fig. 7.13**  Main ferrule details.

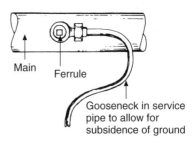

Main

Ferrule

Gooseneck in service pipe to allow for subsidence of ground

**Fig. 7.14**  Connection of service pipe to water main.

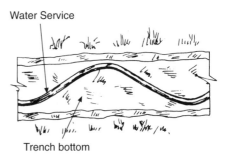

Water Service

Trench bottom

**Fig. 7.15**  Running service pipes in trenches. Note that subsidence provision in the case of polythene services is at the discretion of the supplier.

rotation being achieved by a ratchet lever fitted to the square on the top of the spindle. When the tapping has been completed the centre of the machine is rotated through 180° so the spindle 'A' lines up with the hole. By rotating the ratchet lever the body of the ferrule can then be screwed securely into the main and at this point the machine may be removed.

Figure 7.13 illustrates a half section of the complete ferrule; the shaded portion is the part that has been screwed into the main. The upper part of the ferrule is sometimes called the 'banjo' due to its shape and is made to swivel on the body of the ferrule, being secured to it by the upper body and made watertight by two rubber washers. It should be fitted so the outlet is in line with the main (see Fig. 7.14) which ensures that a gooseneck bend is made at the connection thus allowing for subsidence or shrinkage of the soil surrounding it. For the same reason service pipes should be 'snaked' from side to side of the pipe trench as shown in Fig. 7.15, although some water companies

do not consider this necessary for polythene service pipes owing to their inherent flexibility.

When the connection is completed, the water supply is turned on at the ferrule by opening the integral plug valve permitting water to flow through the holes in the ferrule body into the banjo service connection. To prevent any seepage of water past the valve and to prevent the ingress of earth into the ferrule, a sealing plug is fitted loosely into the top of the upper body.

The maximum size of a tapping in a main will depend upon its diameter and the diameter of the required service. It will be obvious that a large tapping into a relatively small main would be difficult to make watertight; it could unduly weaken the main and excessive draw-off might well cause starvation of water to other consumers at a higher level. For these reasons tappings exceeding 25 mm diameter are rarely used. In cases where large quantities of water are likely to be required, larger storage cisterns are usually recommended.

*Protection of service pipes*
Some soils are very acidic and metal service pipes should be adequately protected against corrosion. Pipelines can be completely sheathed in PVC or they can be wrapped with an anti-corrosive bandage before laying. Any covering removed for jointing must be suitably wrapped when the joint has been made. An alternative is to duct the service completely which has the added advantage of allowing the service pipe to be withdrawn for repair or renewal without a great deal of excavation, an important advantage when the service passes under a main road. It is important that the joints on these ducts, which may be clayware or PVC pipes, are made watertight. Any service pipe passing through or under the footings of buildings must be ducted in a similar way.

All mains and service pipes should be at least 750 mm beneath the surface of the ground as a precaution against damage by frost. The maximum depth of cover should not exceed 1.350 m as pipes buried at greater depths would not be readily accessible.

A diagrammatic illustration of a service pipe, its connection to the main and entry into the building is shown in Fig. 7.16. The square-headed stop valve

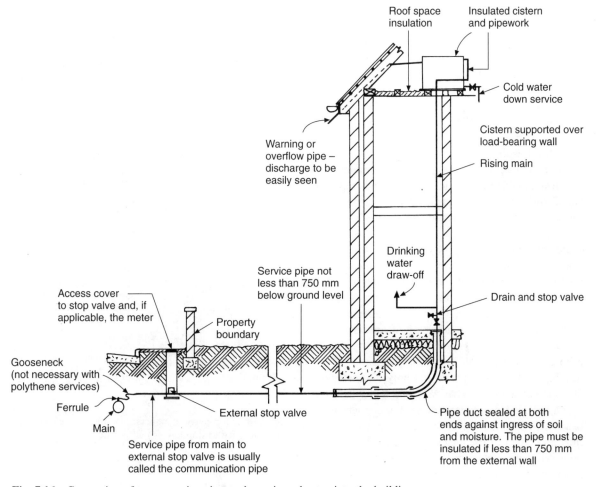

**Fig. 7.16**   Connection of water service pipe to the main and entry into the building.

**Fig. 7.17**  Stop valve ferrule. It should be noted that the square head shown here and in Fig. 7.18(a) is being superseded by crutch-headed stop valve as shown in Fig. 7.18(b).

fitted just outside the boundary line of the building is usually the limit of the water company's responsibility for the service. The only exception to this is where the main is laid under a public footpath and a stop valve ferrule (see Fig. 7.17) is used serving the combined purpose of connecting both the service to the main and as a stop valve.

Access to the stop valve may be obtained via a suitable pipe, 150 mm in diameter, placed over the stop valve, carefully supported on two bricks placed alongside the service. The top of the pipe is fitted with a cast iron cover which should terminate at the same level as the pavement (see Fig. 7.18(a)).

An alternative to the foregoing is a complete unit which is moulded in plastic materials and shown in Fig. 7.18(b). Its height is adjustable and the base is formed to offer full support to the stop valve.

*Water meters*
The metering of water supplies in commercial premises has been usual for a long period of time and is gradually being introduced for domestic properties. Large water meters for commercial premises are installed in a brick-built pit with a cast iron cover, but complete units for domestic properties having small meters have been developed. They incorporate not only the meter but also the service stop valve. Where a meter is fitted into a service having metal pipes, i.e. copper, a permanent earthing bond must be fitted across the inlet and outlet of the meter, so that in the event of removal of the meter earthing continuity is maintained.

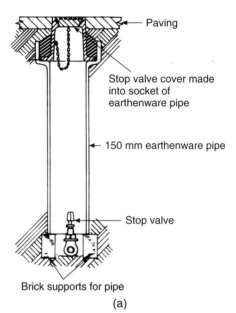

- Paving
- Stop valve cover made into socket of earthenware pipe
- 150 mm earthenware pipe
- Stop valve
- Brick supports for pipe

(a)

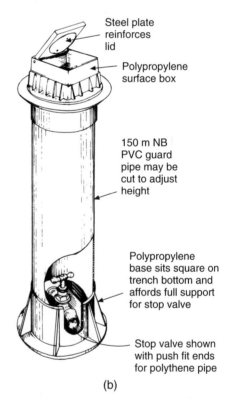

- Steel plate reinforces lid
- Polypropylene surface box
- 150 m NB PVC guard pipe may be cut to adjust height
- Polypropylene base sits square on trench bottom and affords full support for stop valve
- Stop valve shown with push fit ends for polythene pipe

(b)

**Fig. 7.18**  Stop valve chambers: (a) older method of access to outside stop tap; (b) all plastic chamber.

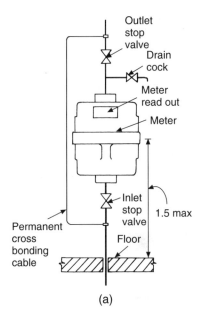

(a)

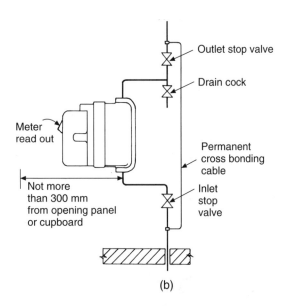

(b)

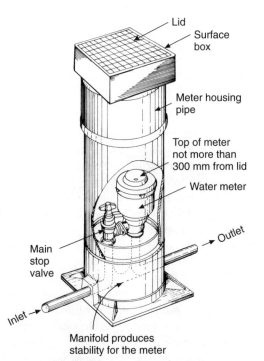

(c) Underground meter installation

**Fig. 7.19**   Water meter – installation details: meters must be positioned so that the dial or readout panel is easily visible; all pipework must be adequately supported and stop valve provided for servicing the meter; the meter must not be situated where it may be affected by frost, vibration or mechanical damage.

CP 6700 lists the following recommendations for the installation of meters for both outside and inside the premises as illustrated in Figs 7.19(a)–(c). In all cases the position of the meter should conform to the measurements shown. This ensures that it can be easily read and serviced when necessary. The special unit shown in Fig. 7.20 is made so that it can be fitted into the bodywork of an existing stop valve conforming to BS 1010 and BS 5433. When the existing stop valve is conveniently situated, its use can save a lot of time and expense.

Prior to the installation of a meter, a careful check should be made to ensure the underground service pipe is sound. Any possible source of water wastage such as defective taps and valves should be serviced or replaced.

### Entry of the service pipe into the building
A consumer's stop valve is fitted where the service pipe enters the building at as low a level as possible with a drain-off tap immediately above it. This enables all the water in the cold water pipes to be drained off if necessary in frosty weather or when repairs are needed. Any low points on the service which cannot be drained by the draw-off tap should have a separate drain tap. Pipes in unheated buildings should be well insulated and turned off and drained during long periods of frost when they are not in use.

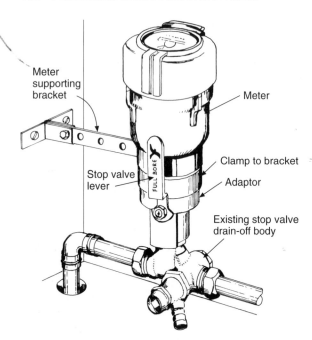

**Fig. 7.20** 'Aquadapt' in-line meter adaptor. This adaptor, marketed by IMI Yorkshire Fittings Ltd, can be fitted to an existing stop valve body after removing the headwork. It can be used in both the horizontal and vertical positions. It is essential that the meter is adequately supported. The adjustable bracket shown is obtainable from the manufacturer. (Courtesy of IMI Yorkshire Fittings Limited.)

### Cold water supply systems in buildings

The two systems used for the supply of cold water in a building are known as *direct* and *indirect* systems. The direct system (Fig. 7.21) has the following advantages: it is cheaper to install as the storage cistern will be smaller and in most cases less pipework is required, drinking water is obtainable from all fittings, and water storage for the hot water supply only is required.

Yet it does have certain disadvantages, the most obvious being a greater possibility of pollution, the causes and prevention of which are fully described in Book 2 of this series.

In areas where many new buildings have been constructed and the supplies are taken from an existing main, it may be found that at periods of peak demand the water supply is insufficient, as the main, which was probably laid many years before, is not big enough. Then again, if the mains are

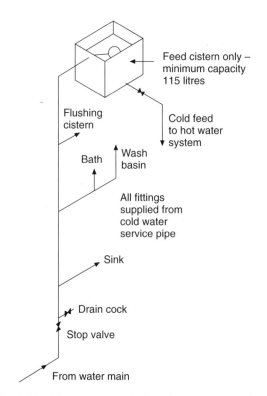

**Fig. 7.21** Direct system. Check and anti-vacuum valves are required by the draft model Water Bylaws (not shown in drawing).

under repair, the premises having no storage facilities may be without cold water for a period of time. Any leaks occurring in the system – the causes ranging from defective taps (i.e. glands) to burst pipes – will result in a greater waste of water. In the case of burst pipes, there will possibly be more damage to household effects because of the high pressures to which the pipes and fittings are subjected. For the same reason, systems of this type have a tendency to be noisy is use.

The indirect system of cold water supply, shown in Fig. 7.22 (do not confuse this with an indirect system of hot water supply), is obviously more expensive to install as extra pipework is required and more storage is necessary to provide water for both hot and cold water services. The minimum storage capacity recommended for such a system is 230 litres. It will be seen that only one draw-off tap is fitted directly to the main, this being for drinking or culinary purposes, and it is almost invariably the sink tap.

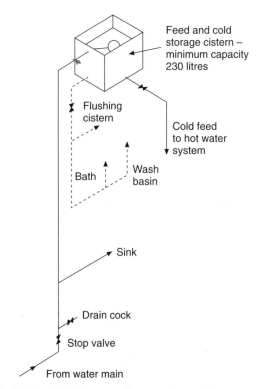

**Fig. 7.22** Indirect system. Larger storage cistern necessary to supply cold water to the bath, basin and flushing cistern; only the sink is supplied with drinking water directly from the service pipe.

Pipes supplying cold water from a storage cistern are called *distribution* pipes. The exception is the supply from a feed cistern to a hot water storage vessel, which is called a *cold feed*. The distribution pipes are represented in the illustration by the broken line. Owing to their relatively low pressure, full-way valves, i.e. gate valves, must be used for the control of these services, as stop valves offer too much resistance to the flow of water.

When fitting float-operated valves in flushing cisterns in a system of this type, always make sure they are either of the low-pressure type or, if the head of water is very low, of the full-way type. In a full-way float-operated valve the orifice through which the water passes is much greater than that in a high-pressure valve; this is to permit the cistern to be filled in a reasonable length of time.

Compared with systems of direct mains supply, indirect supplies are quieter in operation, there is less wear on fittings and some storage is available in case of mains failure.

**Pipe runs**

Pipe runs in roof spaces should be as short as possible, well insulated and never fixed within 2 metres of the eaves to avoid cold draughts, nor should pipes be run near ventilators and outside doors as these are normally very draughty positions. The storage cistern should be sited over a load-bearing wall to provide adequate support for the volume of water it contains. A situation above or near the airing cupboard is an added precaution to prevent frost damage. The storage vessel should be well lagged with non-flammable materials such as expanded polystyrene, glass fibre or vermiculite to a thickness of 75 mm. If pipes are run under suspended ground floors, these should be at oversite level and well insulated to avoid cold draughts from air bricks.

Any pipe supplying cold water must be installed to ensure, as far as practicable, it is not heated above 25 °C. This may require insulation if run is adjacent to hot water services, especially in ducts.

All control valves, except those controlling single appliances, should be labelled to indicate the fittings and section of pipework they isolate. In commercial and public buildings it is also necessary, by means of colour coding, to identify the contents of pipework installations. (See the Appendix in Book 2.)

Generally speaking, pipe runs in a building should be readily accessible. If pipes are run in intermediate suspended floors, it may be necessary to notch or bore holes in the joists. Unless this is carried out as shown in Fig. 7.23 they will be seriously weakened. The formulae shown meet the requirements of the Building Regulations for cutting or boring joists. To give an example, the span of a floor joist which is to be notched 200 mm in depth has a span of 4 m. The maximum distance from the face of the wall is

$$\frac{S}{4}$$

$$\therefore \quad \frac{4}{4} = 1 \text{ metre}$$

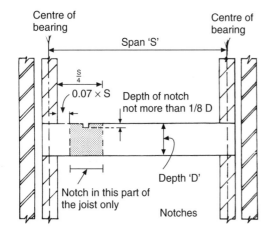

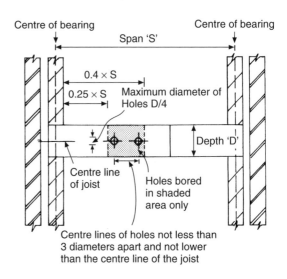

**Fig. 7.23** Notches and holes in joists.

The minimum distance from the wall will be

$$0.07 \times 4 = 0.280 \text{ m}$$

The maximum depth of the notch will be:

$$D = 200$$

$$\therefore \quad \frac{200}{8} = 25 \text{ mm}$$

The methods for calculating safe areas for boring holes are very similar and can be seen from the illustration.

The maximum diameter of any hole must not exceed 25% of its depth. For example, the maximum diameter of a hole bored in a joist 200 mm in depth will be 50 mm.

Notches and holes should be just large enough to give freedom of movement without chafing. If they are too tight expansion and contraction of the pipe may cause noise, especially those pipes carrying hot water. Figure 7.24 shows the recommendation of the notches.

Figure 7.25 illustrates a component which solves many of the problems of pipework installed under suspended floors. Its use ensures that the pipes do not bind on the woodwork, and the steel plate prevents penetration of the pipe by nails or screws when the floor is laid.

Floor boards over pipe runs should be screwed down to permit removal for maintenance purposes. When flooring is removed to facilitate new pipe

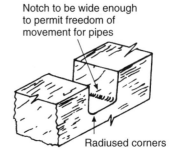

**Fig. 7.24** Notch in joist.

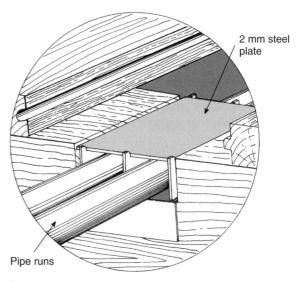

**Fig. 7.25** Pipe guard. The use of these components in notches cut in joists allows for expansion of the pipes and prevents accidental damage from floor fixings.

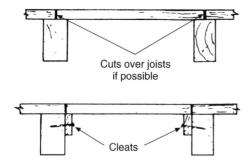

**Fig. 7.26** Replacing wooden flooring.

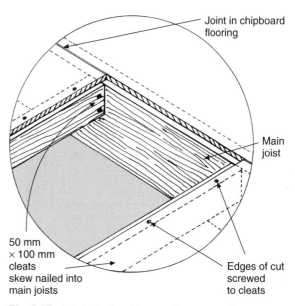

**Fig. 7.27** Method of making good a trap cut in chipboard flooring. Note: As the joints in this type of flooring are normally tongued and grooved the panel may be damaged when it is removed and will require replacement.

runs, the nails should be punched down and in the case of tongued and grooved boards, the tongue should be cut through with a pad- or circular saw to avoid damage to the edges of the board when they are lifted. If possible, board cuts should be over joists but if this is not possible, the edge of the joist may be extended by nailing a stout cleat to it (see Fig. 7.26).

In existing premises where chipboard flooring has been used, it is much more difficult to remove than wooden floor boards in order to gain access to the space below. Because of this problem it may be easier to run the pipework in suitable ducts above the floor level. If this is impracticable a section of chipboard may be removed and replaced as shown in Fig. 7.27. A circular saw with the blade set at the thickness of the flooring is the best way of making the saw cuts – but beware of nails!

Always remember that replaced flooring must be made good in a safe and professional manner.

It is not advisable to run pipes in solid floors where this can be avoided unless they are fitted in properly constructed ducts with removable covers. They are expensive to install and are seldom found in domestic properties. If it is absolutely necessary to run pipes under such floors, a proper channel should be provided by the builder. A little planning is necessary here and the plumber should notify the builder where the channel is to be provided by means of a dimensioned sketch. Pipes should be well insulated to reduce heat loss, allow freedom of movement, and enable them to resist any corrosive effects of the floor screed or finish. Figure 7.28 illustrates some approved

methods of running pipework in the building fabric. Figure 7.29 deals with pipes under suspended floors and Fig. 7.30 shows the recommendations for ducting pipe in solid floors. Because plastic pipes are more flexible than metal ones, the system shown in Fig. 7.31 has been developed. It is essential that access boxes are provided for the inspection of any mechanical joints.

### Protection from frost damage

CP 6700 deals with precautions against frost damage. The Water Regulations also lay specific emphasis on this subject as burst pipes can result in large quantities of water being wasted, quite apart from the damage caused to a householder's property. The best protection against frost is to keep the premises warm, but this may prove difficult when dealing with cisterns in a roof space, supplies fitted outside the building such as standpipes, or cattle feeding troughs. No amount of insulation will prevent freezing, it simply delays it; the thicker and more effective the insulation, the longer the delay. The following deals generally with frost precautions.

Pipe in external wall with access

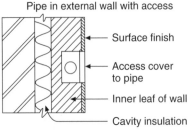

- Surface finish
- Access cover to pipe
- Inner leaf of wall
- Cavity insulation

Pipe in external wall with no access

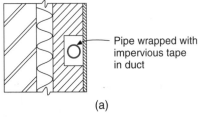

- Pipe wrapped with impervious tape in duct

(a)

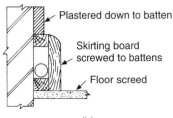

- Plastered down to batten
- Skirting board screwed to battens
- Floor screed

(b)

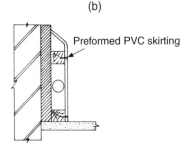

- Preformed PVC skirting

**(c) Accessibility of pipe concealed in or adjacent to solid walls**

Several systems of purpose made PVC skirting are available for both pipe and cable concealment. Most ranges include internal and external corners and special fittings providing for pipe exits through the skirting to permit for example connections to radiators.

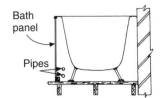

Bath panel

Pipes

**(d) Accessibility of service pipes under baths**

Pipes to be fixed on the panel side of the bath and adequately supported.

**Fig. 7.28** Approved methods of running pipework in the building fabric.

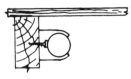

**(a) Pipes under intermediate suspended floors**

Pipes to be suitably supported with provision for access at intervals of not more than 2 m and at every joint. Such pipes are not normally exposed to temperatures of less than 0 °C and in such cases insulation may only be required if they are part of a hot water installation.

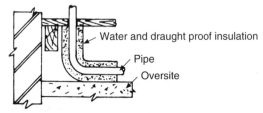

- Water and draught proof insulation
- Pipe
- Oversite

**(b) Pipes under suspended ground floors**

Pipes should be ducted and preferably be laid on the oversite to avoid any cold draughts. Risers must be fitted as far as possible from any air bricks. Pipes in the duct must be insulated if less than 750 mm from an external wall.

**Fig. 7.29**

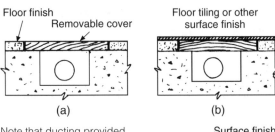

Floor finish / Removable cover

Floor tiling or other surface finish

(a)  (b)

Note that ducting provided with access covers must be carefully levelled to the finish floor level

Only permissible if no joints are enclosed and the pipe can be withdrawn for inspection

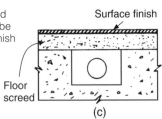

Surface finish

Floor screed

(c)

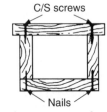

C/S screws

Nails

Purpose-made PVC and galvanised sheet steel ducting is available and recommended if the work is extensive

**(d) Suggested construction of wooden duct. All timber used must be treated against decay and coated externally with bituminous paint. Commercial floor ducts are available**

**Fig. 7.30** Pipe ducts in solid floors.

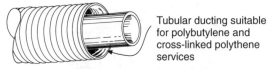

Tubular ducting suitable for polybutylene and cross-linked polythene services

(a) Flexible plastic sleeves

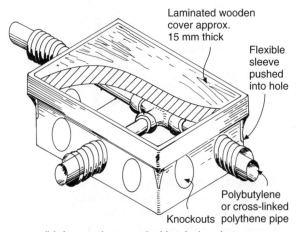

Laminated wooden cover approx. 15 mm thick

Flexible sleeve pushed into hole

Polybutylene or cross-linked polythene pipe

Knockouts

(b) Access box used with tubular sleeves

**Fig. 7.31**

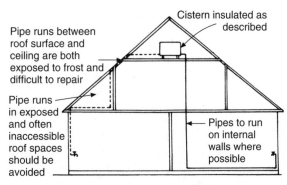

Cistern insulated as described

Pipe runs between roof surface and ceiling are both exposed to frost and difficult to repair

Pipe runs in exposed and often inaccessible roof spaces should be avoided

Pipes to run on internal walls where possible

**Fig. 7.32** Avoiding frost damage to pipework.

Reference should also be made to the section on insulation in Chapter 8.

Storage cisterns in roof spaces should be situated, if possible, over or near the airing cupboard. If this is impossible, siting near a chimney is the next best choice. The underside of the cistern should not be insulated; heat from the rooms below must be allowed to keep it warm. Both cisterns and pipes should be situated away from draughts and well insulated. Pipe runs in the roof should be as short as possible and in no circumstances fitted near the eaves. Never run a pipe immediately under a roof as shown in Fig. 7.32 as it will be very exposed to frost damage and access would be difficult.

All water pipes, both hot and cold, should be fitted in such a way that the whole system can be drained in the event of the building being unoccupied during periods of frost. In those buildings fitted with fully automatic heating, it is suggested that the thermostat is turned down and the time switch overridden so that the heating system is operational continuously during the owner's absence. This ensures that the premises are kept warm and there is no danger of the heating pipes becoming frozen, as in many cases it is almost impossible to drain these systems fully.

Outside taps should be fitted in such a way that they are controlled by a stop valve inside the premises. The pipe should be insulated where it passes through the outside wall as shown in Fig. 7.33(a). Standpipes away from the building

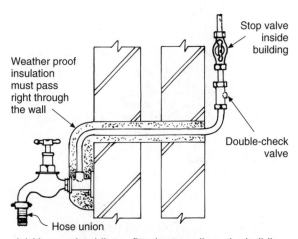

Stop valve inside building

Weather proof insulation must pass right through the wall

Double-check valve

Hose union

(a) Hose union bib tap fitted externally or the building

All new domestic installation employing hose union taps must be protected against contamination by a double-check valve. Replacement of existing taps may have a double-check valve built into the tap or at the connection of the hose.

**Fig. 7.33** Protecting outside water supplies from frost damage.

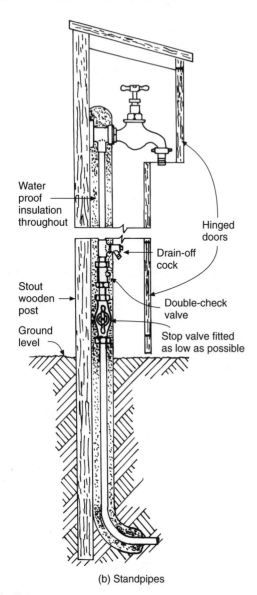

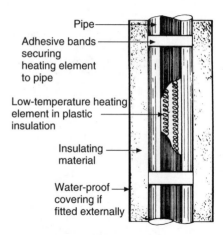

Fig. 7.34 Use of low-temperature heating element for tracing exposed pipework.

**Fig. 7.33** *continued*

(b) Standpipes

should be fitted in an insulated box with a stop valve and draw-off fitted at low level. Figure 7.33(b) illustrates the treatment for outside draw-offs. The same applies to external WC apartments which are exposed to frost damage. Do not forget to instruct the householder or client about the precautions to take during frost.

*Automatic frost protection*

In the case of large or industrial buildings where the water supply must be kept working during severe weather, several alternatives to insulation are available for protection against frost damage.

In cistern or tank rooms of buildings which are centrally heated, a radiator fitted with a thermostatic valve may be used. The sensory element on the valve is set at the frost position and any drop in temperature below this will cause the valve to open and keep the temperature of the room just above freezing point. An alternative is to insulate the cistern with materials containing an electrical heating element, rather similar in fact to an electric blanket. Pipework can be traced with an electrical heating element in tape form (see Fig. 7.34) which is bound to the pipe with adhesive tape, after which the whole is insulated with a suitable material. To avoid the possibility of anyone unwittingly cutting the heating element, warning notices must be attached to the insulation stating that the pipework is electrically traced. Control of the heating element can be manual (i.e. switched on during low temperatures) or automatic by means of a frost thermostat. These operate in the same way as a room thermostat for central heating, but they work over a lower temperature range, usually −10 °C to 20 °C. The position of the frost thermostat should be carefully chosen to ensure that it operates correctly. If, for example, the thermostat is situated on a wall facing east, the heat from the early morning sun may well cause the heating element to be de-energised when freezing conditions still prevail in other parts of the building. A suitable thermostat setting is 2–3 °C.

The Water Regulations permit the installation of electric immersion heaters in cold water storage cisterns to prevent the contents freezing. This form of protection is well worth consideration in situations such as where bulk water storage is installed in unheated areas. Control of these heaters may be effected in the same way as an ordinary heater but having a special thermostat which works, like a frost thermostat, over a lower temperature range. A suitable thermostat setting would be 3–5 °C. This ensures the water temperature would never drop below this point; above this temperature the heater would be inoperative.

In the case of a feed cistern which supplies water to a hot water scheme only, the temperature setting may be up to, but not in excess of, 39 °C.

Finally, do not forget that any insulation fitted outside a building *must* be waterproof. If it is allowed to become wet the result will be worse than having no insulation at all.

### Pipe supports

Selecting the correct type of fixing methods to be used in plumbing installations is a very important part of a plumber's work. Lack of adequate support to pipework and fittings can result in their damage and possibly subsequent failure under service conditions, or to the development of faults such as air locks. In the case of large installations, the type of fixings is specified by the designer, but on smaller jobs the plumber often has to make the decision. The factors to be considered include the type of construction and surfaces into which fixings must be made, the mass of the component to be fixed and whether, in the case of a pipe run, it is fitted horizontally at high level, low level (when it may be prone to damage) or in a vertical position. Pipework fitted in the last position generally requires fewer fixings. The pipework material will also have some influence on the number of fixings required, i.e. plastic pipes require more support than those of copper or steel. To avoid possible corrosion problems, the support or clip should, if possible, be made of the same or similar material as that of the pipe.

Pipe clips and supports vary considerable both in cost and effectiveness, and a variety are illustrated in Fig. 7.35. Clips or other supports in schools, hospitals and factories should be strong, robust, and, to achieve maximum support, should be built into the fabric of the building. The holderbat bracket shown in Fig. 7.35(a) is typical of such a bracket for building into a wall. These brackets are made of cast brass or malleable iron and are suitable for copper and steel pipes. Holes may be made in a wall to accept these brackets by using a large tungsten-tipped drill or by using a chisel of the type shown in Chapter 2, Fig. 2.33(c). In the case of brick walls, if the 'perpends' or upright joints can be located, a fixing is often more easily achieved as these joints are not always filled so solidly with mortar. The techniques for 'making' the brackets into the wall are the same as those described in Chapter 9 for fixing cantilever brackets.

Saddle clips (Fig. 7.35(b)) are made for pipes of steel, copper and semi-rigid plastic. The main objection to these clips is that they clip the pipe back tightly to the wall making painting difficult and allowing an accumulation of dust and dirt to build up between the pipe and the wall surface. For these reasons it is considered good practice when fixing rigid pipes such as steel, copper and semi-rigid plastic pipes such as PVC, to use 'stand-off' or spacing clips, of which many patterns are available. One of the most common of these is shown in Fig. 7.35(c). They are made of polypropylene or PVC and have the advantage of being secured by just one screw which saves fixing time. They are suitable for copper, stainless steel and PVC pipes which 'snap' into the clip after it has been fixed.

A similar clip for securing double-pipe runs is shown in Fig. 7.35(d). The double-pipe clip is made of PVC and is secured by one screw only instead of the two or three normally required. They also ensure both pipes are parallel throughout their length. Extra care is necessary when fitting the pipework as like all pipe fixings made of plastic, this material has insufficient strength to 'pull' a badly fitted pipe into the correct position. A good rule to follow is to carefully set out and fix the clips prior to fitting the pipework.

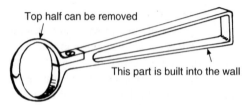

(a) Holderbat built-in bracket

(b) Saddle clip

(c) Thermoplastic 'snap on' clip for copper or plastic tubes

Pipe stands off wall with this type of clip.

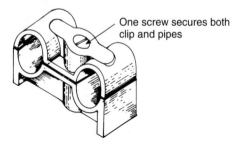

(d) Double pipe clip for copper or PVC pipes

15 and 22 mm OD only.

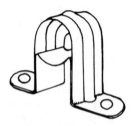

(e) Two-piece spacing clip

(f) School board bracket

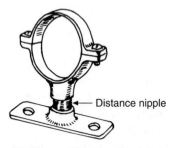

(g) Pipe or Munson ring fixing

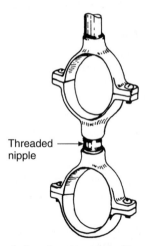

(h) Series of pipe rings supported by a hanger

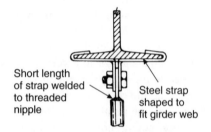

(i) Fabricated girder clamp for pipe rings

**Fig. 7.35**   Pipe clips and supports.

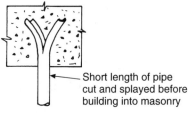

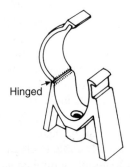

Short length of pipe cut and splayed before building into masonry

(j) Masonry fixing for pipe rings

Hinged

(k) Hinged clips. The type shown is suitable for polybutylene and cross-linked polythene. More robust clips of this type are made for copper tubes

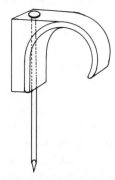

(l) Shows a nail on clip for polybutylene or cross-linked polythene pipes

**Fig. 7.35** *continued*

The clip shown in Fig. 7.35(e) is made of pressed copper and prior to the introduction of the plastic type was the most popular for copper tube fixing in domestic work. None of the 'stand-off' clips shown in Figs 7.35(c), (d) and (e) are very strong and their use is not recommended in public buildings, especially for skirting-level work. A stronger, more rigid type of screw-on bracket is shown in Fig. 7.35(f). With the exception of the base plate, they are similar to built-in holderbat brackets, although the name does vary according to the locality.

Another type of bracket, referred to as 'pipe' or 'Munson' rings, is very adaptable and this bracket can also be used as a floor-mounted support as shown in Fig. 7.35(g), or as a double-ended type as shown in Fig. 7.35(h). The use of a nipple permits a variation of the centre-line distance between the pipe and the floor, or pipe to pipe, and gives a considerable degree of flexibility. In the case of Fig. 7.35(h) a series of pipes can be secured from one fixing such as the girder clamp shown in Fig. 7.35(i). Another alternative for fixing directly into masonry is shown in Fig. 7.35(j).

Pipe or Munson rings are tapped with a $\frac{1}{4}$ in or $\frac{3}{8}$ in BSP thread, the latter size being used for pipes of over 25 mm nominal diameter. As with school board and built-in brackets, those used with steel pipes are made of malleable iron while those used for copper or stainless steel are made of cast or pressed brass. Figures 7.35(k) and (l) show two types of clip for polybutylene pipes. Type (k) is similar to a hinged top clip, sometimes used for copper tubes, and is mainly used for surface fixing. Type (l) is a simple nail-on clip similar to those used for electric cables and is mainly used for concealed work.

Good general recommendations for the distance between pipe supports are given in BS 6700, and are shown in Tables 7.2–4. It will be seen from the tables that fewer supports are needed on vertical pipes. No recommendations are given for extra support for pipes run at low level so the plumber uses his or her own judgement here.

**Table 7.2**  Spacing for copper tube supports.

| Diameter of pipe (mm) | Interval for vertical runs (m) | Interval for horizontal runs (m) |
| --- | --- | --- |
| 15 | 1.8 | 1.2 |
| 22 | 2.4 | 1.8 |
| 28 | 2.4 | 1.8 |
| 35 | 3.0 | 2.4 |
| 42 | 3.0 | 2.4 |
| 54 | 3.0 | 2.7 |
| 67* | 3.6 | 3.06 |
| 76 | 3.6 | 3.0 |
| 108 | 3.6 | 3.0 |

*Note that 67 mm is a non-standard size

**Table 7.3**   Spacing for LCS pipe supports.

| Nom. diameter of pipe (mm) | Interval for vertical runs (m) | Interval for horizontal runs (m) |
|---|---|---|
| 15 | 1.8 | 1.8 |
| 20 | 3.0 | 2.4 |
| 25 | 3.0 | 2.4 |
| 32 | 3.0 | 3.0 |
| 40 | 3.6 | 3.6 |
| 50 | 3.6 | 3.6 |

**Table 7.4**   Spacing for plastic pipework supports.

| Polybutylene and cross-linked polythene complying to BS 5955 Part 8 | | |
|---|---|---|
| Nom. diameter of pipe (mm) | Interval for horizontal runs (m) | Interval for vertical runs (m) |
| 10 | 0.5 | 0.3 |
| 15 | 0.5 | 0.3 |
| 22 | 0.8 | 0.5 |
| 28 | 1.0 | 0.8 |

| Medium-density polyethylene complying with BS 1972 (colour black for above-ground use) | | |
|---|---|---|
| Nom. diameter of pipe (inches) | Interval for vertical runs (m) | Interval for horizontal runs (m) |
| $\frac{1}{2}$ | 0.8 | 0.4 |
| $\frac{3}{4}$ | 0.8 | 0.4 |
| 1 | 0.8 | 0.4 |
| $1\frac{1}{4}$ | 0.9 | 0.45 |
| $1\frac{1}{2}$ | 0.9 | 0.45 |
| 2 | 1.1 | 0.55 |

Copper pipes especially need to be well fixed in this position as they are more easily damaged than pipes made of LCS. Much the same applies to plastic pipe services and some thought should be given to the possibility of fitting such pipes in a position where possible damage to them is minimised. Always remember one clip too many is better than one too few.

Manufacturers of plastic pipes for discharge pipework always include directions for fixings in their freely available technical literature. The tables of fixing distances shown here relate to water supply pipework only.

*Fixing of pipe supports*
Whatever type of anchor, clip or pipe support used, its effectiveness will depend entirely upon how well it is fixed to the wall, ceiling or floor of the building. Fixings have to be made to a variety of structural materials in modern buildings and there are many differing types of equipment available which enable this to be achieved effectively.

Wood is the easiest material to which a fixing can be made, wood screws usually being used for this purpose. Generally they should be of sufficient length and stout enough for the purpose for which they are used. They can be made of brass, LCS or aluminium alloy, the last often being used by the plumber for the fixing of aluminium sheet fixings and cleats. Screws are made with countersunk, round or mushroom heads (see Fig. 7.36).

The mushroom-headed screw is also countersunk and is used for fixing equipment such as towel rails and chromium-plated bathroom ware. Countersunk screws are used where the screw head must be flush with the surface of the work. Typical examples of this type of screw are found in the fixing of floor boards and wooden ducts over pipe runs, and also for fixing down wash basin pedestals and WC pans. Round and dome head screws are used for securing materials which are not thick enough to be countersunk or those in which no countersunk recess has been provided. Brass and alloy screws have less tensile strength than those of LCS and if the hole into which they are screwed is in hardwood or other dense materials they may shear off when screwed home. To prevent this a hole should be drilled in the material equivalent to the root core or minor diameter of the screw (see Fig. 7.37). A little oil or

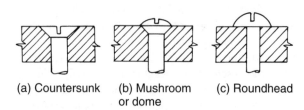

(a) Countersunk   (b) Mushroom or dome   (c) Roundhead

**Fig. 7.36**   Types of screw heads.

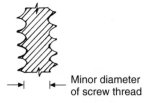

—→| |← Minor diameter
        of screw thread

**Fig. 7.37**   Wood screw details. In certain circumstances it is necessary to drill a hole into the material before inserting a wood screw.

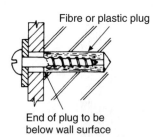

Fibre or plastic plug

End of plug to be
below wall surface

**Fig. 7.38**   Screw fixing in masonry using fibre or plastic plugs.

(a) Spring toggle fixing for steelwork or
hollow blocks

Rubber sleeve

Nut moulded
into rubber

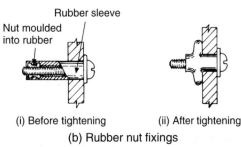

(i) Before tightening            (ii) After tightening

(b) Rubber nut fixings

When the screw is tightened the rubber sleeve is pulled up
to form a flange on the rear of the fixing surface.

**Fig. 7.39**   Fixings in hollow walls.

grease on the screw thread will assist the screwing operations into wood. The use of a lubricant also has the advantage of protecting the screw from corrosion, and enables easy withdrawal if and when required.

The original method of screwing into masonry was to use wooden plugs or wedges. However, as wood shrinks over a period of time owing to loss of moisture, fixings of this type often become loose, and plastic plugs are used as they are much more reliable and save time. If the top of the plug is flush with, or only just below, the wall surface, then the shank of the screw will expand the plug causing it to crack the surface of the work, therefore it is very important, especially when fixings are made to ceramic tile surfaces, to sink the plug into the hole clear of the wall surface. Careless fixing can result in cracking a tile which is time consuming to replace. The correct position of the plug in the hole is shown in Fig. 7.38.

When fixing to materials such as hollow blocks or steelwork, the spring toggle shown in Fig. 7.39(a) is typical of the many types of fixing available which enable the fixing load to be spread over a large area.

For lighter fixings the rubber nut is a cheap convenient alternative. Figure 7.39(b)(i) shows the fixing in position before the screw is tightened, while Fig. 7.39(b)(ii) shows how the rubber is compressed to form a washer at the rear of the surface when the screw is tightened.

It should be recognised that there are limitations on the weight that can be supported by, for instance, plasterboard, which in itself is not a very strong material. In buildings under construction where partition walls are built of hollow blocks or plasterboard on studs, careful planning at an early stage can ensure good fixings. In the case of hollow work, a breeze- or slag-based block may be inserted where fixings are likely to be required for, say, a radiator, at a later stage of the job. In the case of studwork partitions 'noggins' should be fixed across the studs and nailed in appropriate positions as shown in Fig. 7.40 before the plasterboard is fixed.

Where it is necessary to provide fixing for a series of pipes from the floor to the ceiling, i.e. in an airing cupboard, it is good practice to provide a board on which to fix the pipes. Its use will save time and reduce the possibility of damage to the

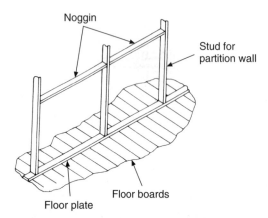

**Fig. 7.40** Noggins in studwork for fixings.

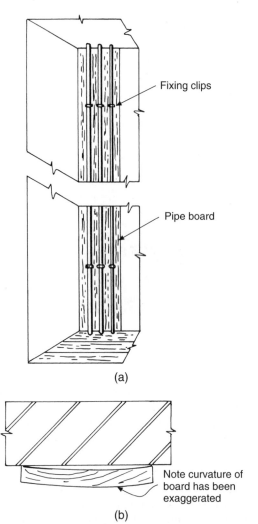

(a)

(b)

Wide boards may have a slight bow and should be fixed as shown. One or two fixings in the centre of the board will pull it back to the wall.

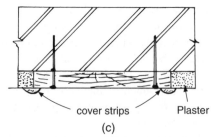

(c)

Section of board showing fixings. Board should be nominally 25 mm thick. Cover strips may be used to seal the shrinkage gaps between the board and plaster where the board is exposed.

**Fig. 7.41** Pipe fixing boards.

finished plaster. The board can be easily fixed to lightweight blockwork using cut nails, but if fixings are to be made to brickwork it is recommended that the wall is plugged and screwed. Figure 7.41 shows details of this type of fixing.

There are two main methods of making heavy fixings in dense building materials such as concrete. The simple rag bolt shown in Fig. 7.42(a) can be positioned and built into the structure as it is being constructed, if sufficient information is available at an early stage in order to locate the rag bolt correctly. Alternatively, they may be built in at a later stage by first drilling a hole in the wall and then cementing or caulking in position using lead wool. The only problem with this type of fixing is that a period of time must be allowed for the cement mortar securing the bolt to dry out or 'cure'.

Another method of fixing which is used in a similar way to the rag bolt is where the fitting and fixing can be made in one operation, the only prerequisites being the correct location of the bolt and an accurate hole in the structure affording a push fit for the bolt. One of the many variations of this type of fixing is shown in Fig. 7.42(b). After the bolt has been pushed into the prepared hole the nut is tightened and the tapered end of the bolt drawn outward, opening up the four keyed segments against the sides of the hole. Not only must the hole be accurate is size, but the structure itself should be fairly dense to use these bolts to their full effect. In old buildings constructed with soft bricks laid in

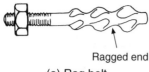

(a) Rag bolt

The ragged end of the bolt is built into concrete, etc.

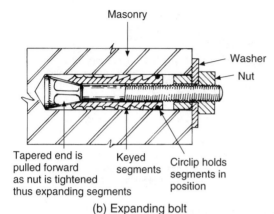

(b) Expanding bolt

**Fig. 7.42** Heavy-duty masonry fixings.

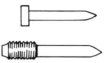

**Fig. 7.43** Typical pins fired by explosive charge into masonry, concrete and steel.

mortar joints, a 'built-in' rag bolt is often a more satisfactory fixing.

In most cases the fixings made by a plumber are either screwed or bolted but occasions do arise when light fixings can be nailed directly into the structure. Nails made of high tensile steel are available, being strong enough to be driven directly into masonry without bending. By using short positive hammer blows, they can be driven to a depth of approximately 20–25 mm into masonry, depending upon its density. These nails may be brittle and could shatter if struck with a glancing blow, therefore care should be taken to ensure that no other person is working in the vicinity where they are used and the operator should always wear suitable goggles.

One of the most versatile and labour-saving devices for fixing is the cartridge tool. These actually fire a pin or bolt into the structure with a force capable of penetrating LCS, and this enables them to be used for fixing into dense brickwork or concrete with the minimum of effort. Two types of pin are used, those having a head similar to that of an ordinary nail and those with a threaded end to which fixings can be bolted; both types are

illustrated in Fig. 7.43. Needless to say, the tool that fires these pins is virtually a firearm and as such it is very important it is used only by a qualified person. Most companies manufacturing this type of equipment also offer short courses in their safe use and the potential user would be well advised to take advantage of such an offer.

**Cold water storage**

In domestic premises where all the cold water supply, except that of drinking water, is supplied by a cistern, the minimum storage capacity should be 114 litres. If the same cistern is also used as a feed cistern for hot water supply, the minimum capacity should be 227 litres. The materials from which these cisterns are constructed are described in Chapter 4.

Some consideration must be given to the siting of cisterns in the roof space where it is not possible to position them over a load-bearing wall. If the mass of water in the cistern is supported over too few joists or trusses, or the bearers are insufficiently strong, the building fabric will be damaged. Cistern platforms should be constructed of a material that will not distort owing to dampness which may cause the water connections to the system to leak. Plywood or a suitable alternative not less than 25 mm thick is suitable for this purpose. Chipboard is unsuitable owing to the fact that it will soften and distort in damp conditions. Figure 7.44 illustrates the main points relating to cistern support on roof timbers or trusses.

On all but the smallest of these cisterns a metal or plastic reinforcing plate is supplied with the cistern to reinforce the area through which the float-operated valve passes giving the side of the cistern more rigidity against the upthrust of the float. Holes in these cisterns should be made with a proper cutter – those shown in Chapter 2, Fig. 2.2

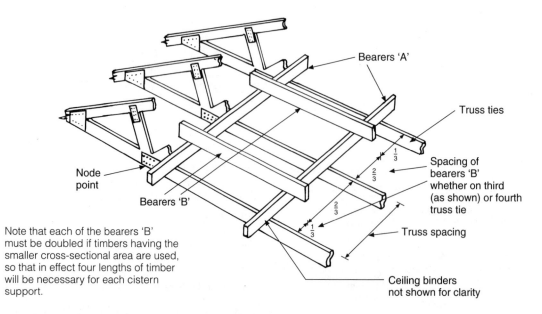

Bearers 'A'

Truss ties

Node
point

$\frac{1}{3}$

$\frac{2}{3}$

Spacing of
bearers 'B'
whether on third
(as shown) or fourth
truss tie

Bearers 'B'

$\frac{2}{3}$

Note that each of the bearers 'B'
must be doubled if timbers having the
smaller cross-sectional area are used,
so that in effect four lengths of timber
will be necessary for each cistern
support.

$\frac{1}{3}$

Truss spacing

Ceiling binders
not shown for clarity

Size of each bearer 'B' relating to truss span and cistern capacity

| Truss span (m) | 300 litres actual capacity | 230 litres actual capacity |
|---|---|---|
| 6.5 | Two* 38 × 100 mm or One 50 × 75 mm | One 50 × 100 mm |
| 9.0 | Two 38 × 100 mm or One 50 × 150 mm | Two 38 × 100 mm or One 50 × 125 mm |
| 12.0 | Two 38 × 150 mm | Two 38 × 125 mm or One 50 × 150 mm |

*This indicates that two lengths of timber must be laid for *each* bearer

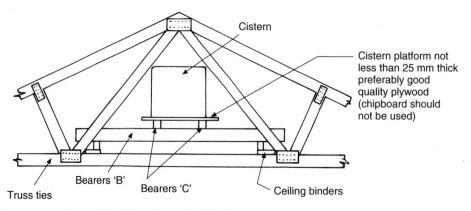

Cistern

Cistern platform not
less than 25 mm thick
preferably good
quality plywood
(chipboard should
not be used)

Bearers 'B'

Bearers 'C'

Ceiling binders

Truss ties

**Fig. 7.44**  Recommendations of Building Research Establishment for supporting storage cisterns on roof trusses. Cisterns of 300 litres actual capacity must be carried on not less than four trusses. Those of 230 litres may be supported on three as shown. Bearers 'A' must be fixed as close as possible to the 'node' points. Bearers 'A' and 'C' should be constructed of timber not less than 50 × 75 mm. The trussed rafter designer should be informed of the cistern capacity and its location in the roof space.

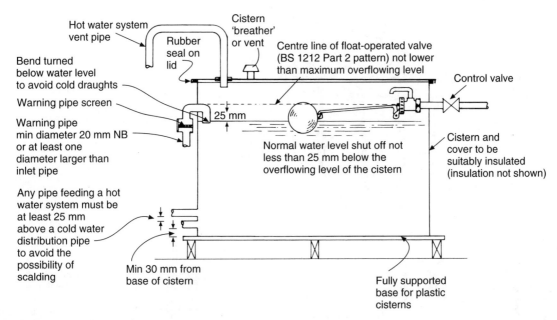

**Fig. 7.45** Connections to cisterns.

are satisfactory for this purpose and ensure a hole of the correct size to accept a standard BSP thread. The practice of heating a piece of copper tube to melt out the hole is most unsatisfactory as it can give rise to subsequent cracking at its edges due to possible overheating. This has been known to cause complete failure of the cistern after a very short period of time. Care must also be taken to use a suitable packing material on any connection made in the cistern. Traditional materials such as paint, putty or paste jointing must not be used as the oil they contain has the effect of softening most thermoplastics and early failure of the material may occur with disastrous results. It has been suggested that polythene washers should be used but unless they are of a very good fit, a slight dampness or drip may be discernible around the joint when it is tested. Satisfactory joints can usually be obtained by the use of PTFE tape in conjunction with polythene washers or rubber seals which can be cut from suitable sheet.

Where holes are cut in galvanised cisterns, it is important to remove all the swarf as failure to do this can often result in serious corrosion of the cistern due to electrolytic action. Cisterns should be tested before any holes are cut as manufacturers will not normally entertain claims made for defects after they have been fitted.

Glass-fibre-reinforced plastic (GRP) has a greater degree of rigidity than those made of thermoplastics and is suitable for the construction of large cisterns that are built up by using bolted sections. The remarks relating to the support and fitting of thermoplastic cisterns apply equally to those constructed of GRP.

Figure 7.45 illustrates the requirement of the Water Regulations regarding storage of water and it should be noted they are much more stringent than hitherto.

It is not uncommon in old installations to find the water polluted with dust from the atmosphere, flies and other small insects and even dead birds and vermin. In many cases in the past, bylaws regarding the use of a cistern cover have been ignored.

Suitable covers must be used and in many cases they are bolted or screwed down on a rubber seal. Atmospheric pressure in the cistern is maintained by a vent in the cover which, like the overflow pipe, must be screened. These fine mesh screens prevent

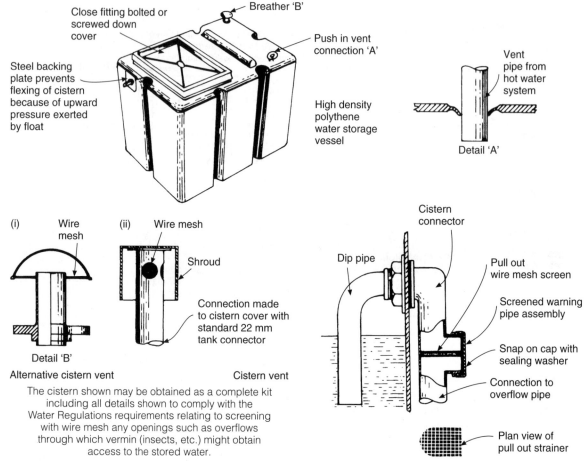

**Fig. 7.46** Cistern and associated components.

access of insects to the stored water. They should be made of stainless steel or other equally suitable material. Building owners should be made aware that these screens should be cleaned regularly, because if they become blocked, especially the overflow, serious flooding could take place. Figure 7.46 illustrates a cistern and its associated components which is available as a complete unit complying with the Regulations. Any replacement of existing cisterns must also comply with the foregoing.

The term 'nominal capacity' of a cistern relates to the quantity of water it contains when it is full to the brim. When the float-operated valve and overflow are fitted, the amount of water is reduced to 'actual capacity'. To ensure the maximum actual capacity, the valve and overflow must be carefully positioned both in relation to each other and in relation to the top of the cistern (see Fig. 7.45). This ensures that the valve cannot be submerged if the cistern is overflowing causing a health hazard due to the possibility of back siphonage.

*Access to float-operated valves* The requirements of BS 6700 relating to working space above cisterns are shown in Fig. 7.47. Sufficient room must be allowed to enable maintenance of the valve, inspection, and means of cleaning to be carried out. The measurements shown are the minimum and should be increased where possible.

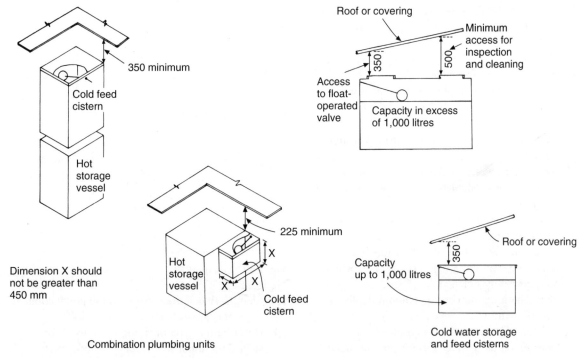

**Fig. 7.47** Recommendations for providing access to float-operated valves in cisterns and combination units.

## Further reading

Useful information can be obtained from the following sources:

BRE Bookshop, Building Research Establishment, Garston, Watford, Herts, WD2 7JR. Tel.: 01923 664000.

Water conservation, BRE 1P 8/97.

Water conservation, design, installation requirements for the use of WCs flushed siphonically or by valves, BRE 326 1997.

*Codes of Practice and British Standard*

BS 6700 Specification for design, installation, testing and maintenance of services supplying water for domestic use within buildings and their curtilages.

*Guide to the new Water Regulations and Bylaws (Scotland)*

Published by the Water Regulation Advisory Services, Fern Close, Pen-Y-Fan Industrial Estate, Oakdale, NP11 3CH. Tel.: 01495 248454.

*Water storage vessels*

Poly Tank Ltd, Naze Lane East, Frickleton, Preston, PR4 1UN. Tel.: 01772 632850.

Tanks and Drums Ltd, Bowling Iron Works, Bradford, BD4 8SX. Tel.: 01274 707000.

*Water services*

FW Talbot & Co. Ltd, Winnall Valley Road, Winchester, Hants, SO23 8LL. Tel.: 01962 705200.

Wavin Plastic Ltd, Parsonage Way, Chippenham, Wiltshire, SN15 5PN. Tel.: 01249 654121.

*Water meters*

Kent Meter Ltd, Pondwick Road, Luton, Beds, LUI 3LJ. Tel.: 01582 438054.

## Self-testing questions

1. State the requirements for the installation of a water meter in a cupboard under the sink unit.
2. Describe the effects of water containing dissolved carbon dioxide on chalk and limestone.

3. What will be the effect on a volume of water if its temperature is raised above or lowered below 4 °C?

4. State the cause of burst pipes in frosty weather.

5. (a) List two types of hardness in water.
   (b) Describe the problems that can arise from using hard water in hot water systems.

6. State the numerical values of (a) a pascal and (b) a bar.

7. The water level in a cistern is 8 metres above a draw-off tap. Calculate the pressure on the tap in kPa/m².

8. State the pressure in bars on a float-operated valve under a pressure of 350 kPa/m².

9. Calculate the section of a floor joist having a span of 3.2 m and a depth of 200 mm which may be safely notched.

10. State the minimum and maximum depths of underground water services.

11. Explain the reason for a gooseneck connection on a water service pipe to the main ferrule.

12. Identity four causes of weakening the floor joist shown in Fig. 7.48 where holes have been bored for pipe runs.

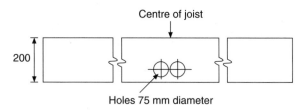

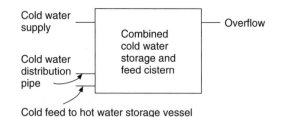

**Fig. 7.48**   Question 12.

**Fig. 7.49**   Question 13.

13. Identify two serious errors in the connections to the storage cistern shown in Fig. 7.49.

14. Describe the type of fixing to be used for a steel pipe run at low level in a horizontal position in a hospital or school.

15. State the minimum working space between the top of a cistern (not exceeding 1,000 litres) and the ceiling or underside of a roof.

# 8 The hot water supply

After completing this chapter the reader should be able to:

1. Identify the effects of heat and understand how heat quantity is expressed.
2. Demonstrate a knowledge of the forms of heat transmission and how they apply to hot water systems.
3. Calculate the boiler power necessary to heat a quantity of water through a given temperature rise.
4. Recognise possible causes of defects and faults in hot water systems and explain how they should be avoided or remedied.
5. Recognise possible dangers in connection with hot water supplies.
6. Understand the working principles of hot water supplies.
7. Identify by name the components and pipes in hot water systems and describe their function.
8. State the purpose of secondary circulations, recognise where they are necessary, and describe methods of control and limiting heat losses.
9. Understand the need for effective insulation of hot water pipes and components.
10. Select and apply various types of insulation for a given purpose.

The introduction of hot water supply in domestic premises is a comparatively recent innovation. Prior to the period of the First World War, 1914–1918, only the very wealthy were able to afford piped hot water in their homes, while one of the most attractive selling features of houses built for speculation during the period 1920–1930 was the introduction of proper bathrooms and hot water supply. One of the problems encountered with houses built during this period was the poor design of the schemes that were installed, often using long exposed pipe runs in uninsulated roof spaces resulting in frozen feed pipes and cisterns. In houses built since the 1945 period the subject of domestic hot water has been given greater thought, with the result that in many cases bathrooms are generally built adjacent to or over the kitchen, enabling sanitary fittings to be grouped together with shorter pipe runs. It has been realised too that if the hot storage vessel is situated in a position close to the boiler, the circulating pipes will be shorter with correspondingly less heat loss.

The design of hot water systems has also improved owing to a better understanding of the basic physical principles involved and it is essential for the plumber to have a thorough grasp of these principles to be able to install hot water systems efficiently.

### Heat

It is difficult to describe heat. Like electricity it cannot be seen, although its effects can be appreciated. It can also be felt, as anyone will know who has accidentally touched, for example, a newly soldered joint. The simplest definition of heat is that it is a form of energy, usually created by chemical change or friction.

## Temperature

The *intensity* of heat is termed temperature, and it can be measured with a variety of instruments. As the temperature of domestic hot water and heating systems seldom exceeds 85 °C, an ordinary thermometer is satisfactory for most purposes in connection with hot water supply.

One of the effects of heat is that as the temperature is increased almost all forms of matter, i.e. solids, liquids and gases, expand. This is a very important principle because it has many differing applications to the work and equipment used in the plumbing industry.

A simple example which proves the fact that liquids expand when heated is shown in Fig. 8.1. It also demonstrates the basic working principles of mercurial and alcohol thermometers. A flask is filled with a coloured, heat-sensitive liquid, usually alcohol, which expands rapidly on a rise in temperature. The flask is closed with a rubber bung through which is passed a capillary glass tube. When the hands are placed on the flask, heat from the body raises the temperature of the fluid, causing it to expand up the tube. So that the effects can be clearly seen, a piece of white card is placed behind the tube.

A thermometer works in exactly the same way, the liquid used being mercury or alcohol, both these materials being very sensitive to temperature change. A fixed scale is used in a thermometer, the boiling and freezing points of water being indicated as fixed points with suitable degree graduations between the two.

## Expansion of hot water and heating pipes

Long lengths of hot water pipe are affected to a considerable degree by increased temperature and the accompanying expansion so precautions must be taken to prevent the effects of the expansion causing a fracture or breakdown in the joints and fittings. Additionally, one of the most common sources of noise in pipework systems is due to the expansion and contraction of hot water pipes laid in wooden joists or passing through floors without sufficient space for thermal movement.

### Coefficient of linear expansion

If a piece of material 1 m long expands by 1 mm when its temperature is raised through 1 °C, it can be said to have expanded by one-thousandth of its length. This number written as a decimal would read 0.001 and relates to the expansion of that material per degree centigrade temperature rise. This fraction is referred to as the *coefficient of linear expansion*.

The foregoing has only been used as an example but reference to the materials listed in Table 8.1 will show the coefficients of linear expansion of many of the materials with which the plumber is commonly in contact.

To calculate the increase in length of a given piece of material due to thermal expansion, three facts must be known.

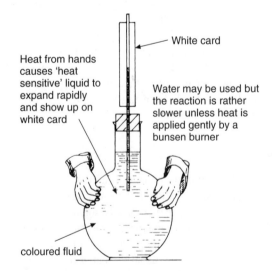

Heat from hands causes 'heat sensitive' liquid to expand rapidly and show up on white card

White card

Water may be used but the reaction is rather slower unless heat is applied gently by a bunsen burner

coloured fluid

**Fig. 8.1** Expansion of liquids.

**Table 8.1**  Coefficient of linear expansion.

| Material | Coefficient ($°C^{-1}$) |
| --- | --- |
| Brass 66% copper; 34% zinc | 0.0000189 |
| Cast iron | 0.0000102 |
| Copper | 0.0000167 |
| Invar steel | 0.0000009 |
| Lead | 0.0000291 |
| Low carbon steel | 0.000011 |
| Polythene | 0.00018 |
| PVC | 0.00005–0.00006 |
| Stainless steel | 0.00001 |
| Zinc | 0.0000258 |

(a) The original or normal length of the material.

(b) The increase in temperature through which it is raised.

(c) The coefficient of expansion for that material.

For instance, if the temperature of a 6 m length of copper pipe used on a central heating system is raised through 70 °C, it will expand by approximately 7 mm. The calculation is made using the following simple formula.

$$\text{Coefficient} \times \text{Original length of material}$$
$$\times \text{Temperature rise}$$
$$= \text{Increase in length}$$
$$= 0.0000167 \times 6\ \text{m} \times 70$$
$$= 0.007014\ \text{m, or approximately 7 mm}$$
$$\text{increase in length}$$

It will be seen that the materials listed have widely differing rates of expansion. It is worth noting that invar steel has a very low increase in length when it is raised in temperature and for this reason it has wide applications in the field of thermostatic devices.

*Expansion joints*

Provision is made for the effects of expansion in pipework by using expansion joints or bends. Predetermined points on the pipe run are 'anchored' or firmly fixed so that any increase in length must be taken up by the expansion joints. A typical example of this is shown in Fig. 8.2 which illustrates a run of pipe between two rooms with an expansion loop formed in the dividing wall.

A typical method of anchoring pipes in ducts is shown in Fig. 8.3. It is important to realise that any pipe clip or support between the anchors must permit some degree of movement to the pipe. Two types of purpose-made expansion joints are shown in Fig. 8.4.

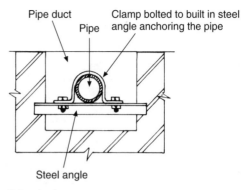

Fig. 8.3 Anchors.

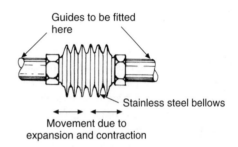

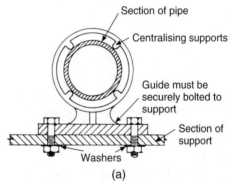

(a)

This illustrates a typical guide which must be used with bellows type expansion joints to prevent damage due to lateral movement. Two brackets, one each side of the bellows unit must be used.

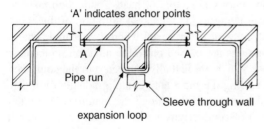

Fig. 8.2 Expansion loop.

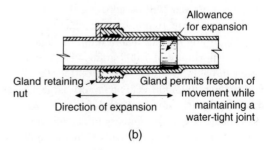

(b)

Fig. 8.4 Expansion joints.

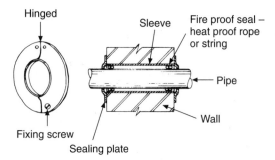

**Fig. 8.5** Sleeving through walls to allow for expansion. Detail of sealing plate which provides a neat finish to sleeves in walls and floors.

To allow for continuity of movement, pipes should not be 'built in' to walls or solid floors, but sleeved as shown in Fig. 8.5. A short length of pipe of larger diameter is used as a sleeve, any gaps between it and the pipe passing through being sealed with heat-resistant rope or similar fireproof material. To provide a neat finish on the face of the wall or floor, a sealing plate is used, made of metal or plastic, usually in two halves to enable them to be fitted after the pipe is installed.

### Quantity of heat
The difference between the intensity of heat (temperature) and its quantity is not always understood and it is often wrongly assumed that the higher the temperature of a substance the greater the quantity of heat it contains. This is not necessarily so as the volume of the heated substance must also be considered.

The basic metric unit of heat quantity is the *joule* (J). It is a very small unit, only 4.186 J (usually approximated to 4.2) being required to raise the temperature of 1 gram of water through 1 °C. A gram of water is also a very small quantity; some idea of its size will be apparent if a cube, each side measuring 10 mm, can be imagined – something a little smaller than a lump of sugar! To raise the temperature of 1,000 grams (1 kilogram) of water through 50 °C, i.e. $42 \times 1,000 \times 50$, would require 209,300 J. This is a very large number for a fairly small amount of heat, so to avoid ridiculously large numbers in calculations the usual practice is to work with *kilo*joules and *kilo*grams:

1,000 grams = 1 kg
1,000 joules = 1 kJ

(Note: 1 litre of water is equivalent to 1 kg of water.)

The following simple calculation will show the difference between intensity of heat (temperature) and quantity. Assume two vessels are taken, one containing 1 litre of water which is heated from 10 °C to 60 °C (a 50 °C increase in temperature), the other containing 10 litres which is heated from 10 °C to 20 °C (only a 10 °C increase). The following calculations show that the former, although hotter at 60 °C, contains less heat than the 10 litres which is at 20 °C.

1 litre $\times 50 \times 4.2 = 210$ kJ heat content
10 litres $\times 10 \times 4.2 = 420$ kJ heat content

From this it will be seen that temperature alone does not indicate the quantity of heat in a substance, but its volume must also be taken into consideration.

### Transfer of heat

Some knowledge of the ways in which heat is transferred is necessary to understand fully the working principles of heating and hot water systems. There are three methods of heat transfer: conduction, convection and radiation. Each of these and its applications will be discussed separately.

### Conduction
If a soldered capillary fitting is made on a piece of copper tube, it will be noticed that although only the area adjacent to the joint is heated, the tube will be hot some distance on either side of the joint, the heat having travelled along the pipe. This heat transference is known as conduction and it occurs in all solids, liquids and gases, the two latter generally being poor conductors of heat. Conduction takes place owing to the increased vibration of the molecules of the material when it is heated. The vibrations are passed on to the adjacent molecules, those which are actually subjected to the source of heat being the most active, and gradually diminish as they move further along the material. This rate or speed of conductivity varies with different materials. Most metals are good conductors of heat whilst

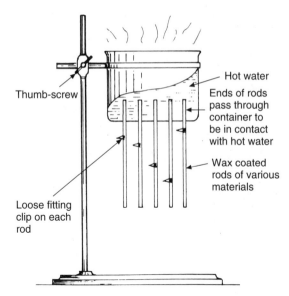

**Thumb-screw**

**Hot water**

**Ends of rods pass through container to be in contact with hot water**

**Wax coated rods of various materials**

**Loose fitting clip on each rod**

**Fig. 8.6** Experiment to show conduction rate through various solids.

plastic, ceramics and timber have a poor conductivity rate, these latter often being the basic materials from which thermal insulation is manufactured. Where it is necessary to reduce heat losses from pipes and vessels made of metal, they should be insulated with a poor conductor such as fibre glass, synthetic foam rubber or a PVC sleeve.

Metals themselves have differing rates of thermal conductivity, and a simple experiment as shown in Fig. 8.6 gives a good guide to the conductivity rates of some of the more common plumbing materials. It consists of an open-topped container into which are fixed a series of rods of differing materials. The upper ends of these rods are inside the vessel while the major lengths of the rods are outside. Loose-fitting clips are placed over the rods all at the same level, and to prevent these from falling by their own weight, each rod is coated with wax – candle wax is suitable for this purpose. The container is then filled with hot water which heats the upper end of the rods. As heat is conducted through the rods, so the wax is melted, causing the clips to slide slowly downwards. Assuming one of the rods is made of copper, it would be seen that it has a high conductivity rate, as the clip secured to it would fall over a longer distance in comparison to the clips on the rods made of iron, lead or plastic. The distance

through which the clips fall gives a general idea of the thermal conductivity of a particular material. There are, of course, precise figures for the thermal conductivity values of different materials.

A practical application of conductivity is the use of the copper coil or annulus in an indirect cylinder, as copper has a higher thermal conductivity rate than, for example, steel.

In years gone by boilers were often made of copper, not only because it has a high conductivity rate, but also because of its resistance to corrosion, especially in soft-water areas. However, owing to the high cost of copper and the fact that corrosion problems can be overcome by using indirect systems of hot water supply, iron or steel is now used almost exclusively as a boiler material. A hot water storage vessel made of copper needs to be well insulated otherwise the high conduction rate of the copper would result in considerable heat loss.

*Convection*
This form of heat transference arises from the physical movement of the molecules within a material. As fluids and gases are the only substances which are able to move freely at ordinary temperatures, this form of heat transference takes place only in these types of materials.

Liquids and gases expand when heated, and the molecules nearest the source of heat expand more quickly than those further away. On expansion they become lighter or less dense and are consequently pushed upward by the colder and heavier molecules surrounding them, these sinking to lower levels thereby creating circulation of the liquid. A simple illustration of what happens when a pot or kettle containing water is heated is shown in Fig. 8.7. The arrows indicate the upward flow of molecules as the water is heated. This movement is maintained throughout the period of heating, the water becoming progressively hotter.

The effect of heat on gases is precisely the same, illustrated in Fig. 8.8. This shows the effect of heating air in a sealed glass-fronted box. The heat emitted by the electric lamp heats the air which surrounds it. The heated air rises and passes out of the box through chimney 'B', causing further air to be drawn into the box via chimney 'A'. The air movement will be seen very clearly if a little smoke

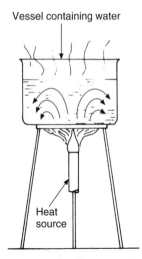

Fig. 8.7 Convection currents in fluids. Heated molecules rise and cooler molecules sink to lower level in vessel.

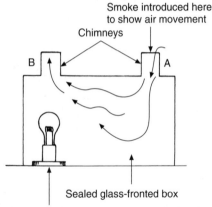

Fig. 8.8 Convection of gases.

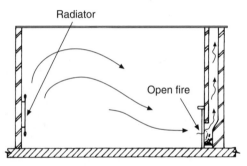

Fig. 8.9 Loss of warm air in room via open fireplace. The air warmed by the radiator may be drawn up the flue by the updraught of the open fire. This could result in a serious loss of heat.

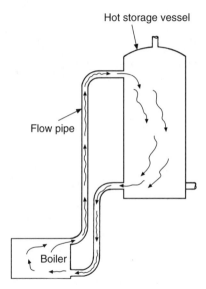

Fig. 8.10 Primary circulation in domestic hot water systems. The circulation of water between the boiler and hot storage vessel is due to convection currents.

is introduced to chimney 'A'. This experiment illustrates how the air in a room with an open fireplace will be heated by the fire and pass up the chimney, and how heat from a radiator causes the circulation of warm air in a room. Figure 8.9 shows how the heated air from the radiator may well be drawn up the chimney and so wasted. It may also be noticed that the wall above a radiator is darker than the surrounding areas. This is caused by dust particles adhering to the wall as the air, due to convection, moves upward. This disfiguration of the decorations is called pattern staining, and when a

radiator is fitted on a wall as illustrated, it should have a shelf fixed above it to deflect the warm air into the room.

Convection is the principle which causes water, when heated in a boiler, to flow upward into the hot storage vessels. The same movement of water occurs as was shown in Fig. 8.7, but in this case the top of the vessel is closed and the hot water rises upward through the flow pipe being replaced by colder and heavier water moving downward in the return (see Fig. 8.10). These two pipes are called the primary circulators.

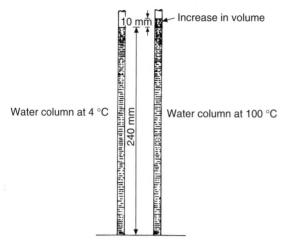

**Fig. 8.11** Difference in height of water columns at 4 °C and 100 °C.

Water has a unique property in that when it falls below or is raised above the temperature of 4 °C, it expands. It therefore occupies the least space (i.e. is at maximum density) at 4 °C. A rapid and considerable expansion takes place when the temperature is lowered, but expansion does not take place so quickly or by so much when the temperature is raised. In fact, water only expands by approximately 1/24th of its volume through a range of 4 °C to 100 °C (shown pictorially in Fig. 8.11). (The fraction is often changed to 1/25 for easy calculation purposes.) This very small amount of expansion means that the hotter water has less mass (i.e. is lighter) than the cooler water which creates the pressure head and causes circulation of the water by gravity. This is called the 'circulating pressure'. To express this in another way, the mass of 1 m³ of water at 4 °C is 1,000 kg while at a temperature of 100 °C the same quantity has a mass of only 958.5 kg.

In the average hot water scheme the difference in densities is much less. Normally water enters a system at approximately 10 °C having a density of 999.7 kg and assuming its temperature is raised to 70 °C, its density is then reduced to 977.7 kg. The difference in mass is only 22 kg at these widely varying temperatures, and as the temperature difference becomes progressively less, so the available circulating pressure head that causes circulation by gravity is reduced even more. For

this reason the primary circulation pipes must assist the circulation as much as possible by being of adequate size, as short as possible, and any changes of direction made with easy bends.

It is often thought that the pressure exerted by the head of water in the feed cistern causes the circulation of water in the primary flow and return. This is quite wrong as the pressure head exerted from this source is the same on both pipes.

*Radiation*

Heat transfer by both conduction and convection requires a medium through which to travel; radiant heat requires none. It may be defined as a form of energy which travels in straight lines through both air and space. The heat from the sun is a good example of radiant heat as it travels through millions of kilometres of space to reach the earth. It will pass through the air without appreciably warming it, but any solid object obstructing the rays will become warmed by them. Dark matt surfaces tend to absorb more radiant heat than those of a light colour, which reflect the heat. For this reason people living in hot climates often wear white clothing and live in buildings which have a light-coloured exterior.

One of the most promising developments in the field of energy conservation is the possibility of using heat from the sun's rays for water heating. Collecting plates consisting of a series of pipes fixed upon a dark heat absorbent surface are situated in a position, usually a roof, where they can receive the maximum exposure to the sun's rays. The dark surface absorbs the heat from the sun's rays and transmits it by conduction through the pipes to the water they contain, the heated water then being pumped to a suitable heat exchanger. Heat transmitted in this way is used to augment that provided by traditional heating appliances. Although solar heating is relatively expensive to install, the continuing shortage of fossil fuels and their alarming increase in cost in recent years has rapidly narrowed the gap between what is, and what is not, an economic proposition.

A simple experiment to illustrate the effect of radiant heat on differing surfaces in shown in Fig. 8.12. Two pieces of sheet metal of the same size are taken, one being highly polished, the other

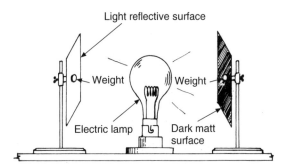

**Fig. 8.12** Effect of radiant heat on light reflective and dark matt surfaces.

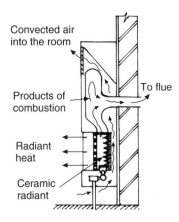

**Fig. 8.13** Diagrammatic section through gas fire convector. Modern gas fires are very efficient, providing heat by both convection and radiation.

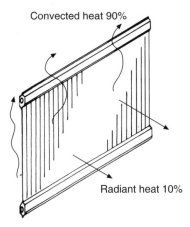

**Fig. 8.14** Heat emission from radiators. The name radiator is misleading as heat is provided mainly by convection creating circulating currents of warmed air.

having a dull matt, preferably black, surface. Into the centre of each plate is secured, by wax, two small pieces of lead, both of the same weight. An electric lamp is used as a heat source and is placed exactly between the two metal plates. The wax securing the weight to the matt surface will quickly melt, releasing its lead weight, clearly indicating that this surface has absorbed more heat. Bright surfaces reflect the rays of heat and for this reason do not absorb it so quickly, a well-known principle used in many forms of insulation.

To reduce heat losses in buildings by radiation plasterboard is obtainable with one surface covered with aluminium foil and, when used with stud work partitioning, radiant heat loss through this material is very low. A further lining of mineral-based felt backing to insulate against conducted heat will result in even further savings. The wall surface behind radiators is sometimes covered with aluminium foil. While this may be effective in preventing radiant heat loss, its use in preventing conduction is limited.

Some heating appliances, especially gas and electric fires, convert the heat generated by a flame or hot element to radiant heat. In the case of electric fires, the heat generated by the glowing element radiates on to a bright shining surface and is reflected out into the room. In the case of a gas fire the ceramic radiants are made having many 'nodules' which increase the surface area when they are heated by the flame and become almost white hot. Heat from these radiants is directed into the room by radiation in the same way as that of an electric fire (see Fig. 8.13). Most modern gas fires are designed in such a way that they produce

convected heat as well. However, it is an interesting fact that the heating equipment we call radiators in fact impart about 90% of their heat into a room by convection and only 10% by radiation (see Fig. 8.14).

## Simple domestic hot water systems

Having obtained some idea of the basic principle of hot water supply, some consideration can now be given to the component parts of the schemes used.

Hot water systems as we know them are the result of many years of careful thought and in some cases, trial and error. Until the last century the only form of hot water supply in many homes was a pot filled with water suspended over an open fire. When cast iron cooking ranges became popular, some enterprising manufacturer fitted a water container with a loose fitting cover on one side of the fire. The cover was removed to top up the container after the heated water had been drawn off by the tap in the base of the container. It was found that the production of hot water sometimes exceeded the requirements of the moment but often fell woefully short so the possibilities of hot water storage were explored, coupled with the need for a supply of hot water directly over the sink. A scheme known as the 'tank system' was eventually evolved, similar in some ways to modern systems, the main difference being that all the hot draw-offs were situated on the flow pipe. This stemmed from the thought that any draw-off should be taken from the hottest part of the system, i.e. near the boiler. Unfortunately what in fact happened was that when the supply of water in the boiler was exhausted, cold water in the return pipe became mixed with that drawn off from the flow. The result was that after a small amount of very hot water was drawn off, the remainder was only lukewarm.

Surprisingly this inefficient system with all its defects survived for many years, only gradually being replaced by that used today known as the 'cylinder' system to identify it from the tank system. This 'cylinder' system owes its name to the cylindrical hot water storage vessel generally used, its main feature being that hot water is drawn from the top of the storage vessel instead of the primary circulation, thus ensuring that water at the maximum temperature is drawn off.

### Boilers

The heart of any water heating system is the boiler, which must be designed to extract as much heat as possible from the fuel it uses. This has become increasingly important, not only because of fuel costs, but also its effects on global warming. When combustion of a fuel takes place, one of its products is carbon dioxide ($CO_2$). Most authorities have now recognized that the vast quantity of this gas now being produced is environmentally harmful. One of the steps to reduce its emission is to ensure that

**Fig. 8.15** Typical logo affixed to appliances approved by the energy saving trust.

heat producing appliances are more efficient and must comply with the provisions of the Building Regulations Part L (2002). Gas and oil boilers are now rated on the "SEDBUK" scale (Seasonable Efficiency of Domestic Boilers). Ratings range from A which are approximately 90% efficient to D which gives an efficiency rating of approximately 78/81%. The one exception to the foregoing are gas fired back boilers which should have a minimum efficiency of 75%. Solid fuel appliances must comply with HETAS (heating equipment, testing and approval scheme). Fig. 8.15 illustrates a typical logo affixed to all energy consuming equipment that complies with the requirements of the Energy saving trust.

*Solid fuel appliances*  Owing to the demise of the coal industry and the availability of gas and oil supplies to most parts of the UK, the market for solid fuel boilers has diminished considerably. Apart from very isolated areas of the country, or where a cheap supply of fuel is available, solid fuel installations are rare. They do have some advantages, however, in that most modern appliances are very efficient and require little or no professional maintenance. They are also popular even in areas where alternative fuels are available because of their aesthetic appearance. One very important point that must always be remembered is that solid fuel boilers, unlike those using oil, gas or electricity, are not on/off. Even when the air inlets and thermostats are closed there is a body of fuel in

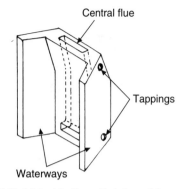

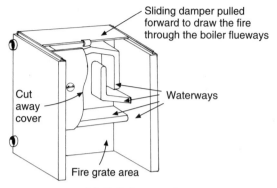

### (a) Solid fuel boiler with integral flue

The waterways in the cheeks of the boiler and those surrounding the flue increase the area in contact with the fire.

### (b) (i) Cut-away view

One of the two plates which cover the flueways is shown cut-away. The periodic removal of these plates enables the flueways to be cleaned.

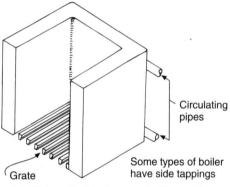

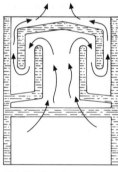

### (c) High-output boiler used with small independent heaters

### (b) (ii) Sectional view

From front of boiler – with the flueway covers removed the path of the hot flue gases is shown.

**Fig. 8.16**  High-output back boilers.

the boiler producing heat. To avoid overheating a gravity circuit must be provided – this is usually called a 'heat leak' to allow for the dissipation of hot water. This is generally achieved by a gravity circulation between the boiler and the hot storage vessel, and if possible a further advantage would be to connect a towel rail or radiator to the same circuit.

*Solid fuel back boilers*  This term is used to identify a boiler heated by a traditional open fire. Small boilers heating water for domestic use are virtually obsolescent and have been superseded by those having a larger heating surface exposed to the heat source. These are far more efficient and are capable of supplying both hot water and space

heating appliances. Typical examples of these boilers are shown in Fig. 8.16(a) (b) & (c). Many modern solid fuel appliances also utilize convection to heat the room in which they are fitted. As with any energy efficient appliance, it is very important to comply with the manufacturers' instructions, especially the flue sizes and connections, as excess air passing up the flue will drastically reduce efficiency.

*Cooking ranges*  Modern solid fuel ranges are very effective and relatively cheap to run as they are used for water heating and cooking, and to some extent space heating. The heavy iron castings used in their construction absorb a considerable amount of heat which is distributed into the room where

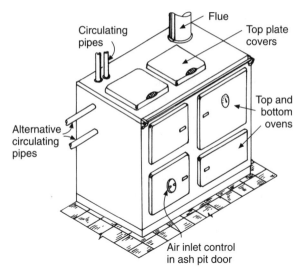

**Fig. 8.17** Solid fuel cooker.

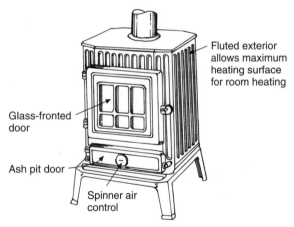

**Fig. 8.19** Small, independent, solid fuel room heater capable of heating water for domestic use and two–three radiators. Larger models are suitable for whole-house heating.

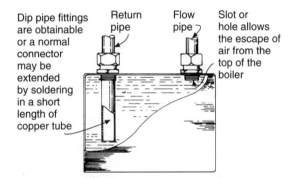

**Fig. 8.18** Boilers with top connections. Some boilers fitted in cooking stoves have top connections. The return pipe must be extended into the boiler to ensure positive circulation.

they are installed. These cookers originally had very small boilers suitable for hot water only, but with larger and more efficient boilers they are able to provide sufficient hot water to supply two or three radiators. It is recommended that the hot water storage vessel contains approximately 180 litres to provide sufficient hot water. This allows longer intervals between the necessity of stoking the fire. Figure 8.17 shows a typical cooking range of this type which has alternative side and top tappings for the boiler. If top tappings are used special arrangements are made to ensure a positive circulation of hot water, see Fig. 8.18. These ranges are also produced for both gas and oil fuels.

*Free-standing boilers* Because they are self-contained and do not have to be built into a chimney opening, these boilers are often called 'independent boilers'. Figure 8.19 shows one example of the many types made with varying heat outputs. As with modern back boilers the larger types of these units are capable of producing sufficient hot water for both domestic and space heating. As with all solid fuel appliances they are used mainly for their appearance, especially in old cottage-type dwellings.

*Controls for small solid fuel appliances* The form of control used for these appliances is by limiting the air for combustion. This is achieved mainly in the following three ways:

(a) **Sliding damper** (see Fig. 8.16(b)). This is a simple device which allows the hot flue gases into and around the boiler flue ways, or effectively closes them off. Used mainly on open fire appliances.
(b) **Spin wheel** (see Fig. 8.20). This controls the volume of air passing into the boiler via the ash pit. The wheel is fitted on a threaded rod and by turning in a clockwise direction it will reduce the gaps through which combustion air can pass.

*Hopper-fed boilers* The main feature of these boilers is an integral fuel store which avoids the necessity

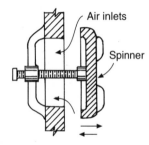

**Fig. 8.20** Detail of combustion air control. By rotating the spinner the volume of air for combustion can be increased or decreased.

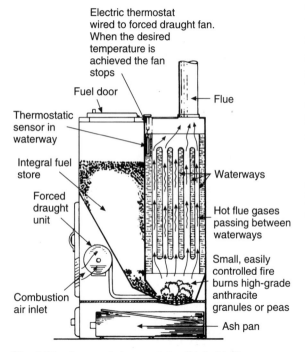

**Fig. 8.21** Automatic hopper-fed solid fuel boiler.

of stoking the boiler every few hours. They are designed to burn small, graded fuel known as anthracite peas. Fuel of this size will flow from the hopper by gravity on to the firebed. An electric blower, which is controlled by a thermostat, provides air for combustion. A typical boiler of this type is illustrated in Fig. 8.21. They are very efficient due to their design, the quality of the fuel they burn, and the forced draught which provides sufficient oxygen for complete combustion of the fuel.

*Gas-fired boilers*   Gas is a very popular and versatile fuel and can be burned in a variety of boiler types all of which embody the same basic principles. The original gas boilers were of the free-standing independent type and very early models required only a supply of gas, all the controls being operated by the gas supply itself, with the exception of the clockwork timer switch which was wound manually. With the advent of electrical control this type of timing device was quickly replaced by those operated electrically, and electrical thermostats and solenoid valves have replaced the older gas thermostats and relay valves to control the burning operation. Gas boilers have also been developed to be fitted behind a fireplace on the front of which is mounted a gas fire that heats the room in which it is situated, normally making the installation of other space-heating appliances unnecessary. Wall-hung boilers are now very popular because of their advantage in saving floor space and being suitable for fitting almost anywhere – in a roof space if necessary – providing access for servicing is available.

*Condensing gas boilers*   These boilers have been developed to increase the efficiency of gas water-heating appliances. They are becoming increasingly used for the replacement of older boilers.

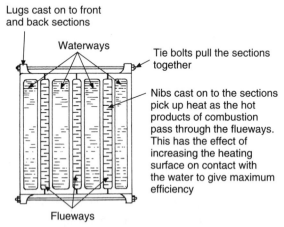

**Fig. 8.22** Plan view of section of gas- or oil-fired boilers showing the flueways. Because these fuels are cleaner the flueways are much smaller than those of solid fuel appliances. It illustrates the need to ensure that fuel/air ratios are carefully adjusted as failure to do so will cause the rapid formation of carbon deposits blocking the flueways.

Many modern gas appliances do not require a chimney, a distinct advantage as building costs can be reduced. These appliances are fitted with a balanced flue and are known as 'room sealed'. They do not take air for combustion from the room in which they are situated, both the air inlet and outlet for the combustion products being situated in a special grille which passes through the wall to the outside. The fact that air for combustion is not taken from the room obviates the cold draughts at low level that are often noticeable with conventional appliances of all types. Both oil and gas fired appliances are dealt with more fully in Book 2 of this series.

*Oil-fired boilers*   The actual boiler construction used with atomising burners is very similar to that used for gas boilers. Figure 8.22 illustrates the flueways of a typical cast iron boiler, which, like many gas boilers, are constructed of a series of sections screwed together with right- and left-hand nipple or tie bolts. Welded LCS plates are also used for the construction of oil-fired boilers. Modern boilers using oil as a fuel employ atomising burners for firing and, like gas, can be adapted for balanced flue operation.

*Electric boilers*   The cost of fuel using electric boilers cannot compete with the cost of oil or gas, but their use compares favourably with other forms of electric heating, such as storage heaters. They do, however, have all the advantages of other types of boilers used in wet heating systems. Apart from running costs they do have many advantages in that installations are very simple, they do not need a flue, they occupy little space and require a minimum amount of maintenance. Outputs vary depending on the model selected and range generally between 6 and 12 kW. They should only be fitted in fully pumped systems of heating; gravity flow installations are unsuitable because of the relatively slow movement of water over the heating elements which would cause constant tripping out of the power supply. All electrical work must be carried out by a qualified electrician and comply with the IEE recommendations. Figure 8.23 illustrates a typical electric boiler.

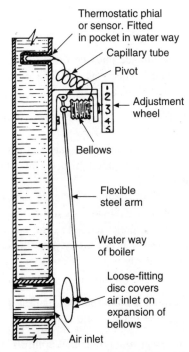

**Fig. 8.24**   Thermostatic control for small solid fuel boilers and room heaters.

(c)   **Thermostatic control** (see Fig. 8.24). This is a simple non-electric thermostatic device, which controls the supply of air for combustion, and is suitable for all small solid fuel appliances except for open fires. It operates by reducing

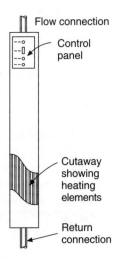

**Fig. 8.23**   Electric boiler.

the air (i.e. the oxygen) supply needed to burn the fuel. The thermostat consists of a loosely fitted metal disc on a flexible steel arm, pivoted at its upper end, allowing it to swing inward so that the disc covers the air inlet. A phial containing a heat-sensitive gas is situated in a pocket in the waterway of the boiler and is connected to the bellows unit by a small-bore copper tube. As the water temperature increases, the gas in the phial expands and opens the bellows which in turn exert pressure on the flexible arm closing the air inlet. As the air inlet closes so the fuel has a reduced air supply, less heat is generated and consequently less hot water is produced. Variations of temperature are achieved by the control which lengthens or shortens movement of the bellows by means of a screw thread.

## Installation

All types of appliance must be fitted to comply with the manufacturer's specifications. The main points to bear in mind are as follows:

(a) Check that sufficient air for combustion is available.
(b) Ensure all seals on the appliance are airtight – this is especially important in connection with convector fires.
(c) The 'throat' of the flue with open fires must comply with the dimensions in the installation guides. It is usually defined as the point at which the appliance actually joins the chimney. It does not apply to appliances connected to the chimney via a flue pipe.

On completion of a solid fuel installation the customer must be made aware that:

(a) The operation of any air inlet controls may vary according to wind conditions causing a variation on the pull of the flue.
(b) It is essential that the recommended grade of fuel is used.
(c) All flueways must be kept clean.

*Capacities of boilers – general considerations*
For hot water supply in domestic premises having a storage capacity of approximately 120 litres, it is usual to allow 3 kW boiler power (i.e. 3,000 watts per hour). If the scheme is combined with space heating a suitable allowance must be added.

It is common practice to oversize solid fuel boilers by about 20% to avoid the necessity of attending to the fire every two or three hours, as having a larger fuel capacity, the air inlet can be adjusted so the fire will keep in for a much longer period. This also makes provision for very cold periods when the air temperature may fall below 0 °C for short periods of time. For the same reason it is advisable to increase the power of both gas and oil boilers by approximately 10%.

Oil- and gas-fired boilers are generally more efficient than those of solid fuel owing to the relative purity of the fuel they burn. A certain amount of deposit from the products of combustion is left on the boiler flueways but provided it is removed when the boiler is serviced, little loss of efficiency will result. These two fuels also have the advantage in that they can be controlled electrically and can therefore be operated automatically to a greater extent than solid fuels. They do, however, require much more specialist attention and unlike solid fuel boilers involve the householder in an annual maintenance charge.

*Boiler sizes*
Boilers are specified by their rated output in watts per hour, this being based on the quantity of heat that will be passed through the boiler plates to the water. Solid fuel boilers can be made to work harder and produce more hot water than their rated output, but this leads to waste of fuel and damage to firebricks and enamelled surfaces. Most boiler manufacturers are aware of this fact and base the output on standard laboratory tests.

*Competency in gas work*  It should be noted that only plumbers and fitters who are qualified in the installation and servicing of gas appliances are allowed to carry out gas fitting, as poor-quality work and lack of knowledge can result in dangerous situations, the only exception to this being where gas work is carried out on a DIY basis. However, all work whether carried out by certified gas fitters or DIY enthusiasts must comply with the Gas

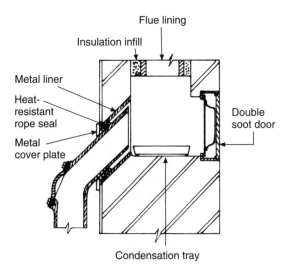

**Fig. 8.25** Connection of flue pipe from independent boiler to chimney.

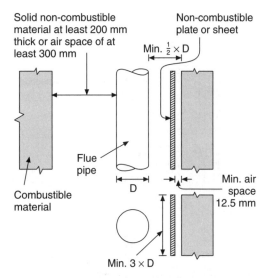

**Fig. 8.26** Recommendations of Building Regulations Approved Document Part J: Flue pipes for solid fuel and oil appliances.

Regulations which are enforceable by law. All plumbers and fitters who undertake gas work must be members of a body known as the Confederation of Registered Gas Installers (CORGI), and to ensure that standards are maintained CORGI representatives make regular inspections of the registered installer's work.

### Flues

The effectiveness of the boiler flue has a considerable bearing on the efficiency of any boiler whatever type of fuel is used. Flues should be lined to prevent condensation forming, a very important consideration when oil or gas appliances are used, owing to the low temperatures of the flue gases.

The formation of the throat of a flue in the case of a solid fuel back boiler must be carried out observing the manufacturer's recommendations in order to obtain the maximum efficiency from the appliance without waste of fuel. The throat is the portion of the flue which is narrowed or reduced in size immediately above the appliance, creating a positive updraught.

The connection of the flue pipe of an independent boiler to the stack should be made as shown in Fig. 8.25. This method allows the flue pipe to expand and avoids cracked plaster where it enters the chimney. A soot door should also be provided

externally to enable the chimney to be swept. Figure 8.26 shows the recommendations for solid fuel flue pipes in relation to combustible materials.

### Hearths

Before fitting a boiler consideration should be given to its position in relation to the suitability of the surrounding building fabric as a boiler support and its fire resistance. Reference must be made to the Building Regulations for details of the requirements for hearths and flues.

### Hot water storage vessels

These must now comply with the updated British Standards BS 1566 and 3198. Those labelled P are only suitable for pumped primary systems. Type G are suitable for both pumped and gravity systems. They are usually cylindrical in shape to enable them to withstand higher internal pressures. The maximum listed capacity is 440 litres. Vessels of larger capacity than this would not be economically viable due to the required increase in the thickness of copper. In commercial buildings where more storage is required, galvanized steel is used for greater strength. Stainless steel vessels complying with BS EN 150 9002 are becoming more popular, especially for unvented systems, as this material is

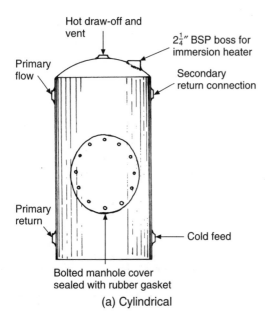

Hot draw-off and vent

$2\frac{1}{4}''$ BSP boss for immersion heater

Primary flow

Secondary return connection

Primary return

Cold feed

Bolted manhole cover sealed with rubber gasket

(a) Cylindrical

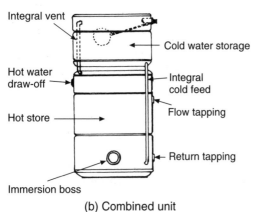

Integral vent

Cold water storage

Hot water draw-off

Integral cold feed

Flow tapping

Hot store

Return tapping

Immersion boss

(b) Combined unit

Reference should be made to Chapter 6 regarding accessibility of the float valve in this unit.

**Fig. 8.27**  Hot water storage vessels.

stronger than copper and can better withstand the higher internal pressures common to such installations. It is not good practice to mix galvanised components and copper in the same system if it can be avoided because of the possibility of electrolytic corrosion. Since 1986 unvented hot water systems have been permitted and because of the higher pressures, stronger storage vessels and more sophisticated controls are necessary with this type of system. Further details

on the subject are dealt with in more detail in Book 2 of this series. Storage vessels should be sited as near as possible to the boiler to reduce heat losses from the circulating pipes and 'dead legs' should be as short as possible. A dead leg is a run of pipe from the hot storage vessel to a drain-off point and most draw-off pipes in small domestic properties are in fact dead legs. Long dead legs will result in waste of water and heat due to the amount of water having to be drawn off before hot water reaches the tap and the subsequent heat losses from the hot water remaining in the pipe gradually cooling when the tap is closed.

It is important to make provision for air movement under the storage vessel by standing it on battens. Failure to do this may result in the formation of condensation, causing damage to wooden or chipboard flooring.

Some manufacturers produce units which combine the cold water storage with that of the hot (see Fig. 8.27(c)). These are suitable for small domestic properties such as flats and pensioners' dwellings where the demand for hot water is not so high. Their main advantage is the limited space they occupy, and the fact that being a unit the vent and feed pipes are built in, which reduces the installation costs. They are also very useful in schemes involving the upgrading of older properties where space is limited.

*Indirect cylinders*  So far all the storage vessels discussed have been of the 'direct' type. This is to say that water heated in the boiler is circulated in the storage vessel and will be eventually drawn off through the taps, resulting in a continuous change of water. This is acceptable when the water is for domestic use only, but if the scheme is combined with central heating, an indirect cylinder must be used (see Fig. 8.28).

The object of using an indirect cylinder is to separate the water in the boiler, radiators and associated pipework, the primary circulation, from that actually drawn off through the taps. The only reason water has to be supplied to the primary circulation after the initial filling of the system is to replace losses due to evaporation. The use of an indirect cylinder avoids the build-up of fur in the boiler and primary flow and return pipes caused by

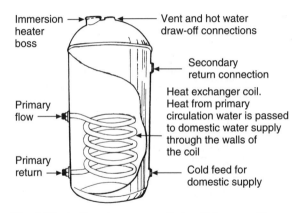

**Fig. 8.28** Indirect hot water cylinder. This type of hot storage vessel is used on indirect systems of hot water supply to reduce corrosion and scale formation in hardwater areas.

**Table 8.2**  Grading of hot storage vessels.

| Vessel | Grade | Test pressure (kN/m²) | Max. working head (m)* |
|---|---|---|---|
| Copper cylinder to | 1 | 365 | 25 |
| BS 3198 and 1566 | 2 | 220 | 15 |
|  | 3 | 145 | 10 |
| Galvanised steel cylinders | A | 483 | 30 |
| to BS 417 | B | 276 | 18 |
|  | C | 138 | 9 |

*The maximum working head is vertical distance between the base of the storage vessel and the water line of the feed cistern.

**Table 8.3**  Hot water storage capacities for small dwellings.

| No. of bedrooms | Storage in litres |
|---|---|
| 2 | 120 |
| 3 | 144 |
| 4 | 166 |

the precipitation of salts from temporary hard water. It also avoids any corrosion problems that could occur with continually changing water coming into contact with radiators made of ferrous metals. Heat is imparted to the water in the storage vessel by a coil or annulus, the coil being preferred owing to the higher pressures developed in fully pumped and pressurised systems of combined hot water and heating supplies. A coiled tube is deemed more efficient than an annulus and so less heating surface is required for the same performance.

*Feed cistern*
This is the term given to the cistern which feeds the domestic hot water supply system to distinguish it from the feed and expansion cisterns used in indirect hot water supplies and from cold water storage cisterns. In most domestic properties, however, where the cistern feeds the cold water services it also serves as a feed cistern for the hot supply. Its minimum nominal capacity, for hot water supply only, should not be less than 114 litres plus the amount required for cold water storage if it is used as a dual-purpose storage vessel.

The materials from which cisterns are constructed will be found described in Chapter 7 on cold water supply.

Hot storage vessels are graded in relation to the thickness of the material from which they are made

and the internal pressure which they can withstand, as can be seen from Table 8.2.

*Hot water storage capacities*  Table 8.3 shows suggested hot water store capacities for domestic properties based on the number of bedrooms. It is assumed here that 3 kW are available for water heating and the temperature of the stored water is 60 °C. The recommendations do not take into account the fact that larger properties may have two bathrooms, in which case extra storage may be necessary, or increased storage temperatures and higher heat input. In special cases where a more accurate assessment of hot water storage requirement is necessary, reference should be made to CP 6700.

*Boiler power for hot water services*  Having decided on the capacity and type of storage vessel, the next step is to ascertain the boiler power required. It used to be thought the boiler should be capable of heating the contents of the storage vessel in 1 hour. This is unnecessary of course as the only time during the day when all its contents are required will be the 'peak' period, this being the time during which the heaviest demands will be made upon the stored water. Two to three hours are allowed for the boiler to heat the stored water, this being termed the firing period. Using this method of boiler sizing, there is still sufficient water at the right temperature to meet the incidental demands of the kitchen sink and wash basin at off-peak times.

To give the reader some idea of calculating the boiler power required to heat a domestic storage vessel containing 120 litres of water, the following example is shown. It is assumed the temperature of cold water entering the system is 10 °C and it is to be raised to 65 °C. Water at 65 °C is hot enough for culinary purposes and bathing, and by keeping the maximum temperature at 65 °C, the loss of heat to the surrounding air is minimised and excessive scale in hard-water areas is reduced. In this example the water has to be increased in temperature by 55 °C and assuming 4.2 kJ are needed to raise the temperature of 1 kg (litre) of water through 1 °C, the following formula is used:

Quantity of water × Temperature rise × Specific heat of water

where the specific heat of water (the heat required to raise 1 litre of water through 1 °C) is 4.2 kJ. Therefore:

$$120 \times 55 \times 4.2 = 27{,}720 \text{ kJ}$$

It should not be forgotten that the time scale on which this calculation has been made is only 1 second, but boilers are rated in kilowatts (kW) per hour so the kilojoules figure has to be divided by 3,600, the number of seconds in 1 hour, to convert it to kilowatts per hour, i.e.

$$27{,}720 \div 3{,}600 = 7.7 \text{ kW}$$

The boiler power required is therefore 7.7 kW per hour, but assuming a firing period of 3 hours:

$$7.7 \div 3 = 2.56 \text{ kW}$$

Just over $2\frac{1}{2}$ kW will be required.

It is seldom necessary to go through these calculations for small domestic systems as all but the smallest boiler will produce sufficient power for this purpose. They do, however, serve to illustrate the method used to calculate boiler power in comparatively simple terms.

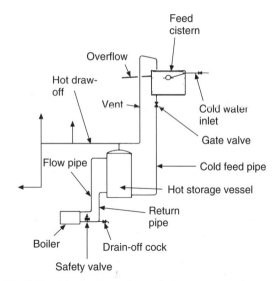

**Fig. 8.29** Simple direct domestic hot water supply system.

### Domestic hot water pipework systems

A simple 'direct' hot water system of this type is illustrated in Fig. 8.29. The primary flow and return pipes should not be less than 25 mm in diameter, except in areas where the water is known to have a high temporary hardness (unless an indirect system is to be used) when one size larger is often used so the build-up of fur and scale does not have immediate effects on the flow of water. The minimum recommended diameter of the cold feed pipe on a system having only three or four draw-off points is also 25 mm. While it cannot be ignored that in many cases a feed of 19 mm diameter appears to function quite well, it should be borne in mind that hot water can only be drawn from the storage vessel as quickly as it is replaced. Thus if two draw-offs are opened simultaneously, it is likely that the supplies to both will be diminished. This can result in air being drawn in through the vent and produce a mixture of air and water at the draw-off, as shown in Fig. 8.30. If a hot water draw-off is fixed at a level too near the feed cistern water level a similar situation occurs, and in such cases the only remedies are either to increase the head pressure on the system by raising the cistern or to increase the diameter of the feed pipe.

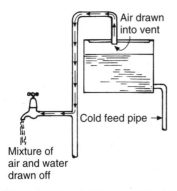

**Fig. 8.30** Effect of fitting hot draw-off too high on vent or fitting cold feed of insufficient diameter.

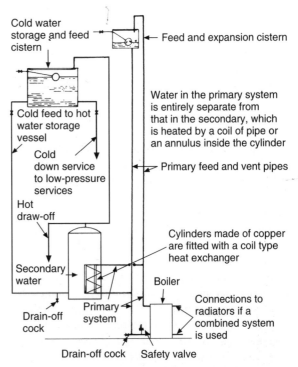

**Fig. 8.31** Indirect system of hot water supply with gravity circuit to domestic hot water.

A valve is fitted on the cold feed to isolate the hot water system for repair work. While in the past stop valves have often been used for this purpose, a gate valve is now used as they are classified as full way and offer little or no frictional resistance to the passage of water, an important point to remember when dealing with the low pressures involved. No other supplies or draw-offs should be connected to the cold feed as this could lead to starvation of the hot water supply by reducing the flow of water into the hot water storage vessel. In a small system supplying only a bath, basin and sink, a draw-off pipe of 19 mm diameter is normally sufficient. This is also the minimum size of the vent, as smaller pipes would be more prone to obstruction. The vent is a very important part of the system, its main functions being to maintain atmospheric conditions in the pipework. It permits the escape of air which has entered the system and, more rarely in the event of the water becoming overheated, allows it to discharge over the feed cistern.

### Indirect systems of hot water supply

An indirect system is designed in such a way that the water in the boiler and primary pipework is never changed, the only loss of water being in the supply cistern through evaporation. When the water is initially heated, any gases it contains are given off, including the oxygen and carbon dioxide, which, as stated earlier, are largely responsible for corrosion and scaling respectively. This is not to say that an indirect system is entirely immune from

corrosion, but its incidence is considerably reduced. A study of Fig. 8.31 shows that the primary water is quite separate from the secondary water, and may be defined as that contained in the boiler, primary flow and return pipes, the annulus or coil in the cylinder, and any radiators that are fitted. The secondary water is that drawn from hot storage vessels to supply the hot taps. It is heated by conduction, being in contact with the annulus or coil.

It is important that indirect cylinders conform to the relevant standards which specify the minimum heating surface area of the annulus or coil. Hot water vessels which do not comply with this standard often have an insufficient primary heating surface which results in slow heating up of the water, wastage of fuel and higher running costs. The cylinders, too, are often made of thinner copper sheet than that specified, resulting in a shorter working life. The moral of this is to ensure that all components carry the British Standard kite mark.

The primary part of the system is fed by a feed and expansion cistern which must be big enough to accommodate the expansion of the water in the system when it is heated. For small indirect systems a cistern of 50–70 litres actual capacity is normally of sufficient size. An accurate method of sizing may be necessary if a *combined system* is fitted. This term applies to systems where one boiler serves both domestic hot water and space-heating appliances such as radiators.

Although hot water expands by approximately only $\frac{1}{24}$ in volume when raised from 4 °C to 100 °C, it is usual to make provision for an expansion of $\frac{1}{20}$ when sizing feed and expansion cisterns; this is to say that a system filled with water containing 100 litres when cold should have provision in the expansion cistern for a further 5 litres. This is a generous allowance as the average temperature of the cold water supply is 10 °C and it is seldom necessary and, indeed, is wasteful to heat domestic hot water to more than 70 °C. Temperatures higher than this give rise to the formation of fur in the secondary part of the system and excessive heat loss from the stored water. Indirect systems of this type are sometimes called 'closed circuits', which can be confusing as, unlike fully pressurised systems which have no outlet to the atmosphere, the system under discussion has a vent and feed pipe open to the atmosphere.

Failure of the boiler to heat the domestic hot water is usually caused by evaporation of the water in the expansion cistern, while the float-operated valve, because it is seldom required to open, becomes corroded and remains in the closed position. When the water level falls below the flow connection on the cylinder, no circulation can take place in the coil, with the result that the secondary water remains cold.

It is rare but not unknown for a leak in the coil to allow the mixing of the primary and secondary water. This may remain undetected for a long time if the surface of the water in the feed and expansion cistern is level with that of the feed cistern. The first indication that something is wrong is the discharge of discoloured water from the hot taps, or worse, corrosion in a radiator causing it to leak. For this reason it is recommended that the cisterns are fitted at different levels so that persistent overflowing of the lower cistern will indicate the defect.

*Single-feed indirect systems*

Single-feed cylinders are no longer installed in new installations and are only available for replacement. The following text has been included so that these installations are recognised and any maintenance necessary can be carried out effectively. This type of system employs the use of a hot storage vessel known as a self-priming cylinder. The difference between this and a normal indirect cylinder is that both the primary and secondary sections of the system are fed with water from a common cold feed, a special heater admitting water to the boiler and any radiators that may be fitted. These cylinders are slightly more expensive than the normal indirect types, the main object in their use being to reduce the amount of pipework involved in the installation and to render a separate feed, vent and expansion cistern unnecessary thus reducing fitting costs. They are, in fact, fitted in the same way as a direct cylinder, the primary and secondary water being kept separate by what is best described as an air lock which is contained in the single-feed heater situated inside the cylinder. This heater contains two hemispheres, under one of which air is entrapped. The movement of air between these two hemispheres allows for the expansion of the primary water when it is heated. A study of Fig. 8.32 shows how both the primary and secondary components are filled with water and charts the movement of the air when the primary water expands. It will be seen that the space containing the air in the hemisphere is limited, and if the volumetric expansion of the water exceeds that of the space available, some air will be forced out and the space it occupied replaced by water. If this continues, a slow but sure change of water in the primary part of the system will take place with all the undesirable characteristics that the use of an indirect system is supposed to prevent. This state of affairs is usually brought about by an excessive quantity of water in the primary circulation which, when it expands, the hemisphere is unable to accommodate. For this reason the heater units are made in more than one size, the volume of the hemispheres being larger for cylinders fitted to schemes having a larger quantity of primary water. In all cases when fitting single-feed cylinders the manufacturer's recommendations

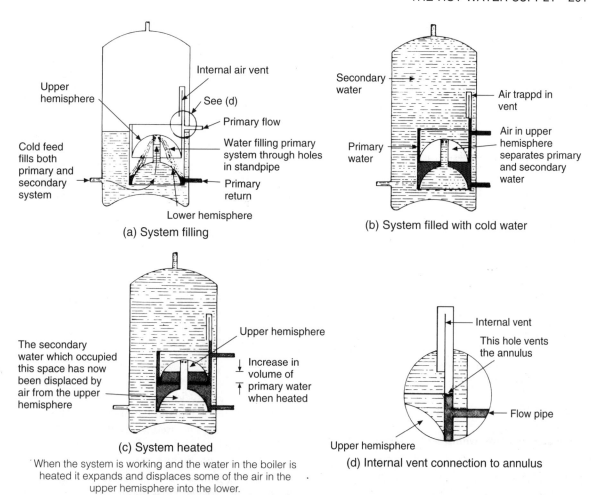

**Fig. 8.32**  Single-feed self-priming cylinders.

regarding the primary capacity of the system must be observed to avoid costly mistakes.

Cylinders of this type must be fitted with a drain-off cock on the cold feed like that of ordinary indirect cylinders, as they can only be partly drained by the boiler drain-off cock. Owing to their design, they must always be installed in the upright position, otherwise no air will be entrapped in the hemispheres.

When self-priming cylinders are drained down and refilled, the heater unit sometimes becomes air locked resulting in the secondary water not heating. This is usually due to a blockage in the inverted U vent fitted to the heater unit. Corrosion products sometimes obstruct the hole where the

vent is connected to the heater unit. If this happens the problem can sometimes be solved by draining the cylinder completely and allowing the feed cistern to fill before fully opening the gate valve on the cold feed. The resulting sudden rush of water into the cylinder often removes the corrosion products allowing the cylinder to function normally. Single-feed cylinders must not be fitted with valves controlling the temperature of the hot water because if the boiler overheats the air seal may be displaced. A hot water system using a single feed cylinder is shown in Fig. 8.33 and it will be seen that the connections are the same as those for a direct installation.

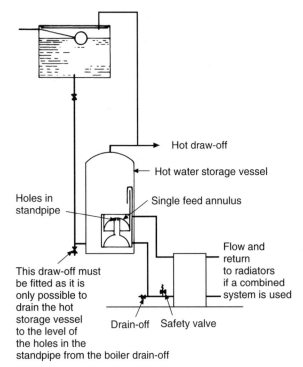

**Fig. 8.33** Indirect hot water scheme using single-feed hot storage vessel.

## Hot water systems with a secondary circulation

As has previously been described, when a hot tap is closed after use, the water that remains in the pipe will cool and when the tap is next opened, this cooled water must be first drawn off before a further supply of hot water reaches the tap. Draw-off pipes of excessive length are therefore a cause of waste of both heat and water. The Water Regulations, unlike previous legislation do not specify the maximum length of dead legs. The pipe lengths and diameters shown in Table 8.4

**Table 8.4** Recommendations of the Water Regulations Guide.

| *Maximum lengths of uninsulated hot water pipes* | |
| --- | --- |
| *Pipe diameter mm* | *Max length m* |
| Up to and including 12 | 20 |
| Over 12 and up to and including 22 | 12 |
| Over 22 and up to and including 28 | 8 |
| Over 28 | 3 |

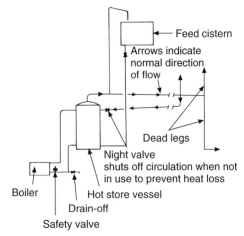

**Fig. 8.34** Domestic hot water system with secondary circulation. This type of system is not normally required in small houses as the draw-offs should be kept as short as possible by careful planning. Note that if a secondary circulation is fitted the hot storage vessel must always be lower than the pipe runs if it is to operate by natural or gravity circulation.

are however almost identical to those shown in BS 6700 as the maximum permissible lengths of dead legs and can be used as such.

Where long draw offs cannot be avoided, consideration should be given to the use of a single point gas or electric water heater. A typical example would be an isolated cloak room, where the maximum length of a dead leg may be exceeded. Secondary circulations should be unnecessary in small dwellings and are only advisable in very large houses, commercial or public buildings where they are absolutely essential. Figure 8.34 shows a direct hot water system incorporating a secondary circulation. It will be seen that its main function of providing an immediate supply of hot water to the draw-offs is accomplished, but the following two disadvantages should be noted. The long lengths of circulating pipe must be well insulated to prevent excessive heat losses and a night valve must be provided to prevent overnight circulation and subsequent cooling of the stored hot water if the boiler is shut down. These used to be manually operated gate valves, but with the availability of motorised valves and time switches, automatic shutdown at predetermined times is now possible.

So far it has been assumed that the secondary circulation operates on a gravity basis. Where such an arrangement is not possible owing to the position of the storage vessel, or head losses are excessive owing to very long pipe runs, then a pump must be provided. Owing to the long lengths of pipe usually associated with secondary circulations, heat loss can make such systems very expensive to run and effective insulation of the circulating pipe is essential.

### Secondary circulations in public and commercial buildings

While secondary circulations were originally designed to function by gravity the design and size of modern buildings often make a pumped circulation necessary. If, for instance, the hot storage vessel is at a high level in the building as shown in Fig. 8.35 a gravity circulation would not operate because of the lack of circulating pressure. In large buildings the length of the secondary circulation would make it extremely sluggish and, in view of the fact that the recommended temperature drop between the flow and return should not exceed 22 °C, a pump would definitely be necessary.

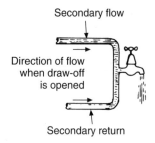

**Fig. 8.36** Effect of opening draw-off on secondary circulation.

Those used for this purpose must be constructed of non-ferrous metals or the working parts will quickly seize up. The system should be arranged to circulate only when the pump is functioning, so that a time switch can be used to determine when the circulation is operative thus avoiding the necessity of fitting a night valve.

When a draw-off on a secondary circulation is opened, water is drawn from both the flow and return (see Fig. 8.36). If the return connection to the hot storage vessel is too low, this will result in cooler water from the base of the storage vessel being drawn off producing a lukewarm supply, especially when large quantities of water are required such as when hot water for a bath is drawn. For this reason the position of the secondary return tapping in the storage vessel must be as high as possible, and certainly not further down the cylinder than one-quarter of its total height.

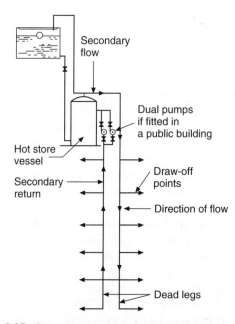

**Fig. 8.35** Pumped secondary circulation. No night valve will be necessary on this system as it will only circulate when the pumps are operating. Pumps can be fitted on the secondary flow if more convenient.

## Defects in hot water systems

Broadly speaking, there are two main causes of faulty hot water supply apart from the effects of frost. There are those that happen naturally and are usually the result either of corrosion caused by the use of dissimilar metals and unsuitable fittings, or scale formation in areas of temporary hard water. There are also the mechanical faults that are often built into a system resulting in air locks and noise. In most cases these defects are the result of lack of knowledge and appreciation of basic principles.

### Air locks

Air locks are a very common cause of trouble with both hot water and heating systems. Most air locks

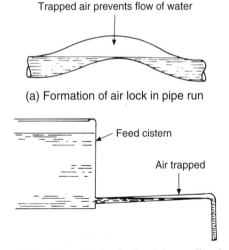

(a) Formation of air lock in pipe run

(b) Air lock caused by badly fitted draw-off or feed pipes. This is possibly the most common cause of air locks in small domestic hot water systems

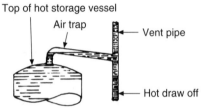

(d) Badly fitted hot draw-off pipe causing air lock

**Fig. 8.37**  Common causes of air locks.

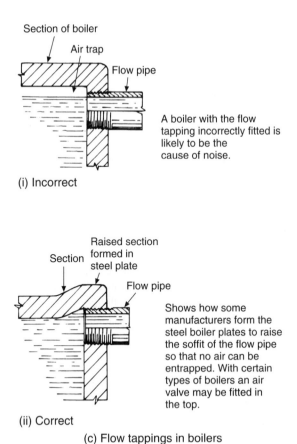

(i) Incorrect

A boiler with the flow tapping incorrectly fitted is likely to be the cause of noise.

(ii) Correct

Shows how some manufacturers form the steel boiler plates to raise the soffit of the flow pipe so that no air can be entrapped. With certain types of boilers an air valve may be fitted in the top.

(c) Flow tappings in boilers

are caused by unventilated arches formed in badly fitted pipework. An air lock is a small quantity of air trapped in a pipe which owing to the very low circulating pressure available prevents water passing through the pipe. Even if an air lock does not completely stop the flow of water it can reduce the flow considerably. They are very often the result of pipes sagging, or not being laid truly horizontally or to appropriate falls. Figure 8.37(a) shows how an air lock occurs in a pipe run, although they are not always as obvious as this.

Another common cause of air locks is shown in Fig. 8.37(b) where the cold feed, instead of falling away from the cistern, rises causing an unventilated arch in the bend.

Air may also be locked in a boiler owing to defective tappings as shown in Fig. 8.37(c)(i) but the boiler manufacturers are aware of this and

usually the flow connection is in the position shown in Fig. 8.37(c)(ii). Probably the worst effect of defective tappings would be a rather noisy boiler, but if this were further aggravated by an air lock in the flow pipe, reversed circulation could take place causing the hot water to be forced up the return pipe instead of the flow. Reversed circulation sometimes corrects itself as the water temperature increases, but the noise caused by this problem is alarming and should be investigated.

Care must be exercised when installing boilers and primary circulating pipes to avoid collections or pockets of air. Yet another common cause of air locking is shown in Fig. 8.37(d) where the hot draw-off has been badly fitted causing it to sag and create an air trap at the cylinder connection.

One of the most common methods of removing air from a hot water system is to fit a hose from a hot draw-off tap to a high-pressure cold supply tap and then turn both taps on, where the high pressure will force out the air. While this method is usually effective, it is not to be recommended as the whole of the hot water system is subjected to water at high pressure and should the vent and feed pipes be obstructed (possibly in freezing weather by ice) it may result in bursting the storage vessel.

Experience has shown that air locks usually occur in the cold feed pipe. The safest method of removing the air is to connect a force pump to the cold feed connection and force water through the air lock. If the air lock is in the hot draw-off, the force pump can be connected by a hose to a hot tap, pumping out the air as described. It must be stressed, however, that before taking any measures likely to exert pressure on the system, ensure that both the vent and feed pipes are not obstructed in any way.

### Boiler explosions

Boiler explosions are fortunately rare, but the result of such a happening can be disastrous, especially when a cast iron boiler is involved. An explosion will only occur if both of the primary circulating pipes are obstructed. This can come about for two reasons, the first being most common in domestic buildings. If a building is left without heat in freezing conditions, the whole of the system including the flow and return pipes may become frozen. If the boiler is fired under these conditions, the heating of the water will cause it to expand and burst. If it is suspected that the system is frozen, no attempt should be made to fire the boiler until it is certain any ice has thawed.

The other cause of boiler explosion is where both primary circulating pipes have been 'valved' for some reason. If both valves are closed the same situation will exist as previously described, with the same result. It should be quite unnecessary for valves to be fitted on domestic installations, although it has been known. On large installations having a series of boilers, it is of course usual to valve both primaries on each boiler to enable servicing or repairs to be made without shutting down the whole system. On installations of this size, however, the valves will be operated by a qualified person who would be aware of the

dangers. Even in these circumstances, when a job is completed on such a system, the valves must be locked in the 'open' position.

### Scale deposits

The formation of scale deposits in temporary hard water areas can obstruct the primary circulating pipes, but this is unlikely to happen to both pipes simultaneously as the flow pipe carrying the hotter water tends to scale up first and the resultant noises in the boiler would prompt an early investigation.

### Safety valves

It has always been considered good practice to fit a safety or pressure relief valve on a hot water system, but in order to cut down on cost they are often omitted. While no legislation exists to enforce their provision, it is good practice to fit them. It is interesting to note that regional gas companies will not undertake to maintain a gas-fired boiler unless it is equipped with a suitable pressure relief valve.

A typical spring-loaded safety valve (pressure relief valve) suitable for domestic use is shown in Fig. 8.38. Any excess pressure in the system lifts the valve against the spring, the excess escaping

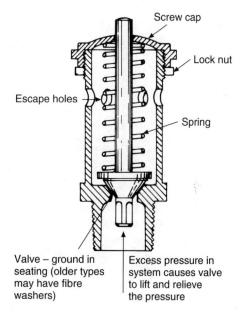

**Fig. 8.38** Spring-loaded safety valve.

through the holes in the valve casing. Safety valves should be fitted directly on the boiler if possible, and some manufacturers provide extra tappings for this purpose. In the case of back boilers this arrangement may not be possible and the usual practice is to fit the valve on the return pipe as close as possible to the boiler. There is less likelihood of scale formation in the return which could cause the valve to adhere to its seating and any water discharged will be at a lower temperature.

Safety valves should be removed periodically and checked to ensure they are in working order. The best practical way of adjusting a safety valve to function satisfactorily is as follows. The cap should be unscrewed, relieving tension on the spring until a drip of water is seen to be escaping from the holes in the casing. It should then be screwed down half a turn and locked in this position with the lock nut. A better type of safety valve is shown in Fig. 8.39. These are factory set between 3 and 6 bar. No adjustment is possible without destroying the cover. It is therefore necessary to assess the pressure at which the system is designed to work prior to its installation.

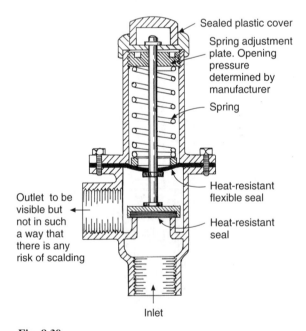

**Fig. 8.39**

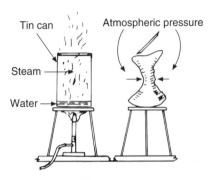

**Fig. 8.40**  Experiment illustrating cylinder collapse.

### Cylinder collapse

Cylinder collapse is due to the pressure on the inside of the cylinder becoming less than that of the atmosphere. A simple experiment is shown in Fig. 8.40 which illustrates quite clearly the effect of atmospheric pressure. A small amount of water is put into a can which has a close-fitting airtight lid. This lid is removed and the water is heated until it boils and gives off steam, when the source of heat is removed and the lid firmly replaced on the can allowing the contents to cool. When the steam condenses and the air in the can cools, it occupies less space, consequently lowering the pressure inside the can. The atmosphere exerts a pressure of approximately 100 kN on an area of 1 square metre, and since this pressure will be far greater than that inside the can, the walls of the can will be forced inward until the can has been completely deformed.

Various situations can occur whereby similar conditions can arise in hot water storage vessels causing their collapse. It almost invariably happens in freezing weather when the vent pipe, often in an exposed position, becomes frozen. There will be no indication that all is not well, as any expansion of the water will take place via the cold feed pipe to the storage vessel. As the water temperature increases, water vapour may form, pushing more water back up the cold feed. In freezing weather the air temperature drops overnight causing the cold feed to freeze as the fire dies down, the system then becomes closed to the atmosphere, and a situation now exists in the cylinder similar to that of the tin filled with steam, and it will have the same result when the boiler is shut down and the water in the storage vessel cools. The cooling water will contract in volume and reduce the pressure inside the cylinder which will slowly deform and partially collapse.

Another cause of cylinder collapse is when a hot water draw-off at a lower level is opened with both the vent and cold feed frozen. Water will flow from the tap but as none can enter the cylinder to replace it, a siphon is started. As the water is withdrawn the pressure inside the cylinder becomes less than that of the atmosphere so the cylinder sides are crushed inwards. In this case the cylinder may collapse fully and become completely flattened.

If only a slight collapse has occurred, the cylinder can often be salvaged by filling it with water and pressurising it with a force pump. Providing the damage is not too severe, the cylinder can be made serviceable again after thoroughly testing for leaks.

The incidence of cylinder collapse is much less in modern homes equipped with central heating systems and where the roofs are made draught-proof by sarking felt under the tiles. It should not be forgotten, however, that many older properties often have very exposed roof spaces where in severe weather it is not unknown for snow to be blown in. Where cisterns and pipework are exposed to such conditions, it is essential they are well insulated and draught-proofed.

*One-pipe circulation*

The positions of the connections on a storage vessel are most important. The hot water draw-off is taken from the top of the vessel as the temperature of the water will be at its maximum at this point. It is for this reason that when a secondary return connection is made it should be at a high level in the storage vessel.

To avoid 'one-pipe' circulation and subsequent waste of heat, the vent and draw-off should be fitted as shown in Fig. 8.41(a). Convection currents will occur in the vent pipe (see Fig. 8.41(b)) if it rises directly from the storage vessel.

*Vent pipe termination in cisterns*

The height of the vent above the water level in the feed cistern depends on the distance between the surface of the water level in the cistern and the base of the hot store vessel and may be calculated as

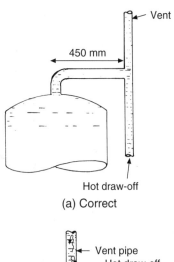

(a) Correct

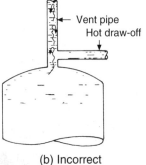

(b) Incorrect

Hot water in contact with pipe walls cools and sinks causing one-pipe circulation to take place.

**Fig. 8.41** Connecting vent and hot draw-off to avoid one-pipe circulation.

follows. Working on the assumption that water will expand by 1/25 from 4 °C to 100 °C, a column of water 1 m high, if raised through this temperature, would expand by 40 mm. On this basis 40 mm is allowed for every 1 metre in height of the system. As it is not desirable for water to be continually discharging from the vent, a distance of 150 mm is added. To give an example, supposing the water level in the cistern is 2.500 m above the base of the storage vessel, the vent should extend above it by 100 mm + 150 mm = 250 mm (see Fig. 8.42). The allowances that have been made are quite adequate, but failure to provide a vent of sufficient height will result in water discharging into the cistern as the water temperature increases. This not only causes a waste of fuel but may also result in lukewarm water at the cold water draw-off points if the cistern is used for hot water feed and cold water storage.

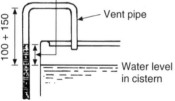

There is a difference between these water levels due to the higher density of the cold water in the full cistern

Vent pipe

100 + 150

Water level in cistern

**Fig. 8.42** Allowance for expansion of water in vent pipe.

The other important factor to observe in relation to vent pipes is to make sure they are trimmed off well above the water level. Should the vent become submerged, cold water could be siphoned out of the cistern when a hot draw-off is opened.

*Corrosion*

The subject of intermetallic or electrolytic corrosion is dealt with in Chapter 4. Surprisingly, many plumbers who are well aware of the disastrous effects of mixing dissimilar metals in a roofwork situation think nothing of using combinations of copper pipes and iron bushes! In hard-water areas a film of scale often protects the iron from any serious corrosion, but this is not the case with soft water when the result is often the rapid deterioration of small ferrous metal fittings such as bushes. To reduce this danger, such fittings should be made of bronze or brass with a high copper content, especially when fitted to direct systems of supply.

Another form of corrosion which arises in both hot and cold water systems is known as 'dezincification'. For many years the manufacturers of copper fittings have used brass with a high zinc content, partly to economise on copper, but mainly to produce an alloy which can easily be pressed or stamped. (These terms relate to a method of manufacture where a component is pressed out into a mould instead of being cast, the latter method being more expensive.) Unfortunately, brasses having a high zinc content are very unstable alloys in the presence of acidic waters and tend to break down into their constituent parts. The zinc content of the alloy is reduced to a basic carbonate and loses all its metallic properties, causing it to

increase in volume and obstruct the waterway of the fitting. The copper content of the alloy, if examined under a microscope, looks like a sponge and does in fact become porous. This form of corrosion is usually identified by a gradual lessening of the supply of water delivered through the draw-offs, and a white growth on the exterior surface of the fitting.

The use of brass as a material for copper pipe fittings has diminished, having been replaced by fittings made of copper or bronze. Quite recently fitting manufacturers have claimed to have overcome the problem of dezincification of brass by subjecting it to a form of heat treatment. Fittings of this type are, however, only available in the compression range, as the effect of soldering capillary joints has damaging effects on these specially heat-treated alloys.

Copper cylinders which were often supplied with brass bosses and brazed with an alloy containing zinc are now specified as being zinc free. Galvanised storage cisterns should on no account be used with waters likely to cause dezincification unless they are suitably protected by painting with a non-coal tar paint approved by the local water authority.

**Prevention of heat loss**

The insulation of plumbing pipework and components is important in terms of (a) conserving heat (energy) and (b) protection from frost damage when fitted in exposed areas. Insulation work carried out by the plumber and heating fitter is normally confined to domestic and small commercial installations. In the case of large contracts an insulation specialist is normally employed. The following text deals mainly with the types of insulation that can be applied with little or no specialist tools and is for general guidance only. For more detailed information reference should be made to BS 5422 and BS 5970 which give comprehensive details on the types and application of all forms of insulation. Manufacturers usually specify their insulation products in W m/K, which means the amount of heat that will pass through 1 cubic metre, per hour, per degree temperature difference, see Fig. 8.43. The reader may be aware

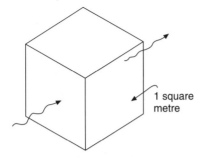

The passage of heat through insulation products is calculated in watts per square metre per kelvin temperature difference and will vary depending on the type of insulating material and its density.

**Fig. 8.43** Heat loss through insulation products.

that the kelvin and the centigrade scales are in effect the same, e.g. 0 K = 0 °C, but because the kelvin scale records absolute zero, its use is more convenient to insulation manufacturers when calculating conductivity. A very important point to consider when specifying insulation is the type of work for which it is to be used, e.g. for hot and cold water and heating services, whether or not it is fitted inside a building, and the ambient temperature (normal air temperature) surrounding it. Pipe diameters and the temperature of the water the pipes contain must also be considered to determine the actual thickness of the insulation necessary. Figure 8.44 shows pictorially some of these considerations. The theory of insulation is very simple and consists of covering the pipework and components of plumbing systems with a material which has a low heat conductivity rate, and which at the same time entraps bubbles or small pockets of air, air being a poor conductor of heat. The reflective properties of a bright surface are also sometimes employed to resist heat loss, a typical example being foil-backed plasterboard which if used for ceilings offers considerable resistance to heat flow into the space above.

Another fact that should also be understood is that the thickness of the insulation often differs as to whether protection against frost damage or the prevention of heat loss (in the case of hot water pipes) is required. The reason for this is that the insulation thickness suitable for preventing heat loss on hot water services may not be sufficient to delay the effects of frost.

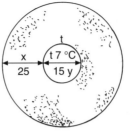
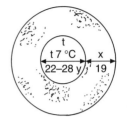

(a) Cold water service     (b) Cold water service

At low temperatures a smaller volume of water contains less heat quantity and will freeze more quickly than that of a larger volume.

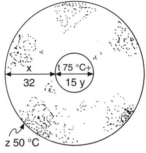
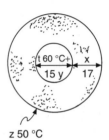

(c) Central heating pipe     (d) Hot water pipe

The insulation shown in (c) and (d) is to conserve heat and it will be seen that the higher temperature of heating pipes requires thicker insulation than those containing hot water at lower temperatures.

Key: t = water temperature in pipe    Conservation of
     x = recommended thickness    energy in heating
       of insulation    and hot water pipe
     y = pipe diameter
     z = ambient air temperature on
       the surface of the insulation

**Fig. 8.44** Frost protection and heat loss insulation, showing the relative thicknesses recommended. See Tables 8.5–8.7. In (a)–(d) the insulation value is 0.035 W/mK.

Table 8.5 shows the recommended thickness of insulation for protection from frost up to approximately 12 hours where the temperature of the water in the pipe is 7 °C and that of the air −6 °C which are considered to be normal conditions. Pipes and fittings fixed externally or in unheated areas inside a building are classified as extreme conditions, and it may be necessary for thicker insulation to be used, or consider the use of trace heating. It will be seen that smaller pipe sizes require thicker insulation than those of a larger

**Table 8.5**  Recommendations of revised BS 6700 Insulation thicknesses for cold water supply pipes.

| Pipe OD (mm) | Min. thickness to delay freezing in unheated areas of occupied buildings | | | |
|---|---|---|---|---|
| | Thermal conductivity (W/mK) | | | |
| | 0.035 | 0.040 | 0.045 | 0.055 |
| | Thickness of insulation (mm) | | | |
| 15 | 25 | 32 | 45 | 70 |
| 22–28 | 19 | 22 | 27 | 32 |
| 35+ | 9 | 13 | 19 | 25 |
| Flat surfaces | 9 | 13 | 19 | 25 |

**Table 8.6**  Recommendations of BS 5422 Insulation thicknesses for hot water supply pipes in unheated areas of occupied buildings.

| OD of copper pipes (mm) | Water temperature +60 °C, ambient temperature −1 °C | | | | |
|---|---|---|---|---|---|
| | Thermal conductivity (W/mK) | | | | |
| | 0.025 | 0.030 | 0.035 | 0.040 | 0.045 |
| | Thickness of insulation (mm) | | | | |
| 15 | 15 | 17 | 17 | 19 | 19 |
| 22 | 16 | 18 | 20 | 20 | 21 |
| 28 | 17 | 19 | 20 | 21 | 30 |
| 35 | 18 | 20 | 21 | 22 | 31 |
| 42 | 19 | 20 | 22 | 23 | 32 |
| Flat surfaces | 23 | 25 | 25 | 29 | 31 |

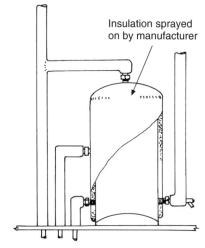

Insulation sprayed on by manufacturer

The Building Regulations specify that pipework connected to storage vessels should be insulated for at least 1 m from their points of connection or to the point at which they become concealed. The exception to this is where the pipe makes an efficient contribution to space heating.

**Fig. 8.45**  Insulation of pipes in heated areas.

**Table 8.7**  Recommendations of BS 5422 Insulation thicknesses for domestic central heating systems in unheated areas.

| OD of copper pipes (mm) | Water temperature +75 °C, ambient temperature −1 °C | | | | |
|---|---|---|---|---|---|
| | Thermal conductivity (W/mK) | | | | |
| | 0.025 | 0.030 | 0.035 | 0.040 | 0.045 |
| | Thickness of insulation (mm) | | | | |
| 15 | 21 | 22 | 32 | 33 | 35 |
| 22 | 22 | 32 | 34 | 35 | 36 |
| 28 | 23 | 34 | 36 | 36 | 36 |
| 35 | 24 | 35 | 37 | 38 | 39 |
| 42 | 25 | 37 | 38 | 39 | 40 |

diameter. This is because the smaller the volume of water in a pipe the more quickly it will lose its heat.

Table 8.6 shows the recommended thickness of insulation for hot water supply pipes in areas such as unheated rooms and storage areas, etc. The insulation thickness shown will not protect pipes from frost at temperatures below −1 °C. If lower temperatures are likely the recommendations of Table 8.5 should be applied. In the past dead legs on hot water systems in heated areas were not insulated, but it is now recommended (see Fig. 8.45) that although the hot water will cool down, by insulating the pipe it will cool more slowly and save wastage of water and energy.

The insulation thicknesses shown in Table 8.7 are designed to prevent heat losses in central heating systems where no heating is required. Note that the insulation thicknesses are greater than those necessary for hot water services owing to the higher temperature difference between the water temperature and the ambient air temperature. Where the heat from the pipework adds to that of other heat emitters, insulation is normally unnecessary. It should also be noted that some slight variations in the insulation thickness vary as to the type of fuel used. In simple terms this is due to the calculated costs of the fuel per megajoule. For smaller pipe sizes in systems working at temperatures less than 100 °C this can be ignored.

Note that the commercial thickness of insulation may vary slightly to those shown in the tables shown.

### Characteristics of good insulating materials

The materials used for insulation vary considerably, some being more suitable for a specific purpose than others, but a good insulating material should be of adequate thickness and have the following characteristics:

(a) it should not be flammable
(b) it should be vermin-proof
(c) it should be draught-proof and impervious to moisture if it is fixed externally or likely to be subject to damp conditions, i.e. under suspended ground floors
(d) it should be sufficiently robust for its purpose

*Pipe insulation*
Hair felt was one of the most commonly used insulating materials, mainly because of its low cost in comparison to alternatives. The original material met only one of the foregoing requirements in that it was a poor conductor of heat. Better qualities of this type of insulation are now manufactured which comply with the vermin and fireproof requirements, but because of its nature cannot be said to be waterproof and should only be used in dry, moisture-proof situations. It can be obtained in both sleeve or strip form, the latter being rolled around the pipe as shown in Fig. 8.46(a). Its performance is considerably improved by overwrapping with a polythene sheet which renders it draught-proof and improves its damp-proofing qualities.

A far superior form of insulation is expanded synthetic cellular sleeving (see Fig. 8.46(b)) which, although expensive, is a very good material. A high-quality insulation of this type should be fire resistant, but it is advisable to check that this is so as some older types of formed plastic and rubber

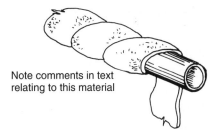

(a) Wrap around hair felt insulation

(b) Foamed cellular insulation

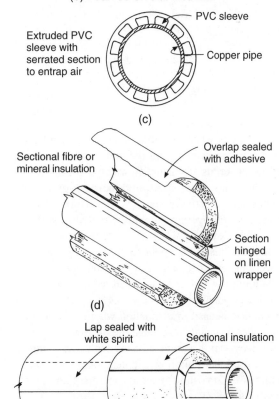

**Fig. 8.46**

are highly flammable and give off toxic fumes when burning. This type of insulation is supplied with a slit throughout its length so that it can be opened up and fitted to existing pipework.

Copper pipes are available with a preformed PVC sleeve which fits tightly around the pipe as shown in Fig. 8.46(c). Like most plastics, PVC is a poor conductor of heat, and the extrusion is made to incorporate the insulating characteristics of still air. Copper tubes fitted with this type of sleeve can be bent with a spring or in a machine with a special former and back guide. It should be noted that sleeved tube of this type is ideal for fitting under solid floors or in aggressive soils, as it also resists attack by acids and alkalis.

Some preformed insulation for pipes is moulded in two halves. Various materials are used, calcium silicate, expanded polystyrene and glass fibre being typical examples. The last is commonly used for hot water work, having a linen wrapper which acts as a hinge allowing it to be fitted over the pipe as in Fig. 8.46(d). The linen overlap may be sealed with an adhesive or stapled, and to clamp it securely on the pipe, black japanned steel bands are fitted at 450 mm intervals.

### Protection of insulation

In situations where external pipework is insulated, it must be protected from the ingress of moisture. Some materials are naturally waterproof, but if not they must be wrapped in a moisture-proof membrane effectively sealed with an appropriate adhesive. All overlaps should be sealed in such a way that they shed water. A modern method of achieving this is the application of polyisobutylene sheet cut into strips allowing for overlaps of 40–50 mm, which is then sealed using white spirit as an adhesive (see Fig. 8.46(e)). If insulation is used in a situation where it is likely to suffer mechanical damage, i.e. at low level in a boiler room, it can be covered with light-gauge galvanised aluminium or stainless steel sheet metal. Figures 8.47(a) and (b) show some methods of its application.

### Fabricating sectional insulation

Fabrication of insulation to cover tees, elbows, valves, etc., can in many cases be accomplished

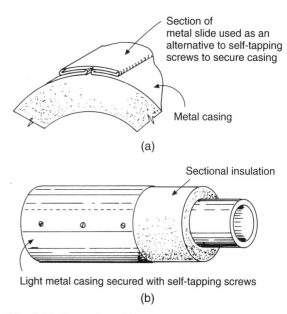

(a)

(b)

**Fig. 8.47**   Protection of insulation.

using offcuts from the material used to cover the pipes. Similar techniques can be employed for larger pipe sizes using sheet material.

Figures 8.48(a)–(d) illustrate some methods of applying cellular polyurethene insulation. Similar methods can be used to set out and prefabricate other materials. The methods shown are based on the *Armstrong Installation Manual*, copies of which may be obtained from the address at the end of the chapter. The company also produces a comprehensive set of templates for a small fee. All cuts are made with a sharp knife having a blade of about 225 mm in length, depending upon the diameter of the insulation used; a small pocket knife will also be useful. The knives must be sharp, so it is **essential every care is taken** to avoid accidents. More rigid materials can be cut with a fine-tooth saw: a hacksaw with a 24 TP1 blade is usually suitable. Marking out is best done with a piece of sharpened chalk, and when repetitive work is required a template will save time. All joints must be sealed with an adhesive recommended by the manufacturer. With a little care a very professional-looking job can be achieved with a little practice.

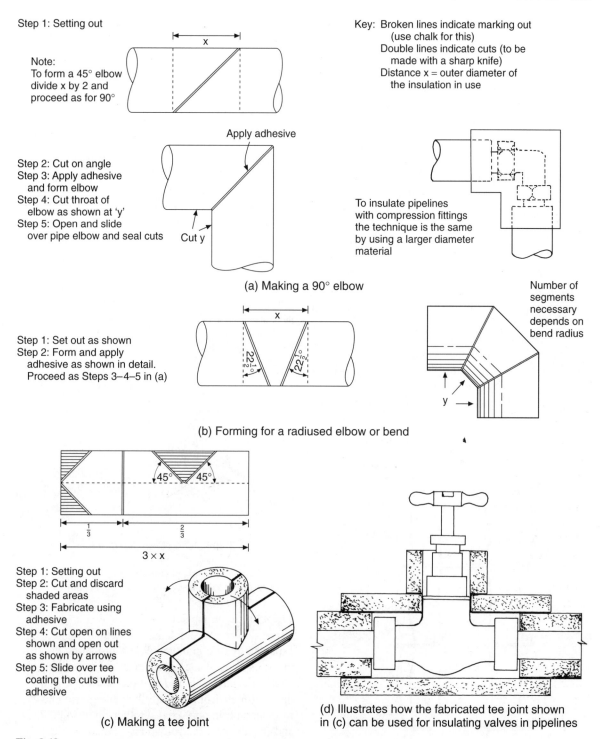

Step 1: Setting out

Note:
To form a 45° elbow
divide x by 2 and
proceed as for 90°

Key:  Broken lines indicate marking out
      (use chalk for this)
      Double lines indicate cuts (to be
      made with a sharp knife)
      Distance x = outer diameter of
      the insulation in use

Apply adhesive

Step 2: Cut on angle
Step 3: Apply adhesive
      and form elbow
Step 4: Cut throat of
      elbow as shown at 'y'
Step 5: Open and slide
      over pipe elbow and seal cuts

Cut y

To insulate pipelines
with compression fittings
the technique is the same
by using a larger diameter
material

(a) Making a 90° elbow

Step 1: Set out as shown
Step 2: Form and apply
      adhesive as shown in detail.
      Proceed as Steps 3–4–5 in (a)

$22\frac{1}{2}°$    $22\frac{1}{2}°$

Number of
segments
necessary
depends on
bend radius

y

(b) Forming for a radiused elbow or bend

45°   45°

$\frac{1}{3}$    $\frac{2}{3}$

$3 \times x$

Step 1: Setting out
Step 2: Cut and discard
      shaded areas
Step 3: Fabricate using
      adhesive
Step 4: Cut open on lines
      shown and open out
      as shown by arrows
Step 5: Slide over tee
      coating the cuts with
      adhesive

(c) Making a tee joint

(d) Illustrates how the fabricated tee joint shown
in (c) can be used for insulating valves in pipelines

**Fig. 8.48**

## Insulation of hot storage vessels and cisterns

Insulation of domestic hot water cylinders is essential to keep fuel costs to a minimum. One of the most effective methods is the use of foamed polyurethane which is sprayed on by the manufacturer. An alternative for existing vessels which are not treated in this way is the use of sectional insulation which is secured by tapes or metal bands, as illustrated in Fig. 8.49.

The sections are made of quilted glass wool or fibre glass with a canvas or PVC cover. When ordering a jacket of this type, the measurements of the storage vessel must be specified.

A similar type of jacket can be obtained for tanks and cisterns. Unfortunately it is not always possible to make a snug fit owing to the pipework connections, which may result in undesirable gaps between the sections being unavoidable. An alternative is to use one of the insulating materials marketed in sheet form, such as rigid sections of mineral wool or foamed rubber. Yet another alternative is to construct a casing out of light timber or hardboard, allowing about 100 mm all around which is filled with a loose fill material such as vermiculite or

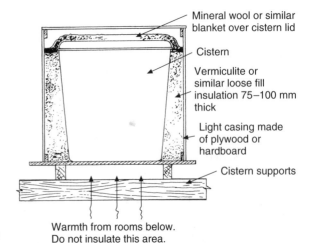

(a) Insulation of cisterns in a roof space

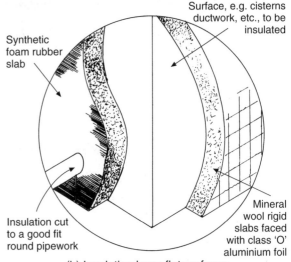

(b) Insulating large flat surfaces

Insulation of the types shown is usually fixed with purpose-made bands or straps or an adhesive. If plastic cisterns are to be insulated check that the adhesive is suitable and will not soften the plastic.

**Fig. 8.50**  Insulating cisterns.

expanded plastic granules. When insulating cisterns in roof spaces, if the cistern rests on the ceiling joists, it is advisable to leave the underside free so that warm air from beneath the ceiling provides a little warmth in very cold weather. Figures 8.50(a) and (b) illustrate the foregoing methods.

### Commercial and industrial premises

The insulation of large installations is usually the province of specialist firms who use modern

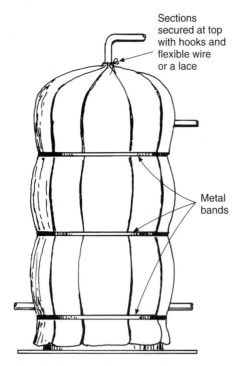

**Fig. 8.49**  Sectional insulation jacket for hot water storage vessel.

techniques to speed up the work of lagging large pipes and surfaces. Asbestos-free mineral fibres mixed with a suitable binder is the usual material used for this type of work and it is often sprayed on. The fuel savings effected with this type of lagging are quoted as almost 50% by one manufacturer and while it may not be possible to show savings of such magnitude on smaller systems, it is a fact that in most cases the cost of insulation pays for itself in a year or two, and is an investment well worth making.

### Further reading

Much useful information can be obtained from the following sources:

*The Building Regulations*
BS 6700 (1987): Design, installation, testing and maintenance of services supplying water for domestic use within buildings and their curtilages.
BS 1566: Specification for double feed indirect cylinders.
BS 699: Specification for copper direct cylinders for domestic purposes.
BS 5422: Use of thermal insulating materials.
All the foregoing are obtainable from HMSO.

*Solid fuel heating appliances*
Aga Rayburn, PO Box 30, Ketley, Telford, Shropshire, TF1 4DD. Tel.: 08457 125858.
Baxi UK Ltd, Brownhedge Road, Bamber Bridge, Preston, Lancs, PR5 6SN. Tel.: 01772 695500.

*Insulation*
Armstrong Insulation Products Ltd, Mars Street, Oldham, Lancs, OL9 6LY. Tel.: 0161 287 7100.
Pilkington Insulation Ltd, PO Box 10, St Helens, Merseyside, WA10 3NS. Tel.: 01744 24022.

*Copper hot water vessels*
Telford Copper Cylinders Ltd, Unit 22, Furrows Business Park, Haybridge Road, Wellington, Telford, TF1 4JF.

### Self-testing questions

1. State the three forms of heat transfer and how they apply to hot water systems.
2. Describe the methods used to accommodate expansion in long runs of hot water pipes.
3. Describe the function of the primary circulation pipes in a hot water system.
4. Assume that a hot water system is to be installed using copper pipework. State the materials you would recommend for the hot storage vessel and feed cistern. Give reasons for your choice.
5. Over a period of 2 hours, 240 litres of water are to be raised by 50 °C. State the boiler power required in kW.
6. State the fundamental requirements of a well-designed domestic hot water scheme.
7. Explain the reasons for fitting a secondary circulation of hot water supply.
8. Explain the causes of air locks and their effect on hot water systems.
9. (a) Describe two sets of conditions under which a boiler explosion could occur.
   (b) Name the component that should be used on all hot water schemes to prevent a build-up of pressure in the system.
10. State the requirements of good insulating materials.
11. Calculate the increase in length of an LCS hot water pipe 12 m long subjected to a temperature rise of 50 °C.
12. Calculate the minimum air gap between a combustible material and a flue pipe 125 mm in diameter.
13. Identify four serious faults in the direct hot water system incorporating a secondary circulation shown in Fig. 8.51.
14. State the recommended storage temperature of hot water for domestic use.
15. From the information given in Table 8.7 specify the recommended thickness of insulation for a central heating pipe in an unheated veranda.

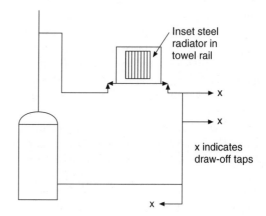

**Fig. 8.51**  Question 13.

# 9 Sanitary appliances

After completing this chapter the reader should be able to:

1. List the materials from which sanitary fittings are made.
2. State the fixing heights of baths, wash basins and sinks.
3. Select appropriate methods of connecting sanitary fitments to various pipework materials.
4. Describe the differences between bidets having an over-rim supply and those fitted with a douche attachment.
5. State the principles of siphonage.
6. Describe the methods of support and fixing for domestic sanitary fittings.
7. Understand the various types of flushing mechanisms in cisterns complying with the 1999 Water Regulations.

## Materials for sanitary appliances

The materials from which appliances are manufactured are dependent upon the type of fitment and the use of the building into which it is to be installed. In general terms the materials must be non-corroding, non-absorbent and easily cleaned. In factories and schools materials must also be capable of withstanding rough treatment, but for domestic use this requirement can be discounted.

### Metal

Baths, shower trays and flushing cisterns were commonly made out of *cast iron*, although this is being superseded to a large extent by the use of plastics. The interior surfaces of cast iron baths and shower trays are *vitreous enamelled* to provide a smooth, hard-wearing surface that is corrosion resistant. Articles manufactured from cast iron are very heavy and brittle and great care must be exercised when handling them. Owing to the high cost of iron sanitaryware, its use is limited to public buildings and commercial use, where its robust characteristics ensure a long working life.

*Vitreous-enamelled pressed steel* is also used for the manufacture of baths and sink units. It is a cheaper material but does not have the long-lasting qualities of cast iron. Manufacturers of *porcelain-enamelled ware* always attach a label to the appliance giving instructions on its care and methods of cleaning. Always make sure that this is intact when the job is complete.

*Stainless steel* is also extensively used for a wide range of sanitary appliances including WCs, sink units, urinal stalls and wash-hand basins. Its popularity derives from the properties it possesses:

(a) clean, pleasant appearance
(b) non-corrosive
(c) hard wearing
(d) has no vitreous-enamelled surface to chip
(e) easily cleaned
(f) very resistant to damage in industrial and public toilet situations

### Ceramic

*Ceramicware* or *potteryware* has always been a popular material for such appliances as WCs, wash-hand basins, urinals, sinks and shower trays. The term ceramic means a substance made by firing clay and includes various forms of pottery such as fireclay, stoneware and vitreous china. Fitments manufactured from fireclay and stoneware would be porous, i.e. would absorb moisture, unless coated with vitreous enamel.

Appliances made from *fireclay* and *stoneware* are strong and heavy and this makes them particularly useful in situations where hard-wearing qualities are essential, i.e. factories and sanitary annexes. Butler's sinks, urinal slabs and stalls, shower trays and WCs are all made from these two materials.

*Vitreous china* is a special type of earthenware which, as result of high firing temperatures, is made impervious, i.e. will not absorb water. Its hard surface coating serves to aid cleaning and to improve its appearance. The strength of vitreous china allows fitments to be manufactured with a very thin section reducing their weight. Wash-hand basins and WCs are the most common appliances made of vitreous china.

### Plastic

Comparatively recently, plastic materials have become increasingly used for sanitary fittings, especially baths and flushing cisterns.

Two types of plastics are used in the manufacture of sanitary appliances: *thermosetting* and *thermoplastic* materials.

Thermosetting plastics are generally harder and have a greater degree of rigidity than thermoplastics.

This fact makes them suitable for the manufacture of such components as WC seats and flushing cistern shells. Conversely a thermoplastic material called methacrylate is one of the most common materials used for the manufacture of baths and shower trays for domestic use. It should also be noted that most of the discharge pipework used for sanitation is made of thermoplastics. Although these materials are cheap and light to handle they have the distinct disadvantage of being softer than metals and thus easily damaged. Baths also suffer from movement when in use because thermoplastics are flexible. This makes it essential to provide a wooden cradle to give the necessary stability.

The specific heat, or heat capacity, of plastics is lower than that of metals, so less heat is absorbed from the hot water by the appliance, which is an advantage especially in relation to baths.

### Storage of sanitary appliances

Arrangements should be made prior to delivery for storage in a clean dry area. Leave all protective wrapping on packages until contents are required for fitting. Stacking should be avoided but if space is limited place sufficient softening material between each appliance, see Fig. 9.1. Wash basins and baths

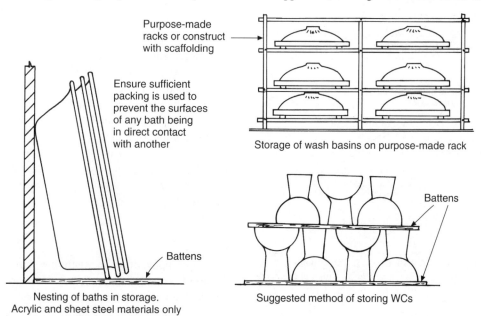

Purpose-made racks or construct with scaffolding

Ensure sufficient packing is used to prevent the surfaces of any bath being in direct contact with another

Battens

Nesting of baths in storage.
Acrylic and sheet steel materials only

Storage of wash basins on purpose-made rack

Battens

Suggested method of storing WCs

**Fig. 9.1** Storage of sanitaryware on site.

can be 'nested' inside each other; WCs should not be stacked more than four high. It is unwise to store other building materials such as bricks, cement and abrasive materials like aggregates in the same store.

### Visual inspection

Where possible all appliances should be checked on delivery for scratches or abrasions of polished surfaces and cracks. A check must also be made to ensure all the materials listed on the delivery note have been delivered, and where possible look for any damage such as cracks in or abrasions of enamelled surfaces that may have occurred during transit.

## Sanitary appliances

### Butler's sinks

These are usually specified as 'London' or 'Belfast' sinks. London sinks were originally less deep than Belfast types, but the only difference nowadays is that the Belfast pattern has an integral overflow while the London has not. Both types are made of glazed fireclay and are capable of very hard wear. They are made in a comprehensive range of sizes, but those for domestic use usually conform to the measurements shown in Fig. 9.2.

Owing to the thickness and the rounded top edge it is difficult to make a watertight joint between such a sink and the wall. The best way of

**Fig. 9.2** London-type butler's sink. The general appearance of both London and Belfast sinks is the same excepting that the Belfast pattern is fitted with a weir overflow. Both types are constructed of glazed fireclay and are made in various sizes. The dimensions shown are representative of the most common size.

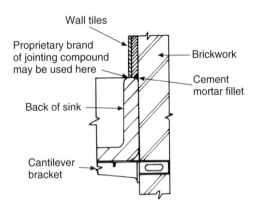

**Fig. 9.3** Making a watertight joint between a butler's sink and the wall surface. Plasterwork must be cut away behind a butler's sink so that the wall tiles finish over the rounded top of the sink. Failure to do this will lead to damp and insanitary conditions behind the sink.

overcoming this problem is to cut away the plaster behind the sink so that it abuts the brickwork. The joint can then be made good with cement mortar and tiled across its top edge to make it watertight (see Fig. 9.3).

The recommended height from the top edge of a butler's sink to the floor is 865 mm.

### Cleaner's sinks

These are seldom necessary in small domestic households, but do differ from ordinary sinks. The main difference is that cleaner's sinks are fitted with either a galvanised or brass-hinged grating on which buckets may be rested while being filled. They are usually fitted at a low level to avoid unnecessary lifting of the bucket and a hardwood face is fitted on the front edge to reduce damage to the glaze. A typical example of a cleaner's sink is illustrated in Fig. 9.4. They are normally supported on cantilever brackets with the upstand at the back screwed to the wall with large brass screws to prevent it sliding forward. Hot and cold taps are fitted at a suitable height over the sink.

### Sink units

This term applies to units where the sink and draining boards are pressed or moulded as a complete component, rendering unnecessary the often insanitary wooden draining boards sometimes associated with butler's sinks.

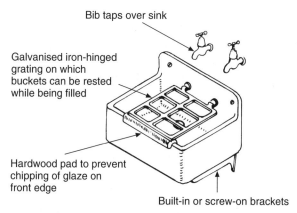

Bib taps over sink

Galvanised iron-hinged grating on which buckets can be rested while being filled

Hardwood pad to prevent chipping of glaze on front edge

Built-in or screw-on brackets

**Fig. 9.4** Cleaner's sink. These sinks are usually fitted between 300 and 450 mm from the floor level to facilitate easy removal of the bucket from the grating.

The most popular materials for construction of sink units are stainless steel and vitreous-enamelled steel. Both these materials are easily pressed into shape by modern manufacturing processes. They are almost invariably fitted to a cabinet which determines the height of the front edge, usually 900–915 mm from the floor. The sink should be a good fit on the cabinet and secured to it firmly by the clips provided. The cabinet should also be screwed to both the wall and floor.

There are many differing combinations of sink units to suit a variety of purposes, some of which are illustrated in Fig. 9.5(a). In situations where a sink or wash basin is inset into a worktop, it is important to seal the edges as shown in Fig. 9.5(b). Worktops are usually made of wood fibres and the absorption of water will cause swelling and distortion quite apart from poor sanitation conditions.

*Baths*
Baths vary little in their basic shape whether they are made of cast iron, steel or acrylic, but refinements in design are available beyond the standard rectangular type to include such features as soap sinkings, hand grips or dropped front edges (see Fig. 9.6(a)).

Baths made for corner fittings are also becoming increasingly popular. The measurements shown in Fig. 9.6(b) apply to standard baths, but longer and wider baths are available if required, the full range of measurements being quoted in manufacturers' handbooks.

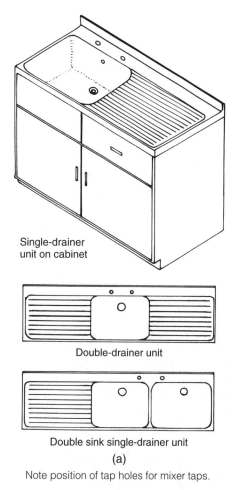

Single-drainer unit on cabinet

Double-drainer unit

Double sink single-drainer unit

(a)

Note position of tap holes for mixer taps.

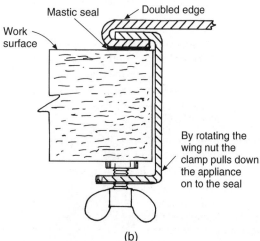

(b)

**Fig. 9.5** (a) Sink units. Standard widths of sinks is 500 or 600 mm, depending on the type of cabinet used. (b) Method of securing and sealing appliances sunk into a worktop.

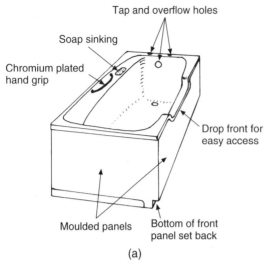

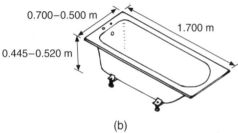

(b)

**Fig. 9.6**   (a) Typical moulded acrylic bath.
(b) Measurements of a standard size cast iron bath. The
height of the bath is variable depending on the adjustable
feet. Generally they should be as low as conveniently
possible.

Baths of cast iron have adjustable feet so that
the bath height can be varied, but it is generally
recommended that they are fixed as low as possible
so they can be used more easily by the elderly and
young children. Most modern baths are designed to
be panelled in, some being supplied complete with
a panel which determines the fixing height. Panels
should be constructed and fitted in such a way that
they are easily removable for maintenance purposes.
Similarly, pipes should not be run in inaccessible
positions behind baths.

*Wash basins*
These are bowl-shaped fittings used for ablutionary
purposes. They are made of glazed earthenware,
vitrified china or, when they are required to be

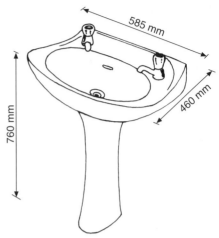

(a) Pedestal wash basin

The height of the basin should not vary substantially from
that shown whether brackets or a pedestal is used.
The measurements shown are generally representative of
this type of basin.

(b) Small hand washing basin suitable for WC
apartments

(c) Angle or corner basin

These basins are useful in bathrooms or toilets where
space is limited.

**Fig. 9.7**   Wash basins.

particularly strong, durable and vandal-proof,
stainless steel or cast iron. Sizes of basins vary
considerably, the standard size conforming to BS
1188 (see Fig. 9.7(a)). Smaller basins for hand
washing only (Fig. 9.7(b)) are often fitted in WC
apartments where space is limited. Angle basins
(see Fig. 9.7(c)) can be fitted in similar
circumstances to save space.

Tap holes on basins vary considerably depending on the choice of taps required. Basins having one centre hole only, or three holes, are available to accommodate the wide variety of mixer taps that are currently produced. To avoid special orders many basins are made now with provision for three 'knockout' holes. Having selected the holes necessary to accommodate the appropriate taps, the glaze covering the hole *must* be tapped out with a ball pein hammer from the 'face' side of the appliance, not the underside. Failure to do this may result in chipping the surface glaze in such a way that it is not covered by the flange on the tap.

Wash basins of all types are supported by a variety of methods which are dealt with later in this chapter. The pedestal shown in Fig. 9.7(a) is most popular in domestic dwellings for the larger type of basins. The basin should be well secured to the pedestal to prevent any movement, and for the same reason, the pedestal must be firmly screwed to the floor via the holes in its base. Two alternative methods of securing the basin to the pedestal are illustrated in Figs 9.8(a) and (b). One shows a bracket which fits round the exposed waste fitting below the basin which is adjustable to meet any variation in the centre line of the waste to the wall. The other is a flat metal flange which is positioned over the waste fitting between the bottom of the basin and the back nut. The two metal screws passing through the flange and the legs in the pedestal are secured by wing nuts. Always grease the threads so they can easily be removed at a later date if necessary.

The basin must also be fixed to the wall to avoid any movement. One method is to use a pair of adjustable brackets as shown in Fig. 9.9(a). The other is to screw the basin to the wall as shown in Fig. 9.9(b). This is not an easy job; the best approach is first to ensure the basin is level, then using a suitable nail as a centre punch, mark the wall or tiles through the holes which are at an angle. Remove the basin and drill the wall at the correct angle (easier said than done). Insert suitable plugs well below the wall surface as this will give more flexibility of angle when the basin is screwed home. Brass countersunk no. 8 screws 38 mm long are usually about right for this job. When secured the top and side edges of the basin should be made

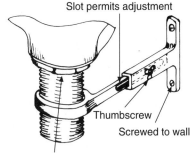

(a) Adjustable basin waste bracket for pedestal basins

This type of bracket is used to prevent movement of basins fitted on pedestals.

(b) Flange method

**Fig. 9.8** Fixing basins to pedestal.

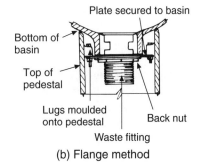

(a) Wall fixing for pedestal basins

This raised portion of the bolt is adjustable and is made to engage into two holes under the side edges of the basin. It effectively ties the basin back to the wall.

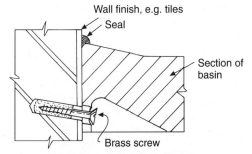

(b) Fixings for pedestal basins

Some basins are provided with fixing holes as shown.

**Fig. 9.9** (a) Fixing pedestal basins to a wall. (b) Fixings for pedestal basins.

watertight using a suitable mastic where it abuts the wall.

### Vanitory units

Although used for the same purpose, a vanitory unit is not a wash basin in the strict sense of the word. It is simply a bowl, made of vitreous china, stainless steel or vitreous-enamelled steel, and fitted into a prepared cabinet with a wooden top, this being cut away to admit the bowl. The main consideration when dealing with these units is to ensure an absolutely watertight joint between the woodwork and the bowl to prevent seepage of water into the cabinet. Manufacturers supply the bowl with the necessary jointing compound and sufficient clamps to fix the bowl securely into the woodwork.

### WCs

Modern WCs may be of the wash-down or siphonic type. The former relies on the momentum of the flush of water from the cistern to remove the contents of the trap while in the latter case the contents are removed by siphonic action.

The integral trap has a 50 mm depth of seal and an outgo diameter of approximately 90 mm. Figure 9.10 illustrates a typical WC which can be converted to an 'S' trap or side outlet using a bent WC connector.

Until recently the design of the wash-down closet has not varied a great deal since it was first produced. However, owing to the statutory requirements limiting the flush to 6 litres maximum some changes have been made. Because of this it is unlikely a flushing cistern complying to the Water Regulations 1999 will operate satisfactorily with an existing WC. Unless a new suite is fitted any replacement should be of 7.5 litres capacity complying with the 1986 Water Bylaws.

### Low-level WC suites

Prior to about 1950 it was generally thought that flushing cisterns had to be fitted at high level to give the flushing water the necessary momentum to remove the contents of the WC. They may still be specified in schools and industrial buildings, but because of a change in fashion manufacturers started producing low-level WC suites for domestic use, the top of the flushing cistern being fitted approximately 1 metre from the floor. To ensure sufficient momentum the diameter of the flush pipe is one size large than is necessary for high-level cisterns. Another more recent development is the close-coupled suite, where the cistern is designed to sit on the back of the WC, see Fig. 9.11. Long horizontal lengths of discharge pipe should be avoided as it may be found that the flushing water may have insufficient momentum to clear the contents of the trap. With the lower volume of flushing water of 6 litres now mandatory, manufacturers are likely to use a 'flap' or 'drop' valve instead of a traditional siphon to achieve the necessary momentum of flushing water. The advantage of both low-level and close-coupled suites is that they are generally less noisy than those employing high-level flushing cisterns. When fitting a WC where the outgo is to be connected to a floor-level drain, the drain socket must be carefully set out to ensure its correct position. Failure to do this may result in the WC seat falling back when it is raised, and in the case of close-coupled suites the cistern may be too far away or too close to the wall.

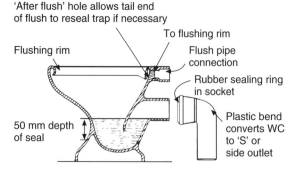

**Fig. 9.10**  Wash-down water closet.

'After flush' hole allows tail end of flush to reseal trap if necessary

Flushing rim

To flushing rim

Flush pipe connection

Rubber sealing ring in socket

50 mm depth of seal

Plastic bend converts WC to 'S' or side outlet

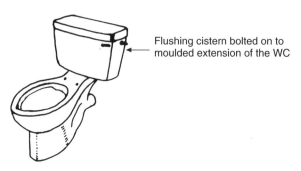

**Fig. 9.11**  Close-coupled WC suite.

Flushing cistern bolted on to moulded extension of the WC

*Siphonic WCs*

Owing to the reduced volume of flushing water permitted by the Water Regulations, this type of WC will no longer be produced, except possibly for replacement. Although modern wash-down closets and the methods used for flushing have improved, the siphonic WC was, to many, the peak of perfection. Unlike the wash-down type which relies on momentum for the removal of the contents of the bowl, a siphonic WC removes them by siphonic action. This is an advantage with low-level and close-coupled suites where the momentum of the flushing water is in some cases insufficient to clear the trap making it necessary to flush the WC twice.

These WCs are very quiet in action and as the water surface in the bowl covers a larger area than that of wash-down WCs, there is less possibility of fouling the bowl. When wash-down closets are flushed, especially when a high-level cistern is used, small quantities of water in the form of a fine spray may be thrown over the edge of the bowl. This is due to the more violent action of the flush and can result in what is, in effect, an unseen bacteria-laden aerosol spray being deposited on the floor adjacent to the WC. The working principles of siphonic closets are dealt with later in this chapter.

Another type of WC, very popular in industrial and public buildings, is the corbel type which has the advantage of leaving the floor area clear for cleaning. Shown in Fig. 9.12 this type is designed for use with a plumbing duct. The water and waste services are all installed behind the panel forming the duct, which is an advantage in appearance and in situations such as hospitals, schools and offices since the wall and floor area around the WC can be easily cleaned. The ducting also reduces the possibility of vandalism in public conveniences. The imposed loads that occur when a corbel-type appliance is in use, e.g. WC or bidet, are taken by the cast iron frame shown in Fig. 9.12 or the steel frame shown in shown in Fig. 9.13. These are now widely used as they are adjustable and can also be adapted for use with flushing cisterns and wash basins. Although the plumber seldom constructs the ducts and partitioning, he or she should ensure that adequate access is available for future servicing of the installation.

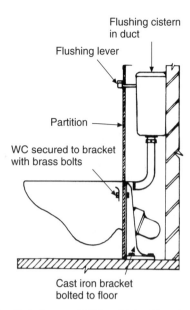

Flushing cistern in duct

Flushing lever

Partition

WC secured to bracket with brass bolts

Cast iron bracket bolted to floor

**Fig. 9.12**  Corbel type of WC and plumbing duct.

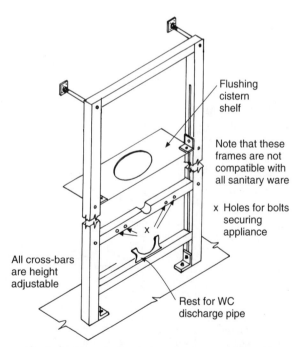

Flushing cistern shelf

Note that these frames are not compatible with all sanitary ware

x  Holes for bolts securing appliance

All cross-bars are height adjustable

Rest for WC discharge pipe

The frame illustrated is designed to accept a WC suite but similar types are available for wash basins and bidets.

**Fig. 9.13**  Prefabricated frames for fixing corbel-type sanitaryware.

Two good tests may be applied to WCs to ensure that the whole of the bowl surface is reached and scoured by the flushing water. First, a suitable water-bound paint (thinned-out emulsion may be used for this purpose) is applied to the inside of the bowl. When the cistern is flushed all the paint should be washed off or at least wetted. The second test relates to the ability of the flush to clear the trap. Six to eight sheets of toilet paper are placed in the bowl and the cistern flushed. If the paper is not removed with one flush, then check the flush pipe joint for obstruction.

*Flush pipe joints*  Flush pipes are joined to the WC by a flexible rubber or plastic connector, the most common type being shown in Fig. 9.14. To be effective they must be tight fitting and as such are sometimes difficult to pull over the WC nozzle. This operation will be simplified if the surfaces are wetted with water or a little washing-up fluid.

Another type of connection between a WC and the flush pipe is shown in Fig. 9.14(b), this type

being usually provided with the WC and flush pipe as a set. They are not suitable for all WCs, being either too tight or too loose.

When fitting flush pipes, make sure they do not enter the nozzle of the WC too far, often resulting in a defective flush.

*WC joints to the drain*  The joint between the WC and the drain will depend upon the material of which the drain is constructed. A very versatile joint marketed under the trade name of 'Multiquick' can be used with nearly all materials, including plain-ended pipes, and is illustrated in Fig. 9.15.

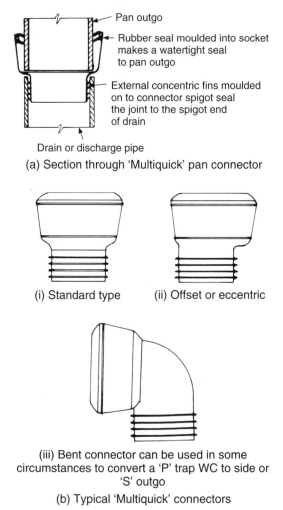

(a) Section through 'Multiquick' pan connector

(i) Standard type    (ii) Offset or eccentric

(iii) Bent connector can be used in some circumstances to convert a 'P' trap WC to side or 'S' outgo

(b) Typical 'Multiquick' connectors

**Fig. 9.15**  'Multiquick' joints.

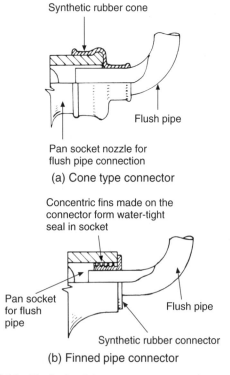

(a) Cone type connector

(b) Finned pipe connector

**Fig. 9.14**  Flush pipe joints.

They are especially useful in cases where a WC connected to a clayware drain is to be replaced. It is often found that the clayware socket is cracked, or if this is not so it is almost impossible to cut away the existing cement joint without cracking it. The easiest way to overcome this problem is to cut the socket away completely until it is flush with the floor. Care must be taken to seal the hole with paper or rags so that chippings are not allowed to enter the drain where they may contribute to a blockage. The 'Multiquick' can now be inserted into the drain.

Connections of WCs to clayware or cast iron socketed pipe are illustrated in Fig. 9.16. If the WC is fitted on a solid floor where there is no likelihood of movement or shrinkage, and assuming the drain pipe to be either cast iron or earthenware, the outlet joint can be made with yarn and cement mortar as shown in Fig. 9.16(a), or by a soft joint using yarn and a proprietary mastic. The advantage of using the latter is that the WC can, if necessary, be removed without breakage (see Fig. 9.16(b)).

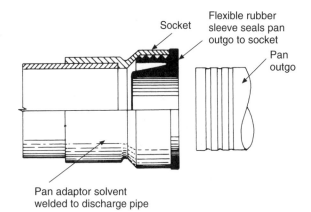

**Fig. 9.17** WC connector for PVC discharge pipe.

When the new WC outlet does not line up with the cast iron pipe, the socket can be cut off and one of the many types of PVC adaptors used, which, like the 'Multiquick', can be pushed into a socketless cast iron pipe. Reference should be made to manufacturers' catalogues to enable the correct adaptor to be selected.

There is a wide range of fittings available for connecting WCs to plastic discharge pipes. Figure 9.17 illustrates a typical example having a synthetic rubber seal, common to all such connections, which requires no other materials to make a sound joint. When making rubber-sealed joints, using a silicone lubricant will make the task much easier. As these joints are inherently flexible, any shrinkage of a timber floor is unlikely to result in a broken WC.

*Fixing WCs to floors* When WCs are fixed to wooden floors, especially new ones, a considerable amount of shrinkage can be expected and the type of joint used to connect it to a drain must always be flexible. A WC must be rigidly fixed on all types of floors with suitable brass screws to facilitate their easy removal. Do not overtighten the screws and make sure they enter into the floor at the same angle as the holes in the WC. Failure to observe these two points often results in a cracked pan.

*Bidets* Bidets are an ablutionary fitting designed for cleansing the excretory organs, or for use as a foot bath. Figure 9.18 shows a bidet without a spray nozzle normally associated with these fittings. It is

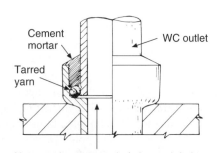

Note gap between end of pipe and drain

(a) Connection to earthenware socketed pipe

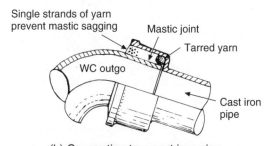

(b) Connection to a cast iron pipe

This type of joint must be used where the WC is sited on a suspended floor. The joint, being soft, will give slightly to accommodate any shrinkage of the timber.

**Fig. 9.16** WC joints to the drain.

**Fig. 9.18** Bidet without spray jet below flood level of the fitting. There is no risk of back siphonage with this type of bidet.

(H & C taps and supplies not shown)

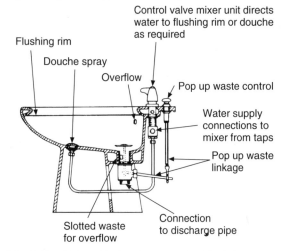

**Fig. 9.19** Bidet with douche spray. As this fitting has a water inlet below its flood level, the cold supply must never be connected directly to a mains water supply, i.e. it must be fed from a cistern.

equipped with pillar taps and a normal waste fitting, the same in fact as those used for wash basins.

The bidet shown in Fig. 9.19 is fitted with a mixing tap set, pop-up waste and spray jet. The mixing valve is constructed so that a mixed supply of hot and cold water can be delivered through the rim so that it is warmed prior to use. The same water fills the bowl for washing purposes. Alternatively, the mixed supply can be diverted to the spray jet for the purpose of douching. Because the spray jet is situated below the flood level of the fitting there is a high risk of water pollution. When fitting this type of bidet the Water

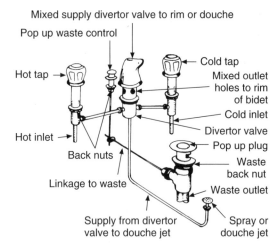

**Fig. 9.20** Integral pipework and waste linkage for bidets.

Regulations must be complied with. More specific information on this subject is given in Book 2 of this series.

Owing to the bowl shape and its use, the waste from a bidet should be treated in the same way as that of a wash basin with the same restrictions on its length, diameter and fall. Discharge pipes to bidets fitted at the ground floor level, if not connected to the main discharge stack, should discharge directly to the drain or be fitted to a back inlet gulley in case of possible misuse.

Diagrammatic details of the integral pipework of a bidet are illustrated in Fig. 9.20 and show how a mixed supply of water is distributed to the rim and douche jet. Two main points should be borne in mind when fitting bidets with a douche attachment. They must be supplied with cold water from a cistern, not from the mains. Both hot and cold supplies must be of equal pressure. Bidets of this type must be installed by a plumber approved by the water supplier, or the water supplier must be informed of the installation in writing. Further details of the Water Regulations relating to bidets are dealt with in Book 2 of this series.

*The principles of siphonage* Siphonic action has many applications in plumbing systems and appliances. However, it is important to understand its basic principles and the part played by atmospheric pressure before studying its practical applications.

Air, like everything else on the earth, is pulled towards the earth's centre by the force of gravity. The pull of gravity on the atmosphere, which we usually refer to as atmospheric pressure, is 101.3 kN/m$^2$ at ground level. Since commercial pressures are normally indicated by the 'bar', it will be seen that the pressure of the atmosphere is approximately 1 bar. To give a practical illustration of atmospheric pressure so that some idea of the force it exerts can be imagined, the pressure created by a column of water approximately 10 m high will be nearly the same as that exerted by the atmosphere. There is a slight variation of pressure due to the contours of the earth. For example, atmospheric pressure will be less great on a mountain than at sea level but this can be ignored as far as siphonage in plumbing is concerned.

The simple diagram shown in Fig. 9.21 shows how a siphon works. In cistern A the pressure of the atmosphere is the same inside the bent tube as that on the surface of the water before siphonic action commences. If, however, air is withdrawn from the tube, the atmospheric pressure exerted on the water in A will be greater than that in the tube and will force the water up and over the bend to discharge into the lower cistern B. This principle has many applications in plumbing, one of the most common being the flushing cistern.

A similar arrangement to that shown is often employed to empty hot storage vessels in hot water systems fitted with defective drain-off cocks.

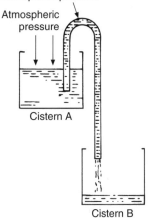

Fig. 9.21 Diagram showing how a siphon works. The basic action of the siphon has many applications in plumbing where, for instance, it may be necessary to empty a hot storage vessel because of a defective or non-existent drain-off cock.

The action of siphonage also affects the working principles of discharge pipe systems as will be seen in Chapter 10.

*Working principles of siphonic WCs* It will be seen that the WC shown in Fig. 9.22 has two traps. Siphonic action is started by a reduction of the air pressure between the two traps. The way in which this is achieved can best be understood by looking

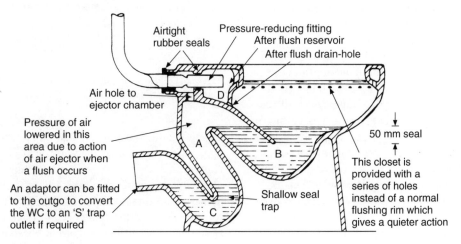

Fig. 9.22 Double-trap siphonic closet.

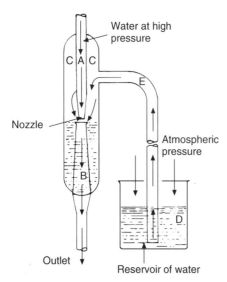

**Fig. 9.23** Principle of a filter pump. Water at high pressure is discharged from the nozzle and spreads out in the shaped tube 'B' abstracting air from the pump body 'C' causing a reduction of pressure. Atmospheric pressure on the surface of the water in the reservoir 'D' forces it upward through 'E' to discharge through the outlet tube 'B'.

at the principle of a filter pump as shown in Fig. 9.23. It embodies the working principles of a siphonic closet. As previously stated, for the siphonic closet shown to work effectively a pressure of less than the atmosphere must be set up in the space between the two traps. Trap B has a full 50 mm seal depth; trap C is shallow, its purpose being simply to retard the discharge of water when the cistern is flushed to allow sufficient time for air to be withdrawn from the space between the two traps shown as A. The principle used to achieve this is almost the same as that of the filter pump: water flowing through the pressure-reducing fitting in the flush pipe abstracts air from area A thus lowering the pressure between the two traps. As in all cases of siphonage, atmospheric pressure acts on the free surface of water (in this case on the water in the WC bowl) forcing it into the outlet and forming a solid plug which removes not only the contents of the bowl, but that of trap B as well. This is replenished by the after-flush water retained in the reservoir (D).

Figure 9.24 shows details of the components used to lower the pressure between the two traps.

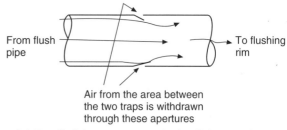

(a) Detail of the pressure-reducing fitting used with WCs having flush pipes as shown in Fig. 9.22

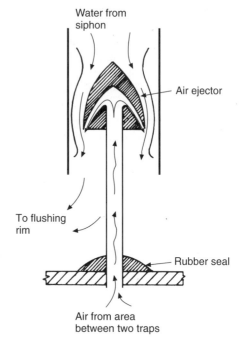

(b) Pressure-reducing fitment used with close-coupled suites. Situated in outer leg of siphon

When the WC is flushed, water passing over the ejector withdraws air from the chamber between the two traps, thus lowering its pressure and setting up siphonic action in the WC.

**Fig. 9.24** Pressure-reducing fitments used with double-trap siphonic closets.

Figure 9.24(a) illustrates the special adaptor for an appliance fitted with a flush pipe, and Fig. 9.24(b) shows a special pressure-reducing fitting used with close-coupled appliances which have no flush pipe. When fitting these components, care must be taken to ensure that all the rubber seals are correctly positioned and airtight. Failure to observe this and any special fitting instructions supplied by the manufacturer will result in inefficiency or failure in the performance of the appliance.

The reader should note that siphonic closets are now obsolescent as they will not function with a 6 litre flush. The foregoing information is given to enable those in existence to be maintained when necessary.

*Flushing cisterns*

These have been used for many years to flush automatically the contents of a WC or similar appliance into a foul water drain quickly and efficiently. Until the Water Regulations took effect the syphon was the only type of mechanism employed in their use. Figure 9.25 illustrates the type of flushing cistern complying with previous Water Bylaws. For years a 9 litre flush was permissible, but in an early attempt to conserve water the 1986 Bylaws specified a maximum of 7.5 litres. Rather than alter the cistern and siphon dimensions most manufacturers chose to limit the volume of water used by fitting a removable plug in the siphon, which by its removal allowed air to break the siphon after 7.5 litres were discharged. The other main difference between these and flushing cisterns manufactured to comply with the Water Regulations is the fact that they are no longer

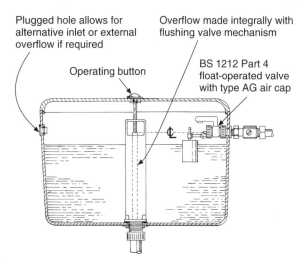

Note that if a siphon is used the invert of the siphon bend must be at the same level as the centre line of the inlet.

**Fig. 9.26** Flushing cistern complying with the Water Regulations' maximum 6 litre working capacity.

classified as waste water preventers (see the figure). This type of cistern, although not permissible for new work, is still manufactured for replacement purposes as existing WCs are unlikely to operate efficiently with the now mandatory 6 litre flush.

A flushing cistern complying with the Water Regulations is shown in Fig. 9.26. The main difference apart from the smaller volume of water it contains is that an overflow pipe is no longer necessary. If an overflow does occur it is now permissible to allow it to discharge into the WC via the flushing mechanism and flush pipe. Some modern flushing cisterns still make provision for a separate overflow pipe, but if it is not required (and this will be the norm) a suitable plug to seal the hole is provided. Some flushing cisterns are now made smaller and will only accept a float-operated valve complying with BS 1212 Part 4. Back flow risks in all flushing cisterns are provided by an AG type air gap.

Many modern flushing mechanisms employ the facility to deliver both a 6 litre full flush and a 3 to 4 litre short flush. In an early attempt to conserve water prior to 1986 many flushing siphons were made to permit this, but because users were insufficiently educated in their use more water was wasted than previously and the 1986 Bylaws

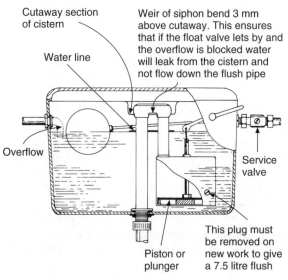

**Fig. 9.25** Flushing cistern with traditional siphon complying with 1986 Water Bylaws.

prohibited their use. The wheel has now turned full circle and the 1999 Water Regulations permit dual flushing. Previous efforts using siphons defaulted on the short flush – hence the waste. The use of valves instead of siphons enables the mechanism to default on the full flush. It should be noted that where dual-flushing apparatus is employed it is important that instructions for its use are permanently displayed adjacent to the cistern.

### Flushing cistern siphons

This type of flushing apparatus has been in use for many years and requires little maintenance. The siphon mechanism of these cisterns was originally made of brass, copper or a lead alloy. They are now made principally of polythene which is both cheap and corrosion resistant. Its working action is very simple. When the handle is depressed it lifts the plunger or piston carrying the water in the cylinder over the siphon bend. As the water falls down the flush pipe it carries away the air in the siphon bend lowering the air pressure in the outlet leg of the siphon. The atmospheric pressure on the surface of the water in the cistern is then greater than that in the siphon causing the water to flow through the piston, which is not solid but spoked like a wheel. The piston simply supports a plastic membrane which gives the effect of being solid only on the upstroke. This is illustrated in Fig. 9.27. When the piston reaches its maximum length of travel in an upward direction, water continues to pass through it, pushing past the piston and the flexible membrane until the cistern is emptied and air is admitted to the underside of the piston, halting the siphonic action. A sketch showing some of the details of this action is shown in Figs 9.28(a) and (b). Defects of this type of cistern are uncommon unless the cistern is

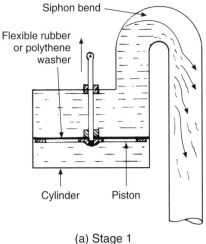

(a) Stage 1

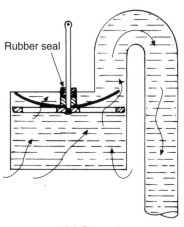

(b) Stage 2

When the piston reaches the end of its travel, water continues to flow through the piston and the flexible washer. As it does so siphonage will continue until the cistern empties and air is admitted under the edge of the cylinder thus breaking the siphon action.

**Fig. 9.28** Flushing action details.

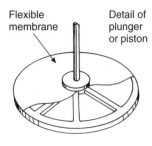

**Fig. 9.27**

misused. The membrane may distort or perish after a long period of time but is easily replaced when the siphon is removed. Until quite recently it was necessary to disconnect the cistern to remove the siphon, but many are now made to avoid this, Fig. 9.29 shows a typical example. The hole through which the piston rod passes will wear in time, when this occurs the cheapest form of repair is to renew the complete unit. Make sure the correct type is replaced, carefully measuring the height and

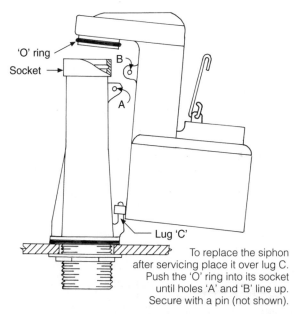

'O' ring

Socket →

B

A

Lug 'C'

To replace the siphon after servicing place it over lug C. Push the 'O' ring into its socket until holes 'A' and 'B' line up. Secure with a pin (not shown).

**Fig. 9.29** Demountable siphon.

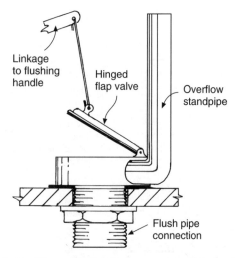

Linkage to flushing handle

Hinged flap valve

Overflow standpipe

Flush pipe connection

**Fig. 9.30** Flap valve flushing mechanism. This is often found on imported sanitaryware and is suitable for full flush only. The flushing handle is made in such a way that the valve remains open until the cistern is empty.

the diameter of the piston. The simplest solution is to take the defective siphon to a builders merchant and match it exactly.

Most flushing cisterns are reversible, which means the inlet can be fitted on the right- or left-hand side. Many are made with bottom entry holes. Unless an overflow pipe is fitted, the spare hole must be closed using the plug supplied. When replacing an existing cistern having a capacity of 7.5 litres, it must be like for like. In such cases it is important that the overflow pipe must be reconnected as failure to do this may result in flooding. The reasons for this can be seen in Fig. 9.25 when the invert of the siphon is above the cut-away section in the back of the cistern.

*Flushing valves*
These are sometimes called drop valves to avoid confusion with the pressure flushing valves described in Book 2. These valves are not new and were commonly used in the UK in the early part of the twentieth century. They were banned by the current bylaws circa 1925 because they often failed to seat, thus resulting in wastage of water. There were several reasons for this and it is claimed that modern materials and technology have overcome this problem.

Although the basic working principle of a siphon is very simple, and they are certainly not obsolete, the drop valve does have the ability to deliver flushing water into the WC more quickly and at a greater velocity. This is advantageous with close-coupled WC suites where the lack of momentum has sometimes been a problem using a 7.5 litre flush.

A very basic type of valve is shown in Fig. 9.30 and is operated by a lever in a similar way to that used with a siphon. They are used both in Europe and North America but are prone to leakage due to the following causes. Because the valve falls through an angle, any scale build-up on the seating is more likely to cause the valve to leak than those dropping vertically. Although made of non-corrosive material the hinge may become stiff to operate over a period of time, which again will prevent the valve closing properly.

The mechanics of some of these valves, especially those incorporating dual flushing, are very complex. They function on combinations of air or water pressure, floatation or springs. Figure 9.31 illustrates some of their working principles. Most are operated by a push button in the cistern lid, but some can be adapted to lever operation, the necessary parts being supplied with the valve. Figure 9.31(a) illustrates a

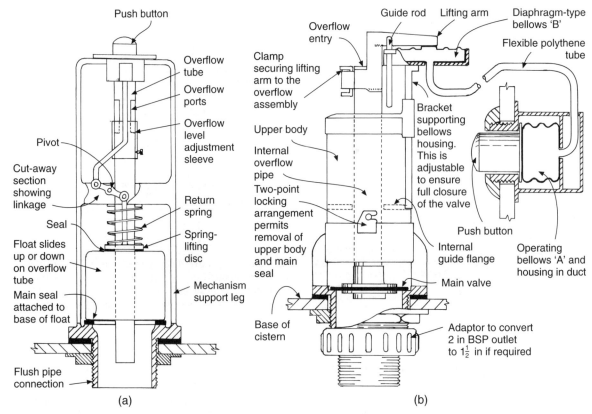

**Fig. 9.31** (a) Basic working principle of a single-flush drop valve. The valve is shown in the closed position. When the push button is pressed the linkage lifts the disc compressing the spring which permits the float to rise and open the main valve. (b) The Dudley pneumatic 'Pinto' valve. To operate, the push button is pressed expelling air from bellows 'A' to bellows 'B'. The subsequent expansion lifts the overflow assembly to which the main valve is attached allowing the contents of the cistern to be discharged.

valve which operates by means of a float and spring. The pneumatic valve shown in Fig. 9.31(b) is designed for cisterns fitted in a duct or in a purpose-made cabinet and is basically very simple with few working parts. Owing to its flexibility the operating button and housing can be fitted within approximately 2 metres of the cistern. A modified arrangement can also be supplied for floor fixing where a 'hands-free' environment is desirable. It should be noted that neither of these two valves incorporates a dual-flush arrangement.

*Electronic flushing* As shown in Fig. 9.32 it will be seen the drop valve principle is employed. These valves are suitable for both new and existing installations and operate in the same way as the electric tap described in Chapter 6. When the hand

is passed within 50 mm of the sensor, a signal is sent to the control box energising the solenoid which lifts the valve allowing water to enter the flush pipe. The volume of water used is governed by the length of time the solenoid is energised and is controlled by a timer in the control box. This must be carefully adjusted when the valve is installed. If it is specified that both 3 and 6 litre flushes are required, two separate sensors will be necessary.

It is always important to comply with the fitting instructions for any appliance, but it is especially so when they are electrically operated in possibly damp conditions.

These valves are mainly used in commercial and public buildings, especially hospitals where a 'no-touch' method of flushing may be important on health grounds. It is essential that frequent checks

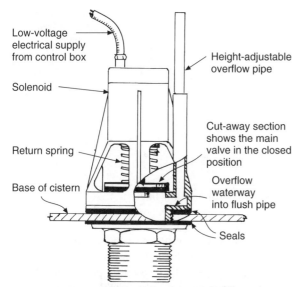

Low-voltage electrical supply from control box

Solenoid

Return spring

Base of cistern

Height-adjustable overflow pipe

Cut-away section shows the main valve in the closed position

Overflow waterway into flush pipe

Seals

Note that operating sensors are made to fit into the hole normally occupied by the cistern flushing handle. Where dual-flush operation is required both sensors are normally fitted into the duct or wall behind the cistern.

**Fig. 9.32**  Electronic flushing valve.

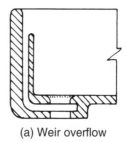

(a) Weir overflow

This type of overflow is usually fitted to butler's sinks and has the advantage of being easily cleaned.

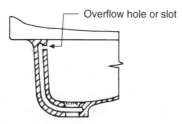

Overflow hole or slot

(b) Basin fitted with secret or slotted overflow

**Fig. 9.33**  Types of overflow.

are made to ensure these valves are working effectively and qualified maintenance staff are available when servicing becomes necessary.

*General maintenance on drop-type flushing valves*
If fitted, electrical components must be isolated by switching off the supply and removing any fuses. Any linkage on the valve should be checked for wear and correct adjustment. If the valve is leaking it may be due to limescale or debris on the seating, or the valve itself may be defective and should be examined for any deterioration of the seals and working parts. On completion of any maintenance work the appliance must be checked for correct operation.

**Overflows to sanitary fittings**

Overflows are provided with most sanitary fittings with the exception of some types of sinks and shower trays. They may be fitted separately, or they may be an integral part of the appliance as in the case of basins and sinks illustrated in Fig. 9.33.

*Weir overflows*
This type of overflow as shown in Fig. 9.33(a) is capable of being cleaned to some extent, and as such is often accepted by sanitary authorities. They are used with Belfast sinks and some types of wash basin. Hospital sinks do not have overflows, thus avoiding inaccessible areas that are difficult to keep clean.

*Slotted overflows*
A type of integral overflow, often referred to as the 'slotted' or 'secret' overflow, is incorporated in the design of most modern wash basins and bidets (see Fig. 9.33(b)). The main objection to its use is the difficulty of effective cleaning.

*Separate overflows*
Few modern sanitary appliances are now made having a separate overflow, as their use always tended to admit cold draughts into the building. The exceptions to this are storage cisterns and some flushing cisterns complying with the 1986 water byelaws. Appliances having an integral overflow

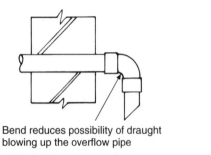

Fig. 9.34   Draught deflection bend for external overflows.

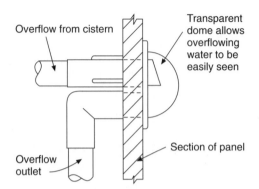

Fig. 9.36   Visual overflow fitting.

are made in such a way that any overflowing water is diverted into the discharge pipe or, in the case of flushing cisterns, the flush pipe. To prevent the entry of cold draughts associated with separate overflows, the inlet should be turned below the water level in the cistern. If this is not possible, a bend may be fitted externally of the building, as shown in Fig. 9.34.

Any overflow pipes terminating outside of the building should extend sufficiently to prevent overflowing water being blown back on to a wall by high winds. This can cause dampness inside the building, especially in older buildings not having cavity walls.

### Standing waste and overflow

The type of overflow shown in Fig. 9.35 is most commonly used in conjunction with hospitals and laboratory sinks, but these can be used with specially designed basins having a space at the back to accommodate the extension of the waste. This

prevents the standpipe being disturbed when the basin is in use. The standpipe is usually made of 'vulcanite', a type of thermosetting plastic. It has a machine taper on one end to enable it to seal the waste effectively in the same manner as a plug. The advantage with this type of overflow is the ease of cleaning and lack of maintenance it requires.

### Visual overflow fitting

This simple device, shown in Fig. 9.36, is designed to enable an overflowing cistern to be detected visually when it is concealed behind a panel and the overflow pipe outside the building cannot easily be seen. These are specially useful in a situation where a common overflow pipe is used, e.g. in a multi-storey building where an overflowing cistern cannot be easily identified. They are only suitable for nominal 20 mm overflow pipes using solvent welded joints.

### Waste fittings

These are all basically of the same type with the exception of their length and diameter which vary depending upon the type of fitting with which they are used. A bath, for example, having no integral overflow and constructed of thinner materials than basins or sinks, requires a waste with only a short length of thread. The other difference is that some wastes are slotted or have a cut-away portion to coincide with entry of integral overflows (see Fig. 9.37(a)). When a waste of this type is fitted it is important to check that the slot in the

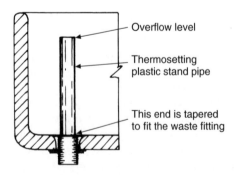

Fig. 9.35   Standing waste and overflow.

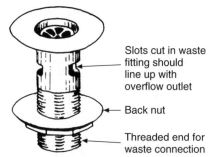

Slots cut in waste fitting should line up with overflow outlet

Back nut

Threaded end for waste connection

(a) Slotted waste fitting for use with appliances having integral overflows

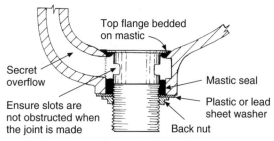

Top flange bedded on mastic

Secret overflow

Mastic seal

Ensure slots are not obstructed when the joint is made

Plastic or lead sheet washer

Back nut

(b) Sealing slotted waste fittings into sanitary appliances

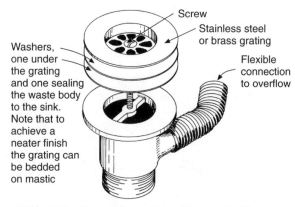

Screw

Stainless steel or brass grating

Washers, one under the grating and one sealing the waste body to the sink. Note that to achieve a neater finish the grating can be bedded on mastic

Flexible connection to overflow

(c) Combination waste and overflow used with metal sinks and baths as an alternative to a slotted waste

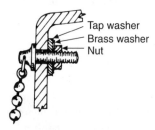

Tap washer
Brass washer
Nut

(d) Method of securing chain and stay to a basin

(e) Special plastic washer used with back nuts to secure pillar taps to sanitary fittings

**Fig. 9.37** Waste fittings.

waste is in line with the entry of the overflow. It is also necessary to ensure that the slots are not obstructed by excess jointing material (see Fig. 9.37(b)). An alternative waste fitting is shown in Fig. 9.37(c) and is sometimes used for baths and sinks. Its main disadvantage is the bar running across the fitting into which the screw securing it is fitted. This type of fitting is hardly the last word in sanitation, as the bar tends to collect solid debris and is difficult to keep clean. The traditional method of making waste fittings into sanitaryware is the use of putty and paint and certain types of mastics have also been used. These methods of jointing are satisfactory for appliances made of ceramic or metal but they are not recommended for those made of plastic materials. Some manufacturers provide rubber washers for making both taps and waste fittings into sanitary appliances but they are not always

satisfactory. Depending on the type of rubber of which they are made, it sometimes distorts and squeezes out of shape causing leakage. This is usually due to the jointing surfaces of some materials becoming distorted during manufacture. Manufacturers' instructions regarding jointing materials must always be strictly adhered to. Generally a silicone-based material is recommended for use with plastic.

It should be noted that manufacturers or suppliers will take no responsibility for the failure of any appliance or component if the wrong jointing materials have been used. Always ensure that after use all surplus jointing media are removed and the appliance is wiped clean.

Basin plugs are normally secured by a stay (or stud) and chain, the stay passing through a hole in the basin. A method of securing the stay and ensuring a watertight joint is shown in Fig. 9.37(d).

Prior to fixing taps into sanitary fittings they should be 'broken' as described in Chapter 6 (see p. 139). The joint between the appliance and the tap must then be made watertight, especially in the case of basins and sink units. To ensure this the taps are bedded down on suitable mastic with the back nuts tightened against a specially flanged washer made of polypropylene (see Fig. 9.37(e)). These washers are usually supplied with the taps as a complete set. The back nut should be tightened sufficiently to stop the tap turning in the hole.

## Brackets and supports for sanitary fittings

The type of support required varies with the type of appliance. Fireclay sinks, for instance, having considerable weight, require stronger fixings than those for wash basins. It is, therefore, important to select the appropriate type of support.

### Brackets
Brackets, a common method of supporting appliances, generally fall into two main groups: they are either 'built-in' or screwed to the wall, the 'built-in' type being chosen for the heavier loads.

The most common support for butler's sinks are cantilever brackets, which are designed to be built into a wall with strong cement mortar. These must be carefully levelled at the correct height, as once the mortar has set their position cannot be altered. One important factor must always be remembered when any built-in fixing is to be made: the hole into which the bracket fits must always be wetted. Failure to do this will cause the moisture in the mortar to be absorbed too quickly by the wall causing it to shrink, with the result that the brackets move. The broken line shown on the bracket illustrated in Fig. 9.38(a) indicates the cut-away portion that is required to accommodate a fitting having an integral overflow, such as a Belfast sink. Ensure the cut-away bracket is built in at the end at which the waste is to be fitted.

A similar type of 'built-in' bracket is available for supporting single wash basins in schools or factories where screwed-on fittings may be pulled off the wall (see Fig. 9.38(b)).

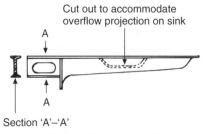

(a) Cantilever bracket

These brackets are built into a wall and provide a strong fixing for heavy earthenware sanitary fittings.

(b) Built-in cantilever towel-rail bracket for wash basins

These brackets are used in schools, hospitals, factories where strong fixings are essential.

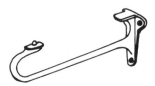

(c) Screw-on towel-rail bracket for wash basins

Used for normal domestic dwellings.

**Fig. 9.38** Brackets for supporting sanitary fittings.

### Pedestals
The usual method used to support wash basins in modern domestic premises is by a pedestal, as illustrated in Fig. 9.7(a). Where it is desirable to keep the floor area clear under a basin, towel-rail brackets which are screwed to the wall surface may be used (see Fig. 9.38(c)). The screws must be of sufficient length to enable secure fixing to be made and purpose-made plastic wall plugs should always be used. The practice of using wooden plugs leaves much to be desired as it is invariably damp on a building site and plugs made of damp wood will eventually shrink causing the brackets to loosen.

All brackets that have so far been mentioned are made of cast iron and protected from corrosion by painting or porcelain enamel. Brackets made of steel are usually galvanised.

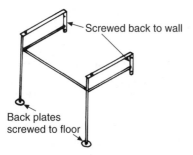

**Fig. 9.39** Chromium-plated tubular basin stand. These stands form a strong attractive support for basins and can be used instead of pedestals or brackets.

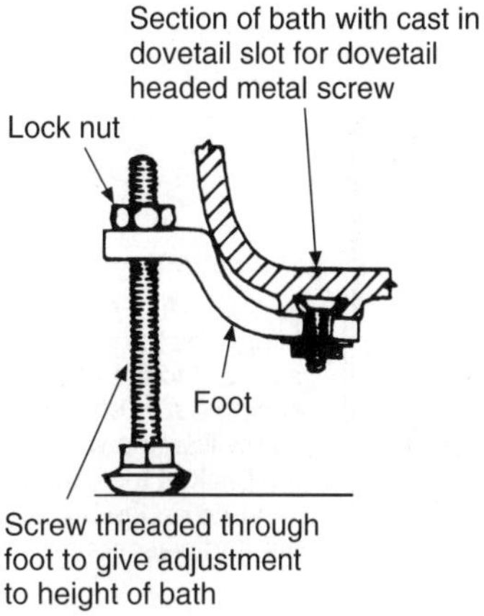

## Stands

When fixings are to be made to soft brickwork walls with mortar joints, a reliable fixing for screws cannot always be obtained. In such cases, either built-in brackets must be considered or, in the case of wash basins, a chromium-plated tubular stand can be used of the type shown in Fig. 9.39. When these stands are used, the floor takes most of the weight, the fixings to the wall only being necessary to prevent the stand from falling forward. A similar type of frame made of cast iron is available for the support of many types of combination sinks and drainers when other types of support are not suitable or desirable.

## Legs

Many sanitary fittings are supplied with adjustable legs or feet so they can be levelled on uneven floors. Figure 9.40 shows such a bracket used to level in and adjust the height of cast iron baths.

## Cradles

Both acrylic and pressed steel baths, being less rigid than those of cast iron, require the use of a special cradle for their support. Figure 9.41 shows a typical cradle with adjustable feet for acrylic baths. Care must be taken to ensure that these cradles are correctly fitted to avoid subsequent damage to the fittings.

## Fixings for flushing cisterns

The fixings for flushing cisterns vary depending on their type. Cisterns made of synthetic plastics vary considerably. Those of good quality are usually

**Fig. 9.40** Adjustable bath foot for cast iron baths.

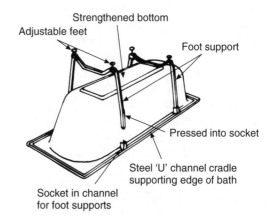

**Fig. 9.41** Cradle for plastic bath.

provided with two back-fixing plates which are screwed to the wall (see Fig. 9.42(a)). The threaded screws passing through the plate are made to fit through the holes in the back of the cistern case, which is then secured to the plates by brass nuts and suitable washers. Very lightweight flushing cisterns are normally screwed through the back directly on to the wall with brass round-head wood screws of sufficient length. Cisterns made of ceramic, being heavier, require more positive support. They are usually supplied with two galvanised steel angle brackets which support the cistern, the top being held back to the wall by two brass screws fixed through the holes provided (see Fig. 9.42(b)).

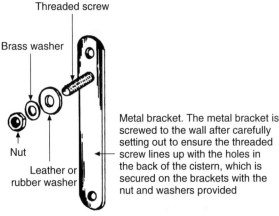

(a) Back fixing plates for light plastic flushing cisterns

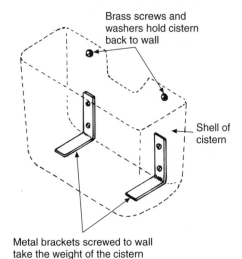

(b) Angle bracket and screw support for cisterns constructed from ceramic materials

**Fig. 9.42** Fixings for flushing cisterns.

## Further reading

Relevant British Standards for sanitary fittings are as follows.

BS 1125: Specification for WC flushing cisterns.
BS 1189: Specification for cast iron baths for domestic purposes.
BS 1206: Specification for fireclay sinks: dimensions and workmanship.
BS 1213: Ceramic wash down WC pans: dimensions and workmanship.

BS 1244: Specification for metal sinks.
    Part 2: Metric units.
BS 1390: Sheet steel baths for domestic purposes.
BS 3380: Specification for wastes for sanitary appliances and bath overflows.
BS 3943: Specification for plastic waste traps.
BS 4305: Specification for baths for domestic purposes made from cast acrylic sheet.
BS EN 12056 Part 1: Sanitary pipework.

Information on sanitary appliances is obtainable from the following companies:

Armitage Shanks Ltd, Rugeley, Staffs, WS15 4BT.
    Tel.: 01543 490253.
Caradon Bathrooms Ltd, Lawton Road, Alsager, Stoke-on-Trent, ST7 2DF. Tel.: 01242 221 221.
Thomas Dudley Ltd, Dauntless Works, PO Box 28, Birmingham New Road, Dudley, West Midlands, DY1 4SN.
    Tel.: 0121 557 5411.
Geberit Ltd, Aylesford, Kent, ME20 7PJ.
    Tel.: 01622 795231.

## Self-testing questions

1. State the essential differences between glazed fireclay ware and vitreous chinaware.
2. Identify the essential difference in design between a Belfast sink and a sink of the London pattern.
3. State the recommended fixing height for a wash basin fixed on towel-rail brackets.
4. State two tests that can be applied for checking the efficient working of a WC.
5. List the advantages of using soft joints for joining a WC either to discharge pipes or to drains.
6. State the two basic factors which are necessary for siphonic action to take place.
7. Explain the action of siphonic closets.
8. State the precautions to be taken when fitting a slotted waste to a wash basin.
9. Explain why special cradles are necessary to support plastic and sheet steel baths.
10. State why purpose-made plugs, instead of those made of wood, are recommended for most types of fixings.

11. State why 'knockout' holes in sanitary ware should be tapped out on the glazed side.

12. Identify the possible cause of a siphon fitted in a flushing cistern failing to operate if the cistern is full.

13. A persistent flow of water is flowing into a WC via the flush pipe where the cistern is fitted with a flushing valve. Assuming the float-operated valve is operating correctly, state two possible causes for this defect.

14. List any checks that should be made when taking delivery of new sanitary ware.

15. State why it is recommended that a filter is fitted in the inlet to a cistern flushed by means of a valve.

# 10 Sanitary pipework

After completing this chapter the reader should be able to:

1. Understand the need for the highest standards of quality in work and materials for sanitary pipework.
2. Explain the causes of seal loss in traps.
3. Understand the evolution of modern above-ground soil- and waste-discharge systems.
4. Identify the main design factors relating to discharge pipe systems.
5. Define the need for means of access in discharge pipe systems.
6. State the special points to be considered when installing washing machine and dishwashing equipment to discharge systems.

## Introduction

The Great Plague of London during the seventeenth century was the direct result of poor sanitation and overcrowded living conditions. It swept large areas of the country and caused many deaths, but it should not be assumed that this was the only outbreak of its type. Such epidemics have occurred throughout history and are still rife in developing countries, though fortunately nowadays they are fairly localised.

It was not until the early Victorian era that forward-thinking people associated the terrible toll of human life with the lack of effective sanitation. In 1875 the London County Council first laid down a system of bylaws or rules to which the installation of sanitary systems had to conform. Other boroughs and local authorities quickly followed, each producing its own local bylaws. While in most cases these bylaws were very similar, any differences which did occur were due largely to their interpretation by individual local authorities. It will be seen that the sanitary pipework systems as we know them today are the product of necessity and legislation by successive governments and local authorities in comparatively recent years.

### Building Regulations

The Building Regulations of 1965 brought a degree of uniformity to the local bylaws and these now apply to all forms of construction throughout the country. They have been modified and amended since they were first introduced to permit the use of new materials and techniques. The latest version of these regulations became operative in 1991. The other important document that applies especially to the plumbing industry in relation to sanitary pipework is BS EN 12056 Part 2 2000. This has superseded BS 5570, the previous code for sanitary pipework. From a practical point of view there is very little difference from the recommendations of BS 5570. One important point is the change in the names of discharge pipe systems, e.g. the single-stack system is now called the primary ventilated stack system. Codes of Practice and British Standards are not legal documents but give a good general guide to the minimum standards of work and materials to be used.

The general arrangement of a well-designed discharge pipe system should embody the following features. It should permit the speedy removal of waste to the underground drainage system and must prevent the ingress of foul air into the building by efficient trapping and maintenance of the trap seals. Adequate provisions should be made for access for the removal of blockages in the system and materials should be chosen to ensure a trouble-free and corrosion-resistant life.

### Terminology

To avoid confusion, some knowledge of the terms relating to sanitary pipework is necessary.

The terms 'soil pipe' and 'waste pipe' are both derived from the original two-pipe system of sanitation where a soil pipe was connected to a WC and a waste pipe to an ablutionary fitting. These terms are not generally used now, both being designated *discharge pipes*.

A *stack* relates to a main vertical pipe whether a discharge 'stack' or a ventilating 'stack'. In certain cases a discharge stack can be used to convey rainwater where a combined system of underground drainage is permitted by the local authority.

A *ventilating* pipe may be a branch or main ventilating pipe, its main function being to maintain atmospheric pressure inside the whole of the discharge and drainage system.

A pipe carrying waste water from a fitting or group of fittings to the main discharge pipe is called a branch discharge pipe.

### Angles of branches (junctions)

The specifications of angles for fittings used in drainage and sanitary pipework in the UK have traditionally been shown as 'obtuse', e.g. angles greater than 90°. In Europe these angles are specified as 'acute', e.g. angles less than 90°. From a practical point of view this makes no difference but care should be taken when ordering.

## Traps

Traps are an integral part of a modern sanitary system, being designed to retain a small quantity of the waste water from the discharge of the fitting to which they are attached as a barrier to prevent foul air entering the building. Traps should be self-cleaning, that is to say, they should be designed so that their walls are scoured by the discharging water. Owing to the necessary bends in their construction, traps are prone to blockage and adequate access for their cleansing must be provided. One of the advantages of modern traps constructed of plastic materials is the ease with which they may be dismantled for cleaning.

### P and S traps

Traps are designated either P or S, the difference being the position of their outlet. The P trap outlets are in an almost horizontal plane, while the outlet leg of an S trap is in a vertical position. A slight angle, usually $2\frac{1}{2}°$, is in fact provided to the outlet of P traps to ensure a slight fall.

Figure 10.1(a) illustrates a tubular trap. These are made of pipe and to reduce the space they occupy, the bends are made to a very small radius. The parts of the trap are indicated and also the depth of seal, which is the measured length from the soffit of the return bend and the weir or invert of the outlet.

A trap of a different design called a bottle trap is shown in Fig. 10.1(b). These are not considered self-cleansing by many but owing to their neat appearance they are commonly used. They are always made with a P outlet, but purpose-made bends are available to convert them to S outlets if required. The effective depth of seal is measured from the base of the dip pipe to the weir.

### Trap seals

The depth of the trap seal varies with the size and usage of the pipework to which it is connected. BS EN 12056 Part 2 specifies that any trap connected to a discharge pipe of 50 mm or less discharging into a main stack or drain should have a seal of 75 mm. It has been found that traps serving flat-bottomed sanitary appliances such as baths are just as effective with a 50 mm pipe which is easier to fit in the limited space. Traps having a diameter in excess 50 mm have always only had a 50 mm seal. The reason for this apparent anomaly is that large-diameter traps are unlikely to receive a full-bore discharge which is often the cause of seal loss. Typical examples of these larger traps may be seen in WCs and gulleys.

Traps with a seal depth of 38 mm are permitted only if the appliance discharges over a trapped gulley, because the gulley provides an air break

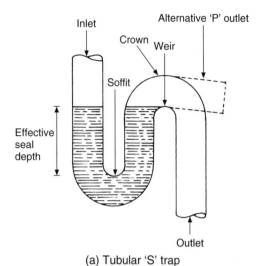

(a) Tubular 'S' trap

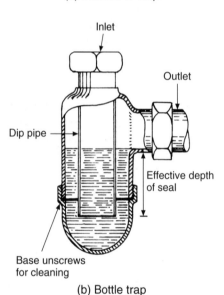

(b) Bottle trap

**Fig. 10.1**  Types of trap.

between the drain and the appliance should the trap seal be destroyed.

### Loss of trap seal

Loss of trap seals will result in objectionable odours entering the building which, if persistent, may cause poor health. For these reasons it will be seen that the water seal of a trap must be retained in all circumstances. The three main reasons for seal loss due to bad design of the discharge pipe system are dealt with fully later in this chapter.

Trap seals, however, are also lost by natural causes, most of which are rare, but of which the plumber should be aware.

*Evaporation*  This is the most common natural cause of seal loss. The warmer and drier the air becomes, the greater will be the rate of evaporation. It is unlikely, however, that traps having a seal depth of 75 mm would evaporate completely, even in very warm weather, providing the appliance was used occasionally. It has been calculated that even during a very hot summer it would take a period of about 10 to 12 weeks to evaporate a trap seal of 75 mm.

*Capillary action*  Capillarity can result in the rapid destruction of a seal due to particles of dish-cloths being deposited over the weir of the trap (see Fig. 10.2(a)). This is, however, a rare occurrence which only happens in S traps. A

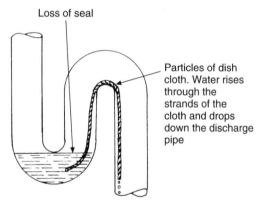

(a) Effect of capillary attraction on tube trap

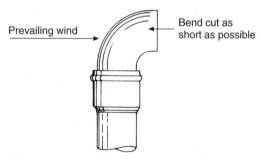

(b) Prevention of seal loss by 'waving out'

**Fig. 10.2**  Loss of trap seal.

similar effect will be seen if a cloth is left hanging over the edge of a bucket of water.

*Wind*   The effects of wind currents and pressures at the vent terminal can cause 'waving' out of traps especially those of larger diameter. Unstable atmospheric conditions in the vent pipe due to this cause sometimes produce a wave movement of the water in the trap allowing it to gradually wash over the weir. This is not a common occurrence and cannot easily be foreseen. It can sometimes be remedied by extending the vent upward or terminating it with a bend as shown in Fig. 10.2(b).

*Momentum*   Momentum is the term used to describe the effect of a sudden discharge of water from a bucket into a WC or gully trap. Owing to the velocity of the discharge, it sometimes also carries away the water which should form the trap seal. When this occurs it can easily be seen and the householder can flush the cistern or empty more water into the trap to reseal it.

### Access

Owing to the low velocity of water flow at which most modern discharge pipe systems operate, access is necessary to remove any silt which may build up and cause a blockage. Access is sometimes provided by an easily dismantled trap, a typical example being shown in Fig. 10.3. The swivel joint also permits some degree of adjustment to be made to accommodate the direction of the discharge pipe connection.

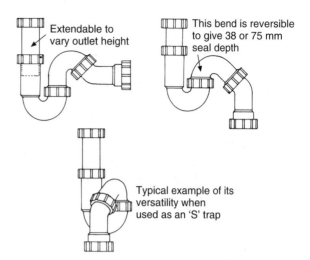

**Fig. 10.4**   Variable outlet trap.

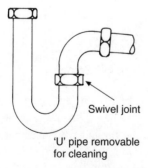

**Fig. 10.3**   Access to traps. Traps made of copper and plastics are made so that they can easily be dismantled for cleaning.

### Special types of trap

*Variable outlet traps*   This type of trap is illustrated in Fig. 10.4. Their unusual design permits them to be connected to a wide variety of plumbing layouts and they are therefore useful in situations where, for instance, a sink unit is to be replaced and the waste outlet on the new unit differs from that of the old. Because of the limit of flexibility with an ordinary two-piece trap, it might be necessary to reposition the waste with all the associated problems of cutting away the fabric of the building and making good. As the illustration shows, a variable outlet trap has a telescopic inlet allowing adjustment for height. The swivel joints enable it to be used as a P or S trap with the outlet at varying angles, while the centre section is reversible giving a choice of seal depth between 40 mm (shallow seal) or 75 mm (deep seal).

*Straight-through trap*   Figure 10.5 illustrates a straight-through trap which can be used in some instances instead of an S trap. Its main disadvantage is the two relatively tight bends which slow up the discharge of water. These traps are especially useful for pedestal basins, as they can be concealed more easily than an S trap behind the pedestal. As they are moulded in one piece and cannot be dismantled easily, adequate provision for clearing is provided.

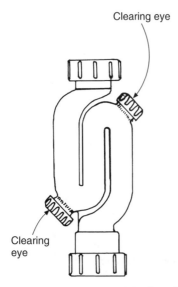

**Fig. 10.5** Straight-through trap (originally called a bag trap). These can usefully be employed instead of an S trap behind a basin pedestal.

**Fig. 10.6** Running trap.

*Running trap* Figure 10.6 shows a running trap which is nowadays only used in situations where it is not possible to use a Ṗ or S trap for some reason. Typical cases would be trapping a sink macerator or a washing machine.

*Combination traps* Another special trap is shown in Fig. 10.7, this being used on a bath waste avoiding the use of a separate overflow. There is a tendency for the waste water to wash back up the overflow pipe, having the effect of depositing soapy waste in its bore, which is a disadvantage with this type of fitting. Most traps of this type, however, are made so that they can easily be dismantled for cleaning.

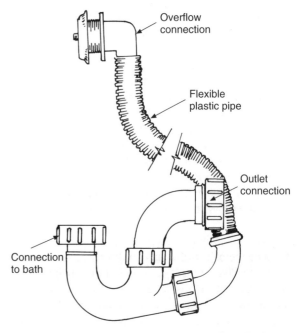

**Fig. 10.7** Combination bath trap and overflow. Bath traps previously required a 75 mm seal, but the revised Code of Practice now allows a 50 mm seal.

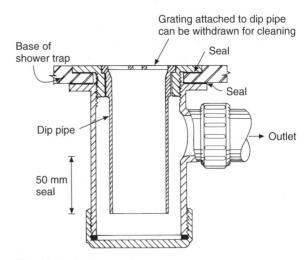

**Fig. 10.8** Bottle-type shower trap.

*Shower traps* Figure 10.8 shows the main feature of a bottle trap having a 50 mm seal and designed to be fitted in the limited space under a shower. Because of the low flow rates, shower traps are prone to blockage from hair and particles of soap and are made so they can be easily cleaned by removing the grating and dip pipe.

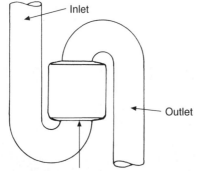

The enlarged bore at this point provides a reservoir of water which drops back and reseals the trap in the event of siphonage taking place

**Fig. 10.9**  Reservoir-type resealing trap.

*Resealing traps*  Later in this chapter the design features which are embodied in a good discharge system to prevent seal losses by siphonage will be discussed. Where these conditions cannot be met, however, a *resealing* trap may be used with advantage. These traps are designed so that a water seal remains as a barrier to foul odours even after siphonage has occurred, this being achieved by the reservoir of water which reseals the trap after conditions causing siphonage have ceased. Figure 10.9 shows a typical example of this type of trap.

Another type of resealing trap is shown in Fig. 10.10. This trap incorporates a bypass arrangement which admits a flow of air into the discharge pipe to maintain equilibrium of pressure inside the trap with that of the atmosphere. Figure 10.10(a) shows the trap with the water at rest; Fig. 10.10(b) shows the effect on the water seal when siphonage is taking place (note the bypass action permitting air to enter the discharge pipe); Fig. 10.10(c) shows the effect on the water seal when siphonage is taking place (note the bypass action permitting air to enter the discharge pipe); Fig. 10.10(c) shows the effect on the water seal after siphonage has occurred. It will be seen that the seal has been maintained but is diminished in depth.

The two main objections to the use of resealing traps of all types is that they are noisy in operation due to the gurgling effect of the air–water mixture as it is drawn through the trap, and the fact that not all the traps are self-cleansing.

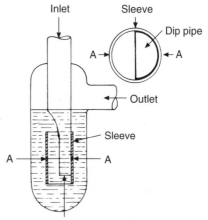

Dip pipe is 'D' shaped at this point and with the sleeve forms a bypass

**(a) Water at rest**

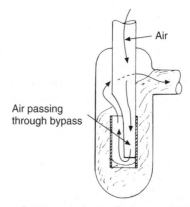

**(b) Resealing trap during a discharge of the fitting**

As the seal of the trap is lost during a discharge the end of the 'D' shaped outlet is exposed and air is allowed to enter via the bypass stabilising the air pressure in the discharge pipe.

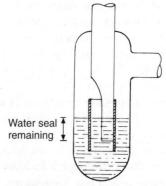

**(c) Shows reduced but effective depth of seal after a discharge has taken place**

**Fig. 10.10**  Resealing bottle trap.

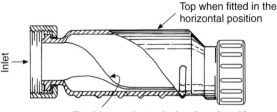

**Fig. 10.11**  Hepworth discharge pipe valve.

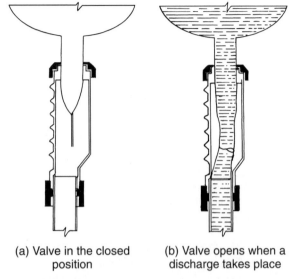

(a) Valve in the closed position

(b) Valve opens when a discharge takes place

**Fig. 10.12**  The operation of the 'Hep O' valve.

Another type of trap which limits the effect of siphonage incorporates a mechanical air valve in its crown. When pressures less than that of the atmosphere occur in the discharge pipe, the valve opens to admit air. These traps are quiet in operation, the only objection being that regular maintenance may be necessary to ensure the correct functioning of the valve.

*Hepworth discharge pipe valves*
Figure 10.11 illustrates the valve which is an entirely new concept to prevent drain air entering a building via appliance discharge pipes. It has been subjected to exhaustive tests by WIMLAS, a testing body recognised throughout Europe, and has been found to meet its requirements. An Agrément Certificate has been applied for and it meets the requirements of British Standards relating to the performance and materials of water-sealed traps. It is, however, suggested that prior to fitting this valve on contracts subject to building control, the Building Inspectorate should be consulted.

Unlike air admittance valves used to maintain equilibrium in discharge pipe systems, it is not mechanical and its working principles are very simple. The plastic membrane is, in effect, a tube which flattens when no discharge is taking place, but is sufficiently flexible to open when water is flowing through it, see Fig. 10.12.

For obvious reasons the manufacturers are reluctant to specify the materials of which the membrane is made, but it is very flexible and tough. It will withstand virtually any materials likely to be discharged through it including fats, gritty substances and various cleaning agents. From a practical point of view this valve has many advantages over the traditional water-sealed trap, especially in awkward areas, i.e. under baths and

behind pedestal wash basins. It can be supplied with bent and straight adaptors enabling it to be used in vitually any situation. It is currently recommended for use in discharge pipe systems complying to BS EN 12056 Part 2, but unlike the water seal in traps, its effectiveness is unlikely to be affected by positive or negative pressure. The use of this valve may result in even simpler discharge pipe systems.

*Flexible discharge pipes*
These are obtainable in short standard lengths fitted to 32 mm and 40 mm plastic discharge pipe. They are also obtainable in 3 m or 6 m coils which can be cut to suit any job requirement. Their use requires special adaptors for making connections to traps and waste fittings, the adaptors being jointed with solvent cement. Because of their flexibility they are useful in awkward situations where, for instance, an existing pipe does not line up with a replacement sanitary appliance, e.g. under a bath. Although the inside bore is smooth and unlikely to collect debris, the outer surface is not easy to keep clean and should only be used where it is out of sight.

## Discharge pipework systems

*The two-pipe system*
The original concept of pipework systems to comply with the London County Council requirements was

the two-pipe system, adaptions of which are still permissible for modern buildings, although it has been largely superseded by more economic installations.

To meet the requirements of the two-pipe system, sanitary fittings are divided into two main groups. Those which are used to dispose excremental matter, such as WCs and urinals, are designated 'soil' fittings. Those used for ablutionary purposes such as baths and basins or those used for culinary purposes such as sinks are termed 'waste' fittings. The main feature of the two-pipe system is the completely separate disposal of the discharges from soil fittings and waste fittings requiring two independent above-ground pipework installations, both eventually discharging into the same underground drainage system. Only soil fittings are allowed to discharge directly into the drain; waste fittings must terminate over a gulley, preferably one of the back inlet type.

This type of system may seem peculiar by modern standards but it should be borne in mind that when it was first devised not all local authorities insisted on the trapping of waste fittings. The gully therefore often provided the only effective trap against drain air entering the building. When the two-pipe system is correctly installed it is an excellent method of soil and waste disposal, but unfortunately it is also an expensive one, due to the double set of pipes required.

What in fact happened during the 1920–40 period – the heyday of speculative building – was the installation of a hopper-head in small domestic buildings. This was used to collect the waste water from the first-floor ablutionary fittings to avoid the cost of a properly constructed main discharge and vent pipe having proper branch connections for the branch wastes. While the use of hopper-heads for this purpose no longer conforms with the Building Regulations, many of these systems are still in use. For this reason an illustration showing a system of this type, with all its defects, has been included (see Fig. 10.13).

A two-pipe system that will meet the requirements of the Building Regulations is shown in Fig. 10.14. The only advantage of this type of installation is its suitability for buildings where, for some reason or other, it is not possible to group the sanitary fittings together around the main discharge

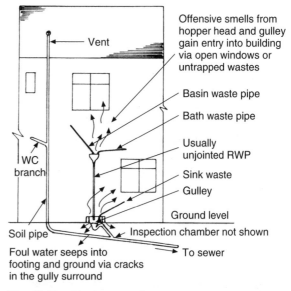

**Fig. 10.13** Obsolete two-pipe system.

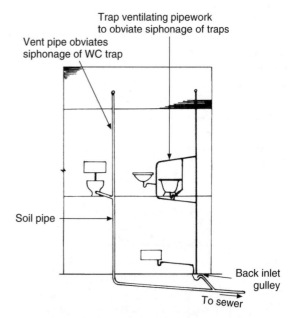

**Fig. 10.14** Two-pipe system conforming to Building Regulations. Note that as with Fig. 10.13 soil and waste water are separated above ground level but discharge into same underground drainage system.

stack. When it is compared with the single-stack system it will be seen how uneconomic it is in terms of labour and materials, coupled with the difficulty of concealing the pipework inside the building.

It is for these two reasons that it is seldom used, or indeed necessary, for most modern sanitary systems.

### The one-pipe system

The one-pipe system, shown in Fig. 10.15, was an early attempt to economise in the labour and materials required for sanitary pipework systems yet maintain the high standards set by the two-pipe system. The main difference between the one- and two-pipe system is that with the former both soil and waste fittings are discharged into a common stack with the addition of some trap-ventilating pipes. Although this was a big advantage economically, it was not accepted by all local authorities for many years. The course of time has proved it to be an effective system and it was used successfully for multi-storey buildings such as blocks of flats in the immediate post-war period. It was seldom used for small buildings, the older two-pipe system incorporating a hopper-head being more popular at this time, not because it was a better system but because it was cheaper to install.

A distinct disadvantage with both the one- and two-pipe systems is the amount of pipework involved making such schemes difficult to conceal. Prior to the 1965 Building Regulations which required all discharge pipework to be fitted inside the building, the use of these systems often resulted in a maze of exposed pipework on the face of the building. The current Building Regulations now permit sanitary pipework to be fitted externally in buildings up to three storeys. This relaxation is helpful for refurbishment work, but where possible it is recommended that discharge pipes are fitted inside the building. Externally fitted discharge pipes have been known to freeze and burst in severe weather as a result of a dripping tap.

### Primary ventilated stack system

With still greater economy in mind, the Building Research Station commenced a series of tests on fully ventilated sanitary pipework. It was found that many of the vent pipes which had previously been insisted upon were in fact unnecessary. The result of these experiments and tests revolutionised the thinking on waste-disposal systems and resulted in what was called the single-stack system, now the primary ventilated stack system, see Fig. 10.16. This system embodies the basic principles of current sanitary pipework systems and the details shown in Fig. 10.17 apply to all its adaptations.

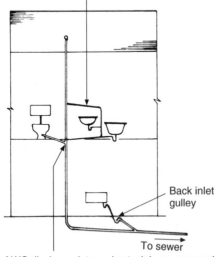

Trap ventilating pipe for basin and bath

Back inlet gulley

To sewer

Entry of WC discharge into main stack is so arranged as to avoid siphoning WC trap

**Fig. 10.15** One-pipe system (*note*: this is almost identical in design to the modern fully ventilated system). With this system both soil and waste discharges are connected to a common discharge pipe. All traps subject to risk of siphonage are fully ventilated.

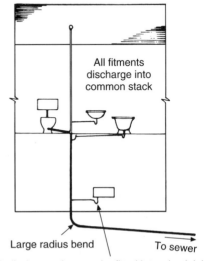

All fitments discharge into common stack

Large radius bend

To sewer

Sink discharge pipe may be fitted into a back inlet gulley instead of connecting to main stack as shown

Diagrammatic layout

**Fig. 10.16** Primary ventilated stack system.

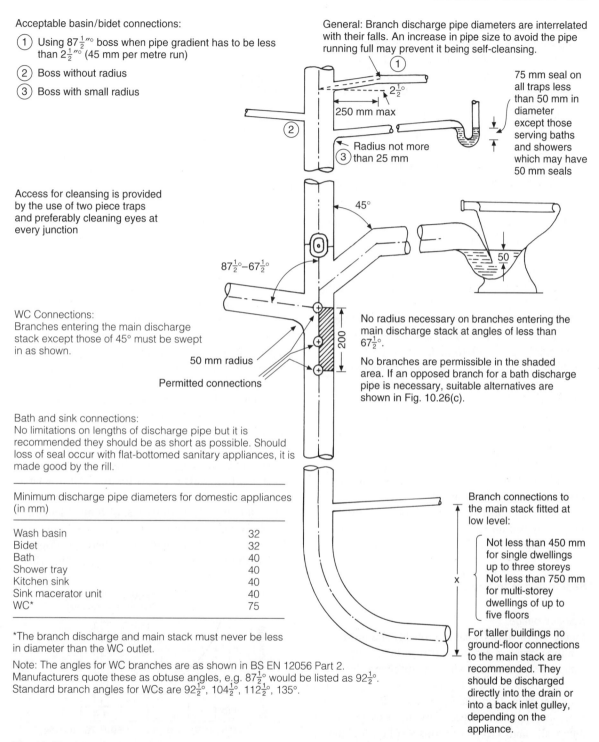

Acceptable basin/bidet connections:

① Using $87\frac{1}{2}$"° boss when pipe gradient has to be less than $2\frac{1}{2}$"° (45 mm per metre run)

② Boss without radius

③ Boss with small radius

Access for cleansing is provided by the use of two piece traps and preferably cleaning eyes at every junction

WC Connections:
Branches entering the main discharge stack except those of 45° must be swept in as shown.

Bath and sink connections:
No limitations on lengths of discharge pipe but it is recommended they should be as short as possible. Should loss of seal occur with flat-bottomed sanitary appliances, it is made good by the rill.

General: Branch discharge pipe diameters are interrelated with their falls. An increase in pipe size to avoid the pipe running full may prevent it being self-cleansing.

$2\frac{1}{2}$°

250 mm max

75 mm seal on all traps less than 50 mm in diameter except those serving baths and showers which may have 50 mm seals

Radius not more than 25 mm

45°

$87\frac{1}{2}$°–$67\frac{1}{2}$°

50

50 mm radius

Permitted connections

No radius necessary on branches entering the main discharge stack at angles of less than $67\frac{1}{2}$°.

No branches are permissible in the shaded area. If an opposed branch for a bath discharge pipe is necessary, suitable alternatives are shown in Fig. 10.26(c).

Minimum discharge pipe diameters for domestic appliances (in mm)

| Wash basin | 32 |
|---|---|
| Bidet | 32 |
| Bath | 40 |
| Shower tray | 40 |
| Kitchen sink | 40 |
| Sink macerator unit | 40 |
| WC* | 75 |

*The branch discharge and main stack must never be less in diameter than the WC outlet.

Note: The angles for WC branches are as shown in BS EN 12056 Part 2. Manufacturers quote these as obtuse angles, e.g. $87\frac{1}{2}$° would be listed as $92\frac{1}{2}$°. Standard branch angles for WCs are $92\frac{1}{2}$°, $104\frac{1}{2}$°, $112\frac{1}{2}$°, 135°.

Branch connections to the main stack fitted at low level:

Not less than 450 mm for single dwellings up to three storeys Not less than 750 mm for multi-storey dwellings of up to five floors

For taller buildings no ground-floor connections to the main stack are recommended. They should be discharged directly into the drain or into a back inlet gulley, depending on the appliance.

**Fig. 10.17** Primary ventilated stack system, sanitary pipework details (previously known as the single-stack system) complying with BS EN 12056 Part 2.

The outstanding feature of this system is the complete absence of any trap-ventilating pipe, except in very special circumstances. This is accomplished by the main factors in the following list:

(a) limitations on the length of waste pipes
(b) limitation on their fall and in some cases an increase in their diameter
(c) the branch entry of a WC into the main stack and the positioning of waste pipes in relation to this

The initial concept of this system was to simplify sanitary pipework systems for both high- and low-rise housing. It has, however, been adapted to the needs of other types of buildings by introducing a limited degree of ventilation to permit its use with ranges of fittings.

Quite apart from economic reasons, this system met the 1965 Building Regulations which required all discharge pipework to be fitted inside the building, the absence of ventilating pipes rendering this system much more easily adaptable to this requirement. The single-stack system and its modified versions are the most likely types to be used nowadays on both new and conversion work. It is, therefore, necessary for the plumber to understand fully the working principles and limitations of the system, not forgetting that trap seals must always be retained.

A thorough understanding of how trap seals are lost by bad design is a prerequisite to a full understanding of the single-stack system.

**Seal loss due to bad design**

*Self-siphonage*
Figure 10.18 shows what is probably the most common cause of seal loss. It is termed 'self-siphonage' and occurs when the discharge from a fitting unseals its own trap in the following manner. As water flows from the fitting down the discharge pipe, its progress is impeded by the resistance of the pipe walls until the water forms a solid plug, the technical term for this being the *hydraulic jump*. As the solid plug of water continues down the pipe, the air between it and the trap seal has to fill an ever

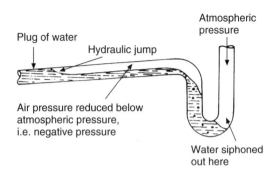

**Fig. 10.18**   Cause and effect of the hydraulic jump.

increasing gap with subsequent loss of pressure. The air in the inlet leg of the trap is at atmospheric pressure and therefore greater than that in the outlet side of the trap, and in an effort to equalise pressure in the pipe, pushes the water in the trap seal out of the trap and down the discharge pipe. This is prevented in one- and two-pipe systems by fitting a vent pipe, sometimes called an anti-siphon pipe. Trap-ventilating pipes are only necessary on modern systems when for some reason the length of a discharge exceeds that recommended by BS EN 12056 Part 2. In small buildings and dwelling houses having a well-designed drainage system this is normally unnecessary.

With the primary ventilated-stack system, vents are used only in exceptional cases, the method of preserving trap seals being as follows. All discharge pipes must be of adequate size and in the case of certain fittings, i.e. wash basins, of limited length, all falls being kept within strictly confined limits. (A graph showing the relationship between slope and length of the pipe is given in Fig. 10.19.) The

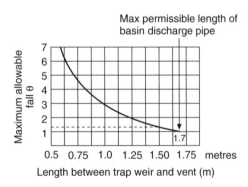

**Fig. 10.19**   Relationship between slope and length of discharge pipes.

length of any discharge pipe should be as short as possible, and one of the most important design features of this system is the close grouping of the sanitary fittings.

The only fittings having a positive restriction on the length of discharge pipes are wash basins and bidets. Both these fittings are very prone to self-siphonage due to their bowl-shaped construction, which means there is no 'rill' or tail-off in the final discharge of these appliances to reseal the trap if the seal is siphoned. If the trap seals of baths and sinks are siphoned they are usually resealed by the tail-off of water due to their flat bottom. The graph in Fig. 10.19 shows the relationship of the length of a basin waste and its fall. To use the graph, the measured length of the waste from the weir of the trap to its junction with the main stack is related to its permissible slope. To give an example of its use, assume the measured length of a basin waste to be 1.500 m. Follow the vertical line upward from 1.50 until it intersects the curved line. From this point, follow the broken horizontal line across to the left and it will be seen to fall between 1° and 2°, in fact a little less than 1.5°. This is the maximum fall that can safely be permitted on a waste of this length.

The accurate measurement of these small angles is difficult under site conditions, but it can be achieved by the use of one of two purpose-made tools. The first is a simple incidence board which can be made by cutting a piece of timber to the required angle (see Fig. 10.20(a)). A similar arrangement is often used for levelling a short length of underground drain to a predetermined fall. When the timber is placed on the pipe and levelled as shown, the pipe will be running at the correct angle. The angle cut on the timber is of course only suitable for one angle of fall, and to avoid cutting a new piece of timber every time a different angle is required, an adjustable incidence level can be made as shown in Fig. 10.20(b). A small bolt and thumbscrew passes through the middle of the two pieces of timber and when the correct angle of fall has been found a protractor is used to set the timbers. The thumbscrew is then tightened and with one timber at 180° (level) the other will indicate the correct slope of the waste.

If it is impossible for a basin discharge pipe to conform to the maximum recommended length,

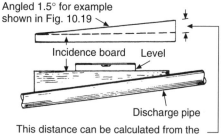

(a) Making and using a simple incidence board
The board should be cut at the angle to which the waste is to fall and used as shown with a level.

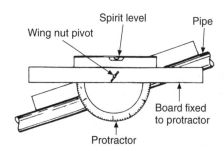

(b) Adjustable incidence level
The advantage of using this tool is that it can be adjusted to any angle.

**Fig. 10.20** Measuring the angle of fall in waste pipes.

three alternatives are possible. Resealing traps can be used, bearing in mind their disadvantages as described earlier, or a trap-ventilating pipe may be used. The third alternative is to increase the size of the discharge pipe, which has the effect of delaying the build-up of the hydraulic jump. Unfortunately, to overcome one problem the measures taken can produce others. One of the biggest problems with all modern systems is the fact that because the water moves through the pipe at a relatively low velocity, it is more likely to leave solid matter such as soap deposits on its invert. The larger the pipe diameter, the lower will be the velocity of the flow, usually resulting in larger deposits. Despite its disadvantages, this latter alternative is often the most viable proposition, but adequate provision must be made for rodding in case of blockages.

### Induced siphonage

The second way in which trap seals may be lost by poor design is known as *induced siphonage*,

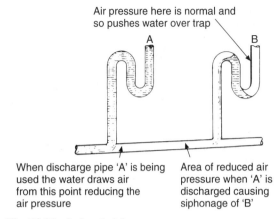

**Fig. 10.21**  Induced siphonage.

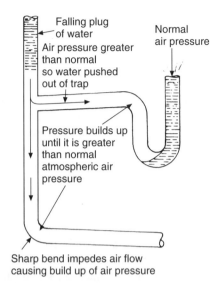

**Fig. 10.22**  Effects of compression.

a typical example being shown in Fig. 10.21. This form of siphonage occurs because of the discharge of one fitting in a range destroying the seal of the trap of another fitting connected to a common discharge pipe. This form of siphonage rarely occurs in domestic properties as it is not normally necessary to connect two fittings to the same discharge pipe. In offices and factories, however, where several fittings are connected to a common waste, some ventilation of the system may be required if induced siphonage is to be prevented.

*Seal loss by compression*

The remaining factor which must be guarded against is the possibility of the trap seal being blown back into the fitting by what is called *compression*, i.e. compression of the air in the discharge pipe. In many cases the water displaced will run back into the trap because of the shape of the appliance it serves, but foul air will have been released inside the building.

This problem only occurs in high-rise buildings where the discharge of sanitary fitments at high level compresses the air in the main stack as it falls (see Fig. 10.22). Sharp-radius bends at the foot of the stack are the most common cause of compression. For this reason it is usual to use a large-radius bend or two 135° bends as shown in Fig. 10.23. For the same reason, offsets or bends should not be used in the wet part of the main stack. If this is unavoidable the offset must be ventilated as shown in Fig. 10.24.

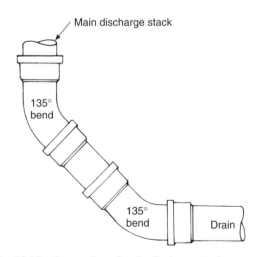

**Fig. 10.23**  Connection of main discharge stack to underground drain. The use of a large-radius bend or two 135° bends at the foot of a main discharge stack prevents 'compression' of air in the stack which would cause trap seals to be blown.

## WC connections to discharge pipe systems

The usual outlet diameter for a wash-down closet is 100 mm, although 85 mm is quite adequate for low-rise buildings. Some types of siphonic closet require an outlet diameter of only 75 mm for their efficient functioning and reference should be made to the manufacturers' fitting instructions when these are

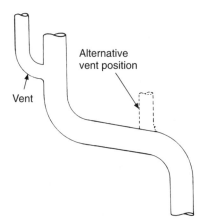

**Fig. 10.24** Ventilated offset used in conjunction with the single-stack system. The use of offsets between discharging fittings should be avoided where possible when the single-stack system is fitted. If an offset must be used it should be vented as shown to avoid the build-up of compression.

installed. Generally speaking, WCs are less likely to be affected by loss of seal due to siphonage than other fitments as their outlet pipe to the main stack seldom runs full of water. The angle at which the WC branch joins the main stack is not critical, but a standard 76° branch is recommended. What is important, however, is the way in which the branch joins the main stack (see Fig. 10.25). The standard 50 mm radius on the invert of the junction leads the discharge from the WC into the main stack in such a way as to avoid induced siphonage of other fitments sharing the same stack.

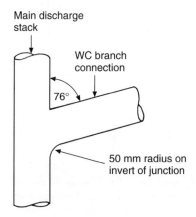

**Fig. 10.25** Recommended WC branch connection to the main discharge stack.

## Cross flow

Another problem encountered in modern discharge pipe systems is that of cross flow. When two opposing branches occur in a stack the discharge from one can shoot across the stack and into the other, causing fouled water to enter the other fittings (see Fig. 10.26(a)). Cross flow is more common with single-stack systems as the branch discharge pipes are not swept into the main stack in the same way as those fitted to one- and two-pipe systems.

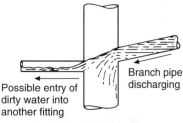

(a) Effect of cross flow

The discharge from opposing branch pipes can often cause cross flow, especially when branch pipes have only a slight fall.

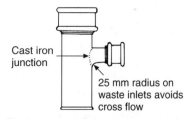

(b) Radiused inlet fitting to avoid cross flow

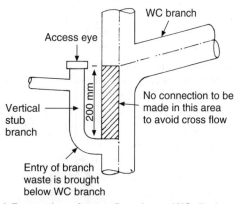

(c) Prevention of cross flow due to WC discharge

Bath and WC are usually at same level so their discharge pipes would most conveniently enter main stack opposite each other, but this would cause cross flow.

**Fig. 10.26** Cross flow.

Swept branches, i.e. those with a large radius, on unventilated wastes can increase the possibility of siphonage. The maximum radius permitted at the junction of a branch discharge pipe and the main stack is 25 mm, as shown in Fig. 10.26(b). It will be found in practice that the WC branch would most conveniently enter the main stack almost directly opposite the bath discharge pipe connection, but as this would encourage cross flow, branches are not made in the main stack within the shaded area shown in Fig. 10.26(c). When necessary a short vertical branch is connected to the main stack below the restricted area in such a way that the maximum slope is not exceeded. The vertical branch is always one size larger than the discharge pipe from the bath, which prevents the formation of a solid plug of water in the vertical stub waste.

## Ground floor appliances

These may be discharged into the underground drain in several different ways. In low-rise dwellings the risk of compression is not great and it is, therefore, suitable to connect the pipe into the main discharge pipe, especially if it is within easy reach. If this is not possible it may discharge into a back inlet gulley.

A ground floor fitting having a small-diameter discharge pipe is sometimes connected directly to a drain using a special reducer (see Fig. 10.27)

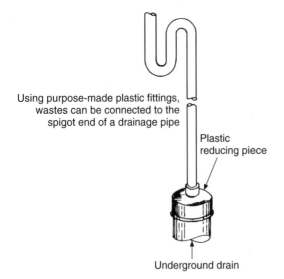

Using purpose-made plastic fittings, wastes can be connected to the spigot end of a drainage pipe

Plastic reducing piece

Underground drain

**Fig. 10.27**  Alternative method of connecting a ground floor fitting directly into a drain.

which is permitted by the Building Regulations. No reference to this method of discharge is made in BS EN 12056 Part 2 and if this method is used a check should be made to ensure a 25 mm trap seal is retained after a discharge. Alternatives to that shown would be to use an external back inlet gulley or use a modified stub stack system as shown in Book 2 of this series.

## Termination of vent pipes

The termination of vent pipes of roof level is the same for all waste-disposal systems. The most important point is to situate the terminal in such a way that foul air cannot enter the building via windows or ventilators. The recommendations given in BS EN 12056 Part 2 are illustrated in Fig. 10.28.

The Building Regulations simply state that the vent should be carried upward to such a height as not to cause a health hazard or nuisance. If, as is usual, the main vent stack is fitted inside the building, the point where the stack passes through the roof should be weathered by a pipe flashing

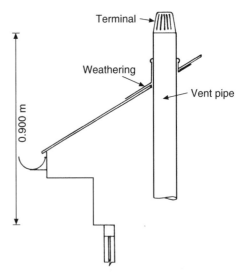

Terminal

Weathering

Vent pipe

0.900 m

**Fig. 10.28**  Recommendations to avoid the entry of drain air to a building. If the vent pipe terminal is situated within 3 m of an opening window or ventilator it must be carried up to sufficient height to prevent drain air entering the building.

as detailed in Chapter 11. When PVC stacks are used, some systems include a special weathering component which avoids the necessity of fabricating metal weatherings. Some form of terminal should also be fitted at the top of the vent to prevent birds nesting in the stack and those constructed of high-density polythene are adaptable for most piping materials.

## Access points

The reason for providing means of access for clearing blockages has already been discussed. It is probably true to say that the traditional fully ventilated systems, especially those fitted to low-rise private buildings, seldom become obstructed, but modern waste-disposal systems are more prone to obstruction owing to the low velocity of flow through the branch discharge pipes and the need for access is often very apparent. Generally speaking, access should be provided at the fitting and at the junction of the branch waste with the main discharge pipe. It was shown earlier in this chapter that easily dismantled traps can be used at the appliance, and clearing eyes at each junction or group of junctions in the pipework system (see Fig. 10.29(a)).

A greater number of access points are required in public and commercial buildings than private dwellings as blockages are more likely to occur from the possibility of misuse. This is especially true of urinals, when owing to the ammonia content of urine a hard surface coating of scale is built up on the inner walls of the pipes. The channel also becomes a depository for cigarette ends, matches and other small items of rubbish which, unless sufficient access is provided, can cause some unpleasant stoppages.

A screw cap type of access point is illustrated in Fig. 10.29(b), these being used at the end of a common waste serving several fittings. Figure 10.29(c) shows a cleaning eye bend suitable for branch discharge pipes while Fig. 10.29(d) illustrates a cleaning door in a length of pipe. These are usually used at ground level on a main discharge stack to permit access to the connection with the underground drain.

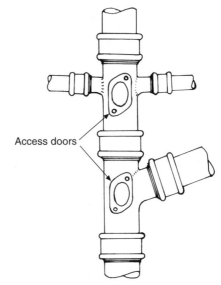

(a) Access to junctions

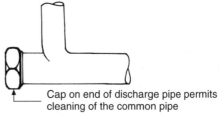

Cap on end of discharge pipe permits cleaning of the common pipe

(b) Access cap to a range of fittings served by a common discharge pipe

Bends in discharge pipes are common causes of obstruction and should be provided with a means of access

(c) Cleaning eye bend

Access door

(d) Access pipe

This is usually fitted at a point where the main discharge stack connects to the underground drain. Some manufacturers of PVC pipe produce an access door which can be solvent welded into a pipe at any required point.

**Fig. 10.29** Access points.

## Branch discharge pipe connections

The method of making additional connections to existing stacks depends upon the type of material in use. In the case of lead stacks, which are very rare except on very old buildings, lead-welded or wiped soldered joints would be used. In an existing copper stack it might be possible to cut in a new junction, but if the plumber is sufficiently competent, an easier method would be to open up a branch hole and braze in a new connection.

If the existing stack is made of cast iron with lead-caulked joints and an additional WC is to be fitted, a new junction must be used. Carefully mark out its position and cut out a section of the existing pipe using a power hacksaw or angle grinder. Do not attempt this from a ladder but provide a secure scaffold. Ensure the cuts are square and insert the junction using time saver joints as shown in Fig. 4.15(b). These are suitable in most cases but when ordering specify the outside diameter of the existing pipe, as the tolerances are approximately only ±2 mm. If the new branch is of small diameter the same technique as described may be used, the alternative being to use the 'Fixicon' bracket as shown in Fig. 10.30 over a previously cut hole in the existing pipe. The hole may be cut with a circular cutter, or using the method shown in Fig. 10.31. It must be stressed that both patience and care are exercised whenever the preceding operations are carried out.

The manufacturers of PVC systems employ several methods of making branch connections. Blanked-off sockets may be moulded on to standard 100 mm junctions, the blank being cut out when the connection is required (see Fig. 10.32(a)).

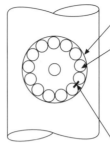

Step 1: Mark out hole to accommodate the branch

Step 2: Drill a series of holes approximately 6 mm in diameter just inside the mark and a centre hole

Step 3: Attach a length of wire or string through the centre hole so that the cut-out section can be retrieved if it falls into the pipe

Step 4: Cut the bridges between the holes and remove the centre section using an adapted hacksaw blade

Step 5: Round off the hole using a half-round bastard file

Detail of ground-down hacksaw blade

24 TPI

**Fig. 10.31** Preparation of a hole to accept a Fixicon bracket.

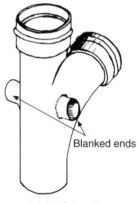

Blanked ends

### (a) PVC junction

This junction has moulded-on blank sockets for waste connections. Blanks are removed as and when needed.

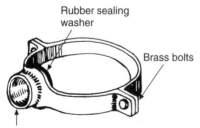

Rubber sealing washer

Brass bolts

BSP thread for connection to galvanised steel pipe or copper and plastic pipe adaptors

These fittings are clamped round the pipe over a precut hole.

**Fig. 10.30** Cast iron Fixicon bracket.

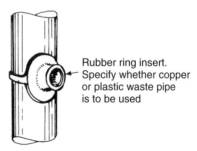

Rubber ring insert. Specify whether copper or plastic waste pipe is to be used

### (b) Branch connection for PVC discharge stacks

**Fig. 10.32** Connections to main stacks.

The other main alternative is the use of a fitting very similar to the 'Fixicon' bracket used with cast iron (see Fig. 10.32(b)). A hole is cut at the point where the branch is required using a suitable hole saw. The area round the hole and the face of the bracket are then treated with cleaning fluid prior to solvent welding the connection in position. The connection illustrated is provided with a moulded clamp which secures it in position until the solvent-welded joint has matured. Others are clamped in position by means of a special tool provided by the manufacturers.

## Mechanical waste-disposal units

These fittings are becoming increasingly popular with householders and the plumber is required to install them in both new and existing properties. They are fitted beneath the sink and are designed to macerate waste materials such as vegetable peelings so they can be discharged into the drains without causing a blockage. They are electrically operated, and unless the plumber has competent electrical knowledge a properly qualified person should be employed to make the necessary electrical connections.

Before an installation is contemplated make sure an electrical supply is available! Such a warning may sound unnecessary, but experience proves otherwise. Another important point to remember is that the waste hole in many existing sinks is not large enough to accept such a unit, and in many cases a new sink will be required – a point the householder may not have considered. A hole of approximately 90 mm in diameter is required, which is far larger than that required for a normal waste fitting. Special cutters are available for enlarging the hole in certain types of existing stainless steel sinks and may be hired from the stockists of the equipment being fitted. Some of these stockists employ specialists who will fit the actual unit and make the necessary electrical connections, leaving only the discharge connections to be fitted by the plumber.

Disposal units do not have an integral trap. A tubular trap must be fitted to the outlet and should be installed so that the invert of its outlet is lower than that of its inlet, which prevents water lying in

the machine, Bottle traps must not be used as they tend to retain waste matter. The maximum length of the discharge pipe is 3 m and its fall should be such that sufficient velocity of flow is maintained to prevent debris being deposited and causing a blockage. The absolute minimum fall is 1 in 10 or approximately 6° but some manufacturers recommend 15°. These relatively sharp falls preclude the direct connection of the equipment to a main discharge stack. They should always be connected to a back inlet gulley as illustrated in Fig. 10.33(a), because if they discharge over an ordinary gulley grating this will quickly become clogged by the waste particles. Suggested arrangements for fitting to a sink on an upper floor are shown in Fig. 10.33(b). The provision of a vent

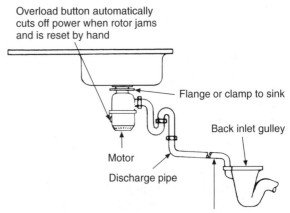

Discharge from sink disposal units must be connected under the gulley grating or via the back inlet to avoid fouling and blocking the grating

(a) Fitted to a ground floor

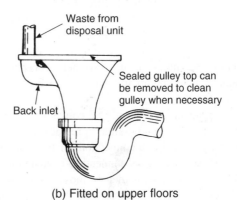

(b) Fitted on upper floors

**Fig. 10.33**  Waste disposal units.

in the gulley is not normally necessary as it is very unlikely the drain will run full.

When this equipment is installed, the client should always be made to understand that a flow of water is required while in use. Failure to observe this will result in constant obstruction in the discharge pipe. Always make sure the client understands how to operate the unit and free it if it jams, a special tool being provided for this purpose. Leave the manufacturer's instructions and any guarantee forms with the client before leaving the job.

## Washing machines and dishwashing equipment

The increasing use of domestic washing machines and dishwashers has also provided the plumber with a new area of work. Before any machine of this type is fitted check that it is approved by the local water company, as some types, especially older models, do not conform with the Water Regulations in respect of back flow.

*Washing machine waste connections*
The discharge outlet on these units is a flexible rubber hose, approximately 20 mm nominal diameter. The waste water is discharged by a pump fitted to the machine. The original method of disposing of the waste was to connect the hose to a 22 mm pipe which terminated over a gulley, but this is now unacceptable for two reasons. It was found that this method often resulted in the pump becoming airlocked, necessitating a service call to the dealer. Secondly, owing to the changeover to modern systems of sanitary pipework, there may not be a gulley available. Figure 10.34(a) shows a suitable arrangement for the connection of the discharge pipe from the machine to the main stack, the rubber waste connection being hooked over the 40 mm riser. The air break so formed prevents siphonage of the pump and waste water back up into the machine.

If the machine is situated at a convenient point near a sink, it is often possible to replace the existing trap with one having a connection for the machine (see Fig. 10.34(b)). This connection is adaptable to a wide range of hose diameters; those not required can be cut off with a hacksaw and the hose secured with an adjustable clamp such as a 'Jubilee' clip.

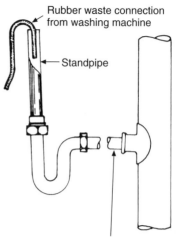

Rubber waste connection from washing machine

Standpipe

Waste may be fitted directly into a main discharge stack as shown or over a back inlet gulley

(a) Waste connection to discharge pipe

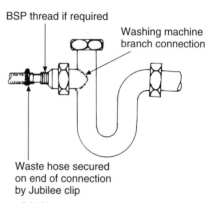

BSP thread if required

Washing machine branch connection

Waste hose secured on end of connection by Jubilee clip

(b) Waste connection to sink trap

The branch connection enables the rubber waste hose to be discharged into the sink trap arrangement by providing a special trap as shown.

**Fig. 10.34** Washing machine waste.

## Further reading

Much useful information can be obtained from the following sources:

BS 4118 1967: Glossary of sanitation terms.
BS EN 12056 Part 2 2000: Sanitation pipework layout and calculation.
BS 1184: Copper and copper alloy traps.
BS 3943: Plastic waste traps.

Design guides are obtainable from the Building
Research Establishment, Garston, Watford,
Herts, WD2 7JR:

DG 248 Sanitary pipework Part 1 Design basis.
DG 249 Sanitary pipework Part 2 Design of pipework.

*Discharge pipe systems and materials*
See manufacturers' lists, Chapter 3.

**Self-testing questions**

1. By means of a sketch show how the 'effective
   seal' of a trap is measured.
2. Name the three causes of seal loss in a trap due
   to the defective design of sanitary pipework.
   Briefly explain the reason for each cause.
3. Explain why a seal depth of 50 mm is
   satisfactory for WC and gulley traps.
4. State the minimum depth of a trap seal for
   baths and showers connected directly to a main
   discharge stack.
5. Describe the circumstances in which a resealing
   trap may be used.
6. Name the systems of above-ground discharge
   pipework.
7. State the minimum diameter of discharge pipes
   for baths, basins and sinks.
8. Describe how 'hydraulic jump' occurs in
   discharge pipes and how this affects the water
   seal in traps.
9. Explain why it is necessary to join a main
   discharge pipe to an underground drain with
   an easy or large-radius bend.
10. Define the term 'cross flow' in discharge pipes
    and give an example.
11. State why the discharge from sink waste-
    disposal units should not terminate over a
    gulley grating. Sketch and describe the correct
    method.
12. Use the graph shown in Fig. 10.19 to determine
    the maximum fall in a basin discharging pipe
    1 m long.

# 11 Lead sheet weatherings

After completing this chapter the reader should be able to:

1. State the main physical and working properties of sheet lead.
2. Describe any preparatory work prior to fixing the weatherings.
3. Understand the preventative and remedial measures to protect lead weatherings from damage by wind, corrosion and thermal movement.
4. State the recommended falls for effective roof drainage.
5. Select suitable jointing and fixing methods for various applications.
6. Understand the basic principles of forming weathering details.
7. Describe the methods of employing bossing and lead welding techniques in weatherings.
8. Calculate the amount of material required for a given job.

## Introduction

Before the introduction of piped water supplies the main activity of the plumber was roof weathering. Lead was the sheet material in general use and some idea of its applications can be gained by visiting historic buildings, many of which still retain their original roof coverings. It should be noted that the plumbing NVQs cover only lead as a weathering material.

In recent years there has been an increase in the use of sheet metal weatherings, especially for large commercial, public and high-quality dwellings, and although the initial costs may be higher, the longer life of these materials over their alternatives has proved to be more economic in the course of time. Modern methods, coupled with the desire of many planning authorities to revert to traditional styles of building, have led to an increase in the use of sheet metal weathering, especially lead. While specialist roofing contractors are mainly involved in large-scale projects, the modern plumber should be competent to weather small canopies, dormers, gutters and chimney weatherings in lead sheet.

## Methods of working

Reference should be made to the section on wood tools, their care and maintenance in Chapter 2. Because lead is a soft metal, any abrasions on the tools will be transferred to the lead, making unnecessary tool marks. Sheet lead components should be fabricated as far as possible on a bench. Two or three clean, reasonably smooth scaffold boards on a pair of trestles is all that is normally required. Some offcuts of 100 mm × 50 mm timber are also useful as forming blocks. When cutting straight lengths of lead, e.g. cap or cover flashings, use a knife to score the sheet, then bend it to and fro through approximately 90° until it breaks. This method is a quick an effective way of cutting long, straight lengths of lead. Tinsnips are, of course, invaluable for making small incisions and cutting out smaller items such as gussets. Figure 11.1 illustrates the method of forming an upstand using a holding-down block. This will produce a clean, straight turn throughout the sheet. The root radius of any angle formed should be approximately 4–5 mm – anything less than this may prevent freedom of movement resulting in fatigue cracks after a short period of service life.

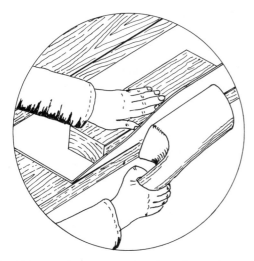

**Fig. 11.1**  Turning an upstand on lead sheet using a dresser and a holding-down block.

### Physical properties of lead

Chemical symbol       : Pb
Colour                : Blue-grey
Density
   Milled           : $11,340 \text{ kg/m}^3$
   Cast             : $11,300 \text{ kg/m}^3$
Melting point        : 327 °C
Coeff. of linear expansion : 0.0000293 °C
Tensile strength     : $18 \text{ MN/m}^3$
BS specification No. 1178
BSCP 143, Part II (Roofing)

Colour and melting point are familiar terms to everyone but density and coefficient of linear expansion will require explanation.

### Density

For commercial purposes density is calculated on the basis of 1 cubic metre. Water is used as the standard of comparison and has a density of $1.000 \text{ kg/m}^3$. The density of lead varies slightly depending on whether it is cast or milled, but it will be seen that lead is a very heavy material being just over 11 times the mass (weight) of water. Note that when a comparison is made between the densities of various materials their volume, i.e. the space they occupy, must always be the same.

When the density of a material is known then the mass of the sheet roof covering made from it may be found by calculation.

*Example*  If lead has a density of $11,340 \text{ kg/m}^3$, calculate the mass of 1 square metre of BS code No. 6 lead sheet having a thickness of 2.5 mm.

First find the mass of 1 square metre of lead sheet which is 1 mm thick:

$$11,340 \div 1,000 = 11.34 \text{ kg}$$

The mass of 1 square metre of sheet lead 2.5 mm thick will therefore be

$$11.34 \times 2.5 = 28.35 \text{ kg}$$

This simple formula can be used to find the mass (weight) of the total area of sheet lead, in square metres, necessary to weather a roof, information which would be needed by a building designer to make provision for sufficiently strong roof supports.

### Coefficient of linear expansion

This subject has been dealt with in Chapter 8 (p. 182) where it was shown that an increase in temperature has the effect of increasing the length of pipe. The same principle also applies to lead sheet coverings for roofing and weathering which will be affected by variations in temperature ranging from snow in winter to heat from the sun in summer. Temperatures of up to 90 °C have been recorded on roofs in the UK in the summer, so the matter of expansion and the subsequent contraction of lead sheet demands careful consideration.

As it is so important to understand the need for restricting the areas and lengths used, a further example of this subject is included here.

*Example*  One bay of a lead roof covering is 2 metres long and has been subjected to a temperature rise of 60 °C. Calculate the increase in its length (coefficient of linear expansion of lead is 0.0000293).

Increase in length = Length × Temp. rise
                   × Coeff. of linear exp.
           = 2 m × 60 °C × 0.0000293
           = 0.003516
       or = 3.5 mm approx.

The bay will be 3.5 mm longer when its temperature has been raised through 60 °C. It must also be realised that this effect is reversed on cooling, i.e. when cooled through 60 °C to the original temperature, the lead will contract by 3.5 mm.

Expansion and contraction due to temperature changes must be allowed for when laying roofing materials or deformation and cracking will occur. This is called metal fatigue and the foremost consideration when laying lead sheet is to ensure freedom of movement by using only approved methods of fixing. These methods are considered later in this chapter.

*Specification and production of lead sheet*
Lead is obtained from the ore galena and is mined and smelted in many different parts of the world, the principal producing countries being Australia, Canada, the USA and the former USSR. Lead is also mined in the UK but the amount obtained nowadays is very small. The plumber's name comes from the Latin name for lead which is·*plumbum*. In Roman times the use of lead in building was restricted mainly to sanitary and ablutionary installations, but during later years, particularly in the Norman period, the use of lead was extended to include roofing and weathering. There are many examples of the use of this material on English churches and cathedrals of all periods, notably Lincoln, Canterbury, York and St Paul's Cathedrals. Sheet material used for these buildings was usually produced on site by pouring molten lead on to a flat bed of fine sand. These cast lead sheets were thicker than most of those used in modern practice and varied between 3.15 and 3.55 mm in thickness. It should be noted that one company producing lead sheet has developed a technique for continuous casting. It is obtainable in roll and strip form like milled sheet and has comparable thicknesses.

As can be seen from the stated densities, cast lead is less dense than the rolled sheet in current use, the latter being compressed during manufacture by being passed between steel rollers to obtain the required thickness. Traditionally cast lead is little used nowadays except for ornamental details such as rainwater pipes and hopper-heads. If cast lead weatherings are renewed using traditional methods the work is usually executed by specialist firms with the necessary equipment and craftworkers.

Lead is ductile, the softest of the common metals and is highly malleable which allows it to be shaped with ease. Lead recrystallises during cold working and this factor means that work hardening does not take place to any great extent. Low tensile strength, softness and lack of elasticity give lead a tendency to 'creep', but this can largely be overcome by the application of correct installation procedures. The term 'creep' is not used to describe the slipping of lead down a pitched roof due to inadequate fixings failing to provide support. Fatigue cracking does occur but is usually caused by bad design or poor work in using oversized sheets and not making sufficient allowance for expansion and contraction.

Rolled lead sheet is available in standard rolls 2.4 m wide in lengths of 3, 6 or 9 m. These sheets are very heavy, difficult to handle and are only recommended where large areas are to be covered. For a small extra charge rolls of smaller width (75 mm up to 600 mm) known as strip can be supplied, these being more economic in use and save time in cutting on site. The thickness of lead is designated by a BS specification code number, by the thickness in millimetres and by an identifying colour; these factors are shown in Table 11.1.

The thicknesses of lead for various situations as recommended by the Lead Development Association are given in Table 11.2.

The choice between the thicknesses of sheet given in Table 11.2 depends on the quality of the building, length of life required, roof design and shape of individual panels. The initial cost is, however, an important consideration so the thinnest lead to suit the fixing position will usually be specified.

*Corrosion*
One of the characteristics of lead is its resistance to corrosion; any failure is far more likely to be due to defective work or poor design. All metals are subject to corrosion where they are in contact with

**Table 11.1**  Identification of sheet lead.

| BS specification Code No. | Colour | Thickness (mm) |
|---|---|---|
| 3 | Green | 1.25 |
| 4 | Blue | 1.80 |
| 5 | Red | 2.24 |
| 6 | Black | 2.50 |
| 7 | White | 3.15 |
| 8 | Orange | 3.55 |

**Table 11.2**  Thickness of sheet lead for various fixing positions.

| Fixing position | BS specification Code No. |
|---|---|
| Small flats with no pedestrian traffic | 4 or 5 |
| Large flats with or without traffic | 5, 6 or 7 |
| Gutters – parapet, box or tapering valley | 5 or 6 |
| Dormer cheeks and roofs | 4 or 5 |
| Chimney flashings | 4 or 5 |
| Soakers | 3 or 4 |
| Cornices | 5 or 6 |
| Valleys, hip, ridge and cover flashings | 4 or 5 |
| Vertical cladding | 4 or 5 |
| Pipe weathering | 4 or 5 |
| Cornice weathering | 4, 5 or 6 |
| Damp-proof courses | 3, 4 or 5 |

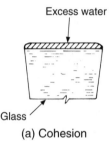

(a) Cohesion

The surface tension of the water molecules is capable of holding a body of water above the edges of the vessel containing it.

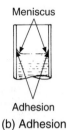

(b) Adhesion

The climbing effect of the water molecules where they meet a solid material such as glass is adhesion.

**Fig. 11.2**  The two forces causing capillarity.

water and the constituents of the atmosphere such as oxygen, carbon dioxide and sulphur dioxide, the latter two forming weak carbonic and sulphurous acid. Unlike ferrous metals, however, the initial attack by these acids forms a protective surface coating called 'patina' which protects the metal from further attack. Rainwater, when it has been in contact with moss or lichen growing on roof coverings, can form an acidic run-off which can damage not only lead but other metals used for weatherings. It is at its worst during periods of light rain or mist where the run-off has been in contact with the tiles or slates for some time. Corrosion can also occur when a metal is in close contact with some timbers, such as oak, teak or cedar, especially cedar as it is sometimes used as a roof covering in the form of shingles. These are very acidic, especially when new, and therefore the metal should be protected with bituminous paint. Electrolytic action between lead and other metals is negligible and can be ignored. Copper, stainless steel and brass are commonly used for metal fixings.

*Capillarity*

Capillarity is responsible for many defects in buildings and can cause serious deterioration of the building fabric. The following text describes the causes of capillarity and its prevention in sheet weathering.

It has been established that every substance is composed of many molecules and it is of interest how these are held together. Although the full scientific explanation is complicated, the force between molecules can be thought of as being similar to the attraction between a magnet and steel. The force is, however, very much more powerful than a magnet and a simple term to describe it is *cohesion*.

An example of cohesion can be seen in Fig. 11.2(a) which shows a glass overfilled with water. The liquid has not overflowed as the molecules of water are attracted to each other and those at the surface are pulled together and downwards, thus creating *surface tension* which acts in the same manner as a surface 'skin'.

There is another force or attraction between the water molecules and the side of the vessel in which they are contained. This is called *adhesion*, an example of which is given in Fig. 11.2(b). If a clean glass tube is placed in the water, the water will adhere to its sides and be raised above the true water level in the tube, thus creating a dish-shaped water level in the tube, known as a *meniscus*.

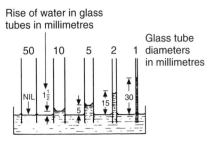

**Fig. 11.3** The effect of capillarity in various tube diameters. Note that the height of the water is greater as the tube bore diminishes.

When reading the water level in a glass tube the lower part of the meniscus is taken as being the correct level.

The two forces of cohesion and adhesion combine when capillarity occurs. The molecules of water by the action of adhesion between the water and the solid 'haul' themselves upwards wherever there is contact. By the simultaneous action of cohesion the main body of water is lifted higher, thereby permitting adhesion to take place again to achieve a yet higher level. These processes are both continuous and simultaneous and cease only when the water column becomes too heavy or when the surface area is too large to be supported by adhesion to the sides of the vessel.

The illustration in Fig. 11.3 shows this quite clearly, the water in the glass tubes of larger diameter having a larger surface area is unable to reach the height attained by that in the tubes of smaller bore. It also indicates the approximate height that a column of water will achieve in a given tube diameter: the smaller the tube bore, the higher will be the column of water capable of support by capillarity.

Water will find its way into buildings owing to capillarity if lapped surfaces are close together, so the most practicable method of prevention is to widen the gap between the surfaces. A typical example of both the problem and its solution can be seen in Fig. 11.4.

*Fatigue cracking*

If a piece of metal is continually bent and straightened as shown in Fig. 11.5, it will crack because of the breakdown of the adhesive forces between its

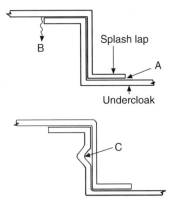

**Fig. 11.4** Preventing the entry of water into a building by capillarity via a drip in a lead gutter. (For further details of drips in lead gutters, see Fig. 11.11.) The formation of an anti-capillary groove shown at C prevents any water rising above it.

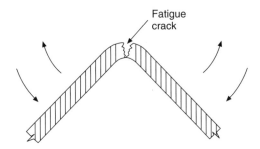

**Fig. 11.5** The effect of fatigue on metals continuously bent in the direction of the arrows.

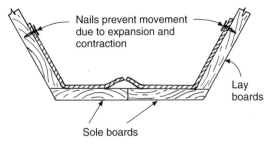

**Fig. 11.6** The effect of fatigue on an improperly fixed valley gutter.

molecules. The effect of expansion and contraction due to a temperature change is the main cause of fatigue in sheet lead. Figure 11.6 shows how incorrect fixings can cause failure in a valley gutter.

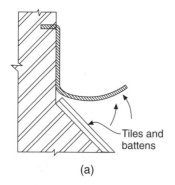

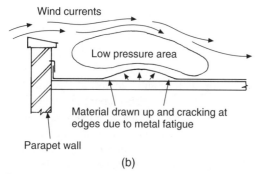

**Fig. 11.7** (a) Strong winds can lift the unsecured edges of apron flashings causing damage to the flashing and permit the entry of driving rain into the building.
(b) Causes of fatigue cracking in areas of high winds. It is uncommon with lead sheet because of its mass (weight).

*Wind effect*

The effect of the wind can cause metal sheets which have a free edge, e.g. cover flashings, to 'flap' about and even continually hit against adjacent surfaces. This 'cold working' resulting from the bending or the blows can eventually lead to the fracture of the metal, see Fig. 11.7(a). Another result of wind effect is the continuous rise and fall of covering materials when subjected to alternating negative and normal air pressures which will also lead to fatigue cracking, see Fig. 11.7(b).

Fatigue cracking can be limited by the use of:

(a) the correct gauges of material as specified in Table 11.2
(b) recommended maximum bay or panel sizes
(c) correct jointing methods
(d) adequate fixings which do not resist controlled movement of the lead

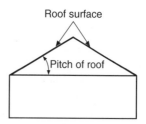

**Fig. 11.8** Angle of roof pitch.

### Roofing terms

Prior to considering the operations involved in applying sheet weathering materials, two important roofing terms, roof pitch and maximum bay sizes, must be clearly appreciated.

*Roof pitch*

Roof pitch is the angle between the sloping roof surface and the horizontal plane as shown in Fig. 11.8. A roof is generally referred to as *flat* when the pitch is less than 10° and *pitched* when it is 10° or more. The minimum satisfactory pitch or fall is $2\frac{1}{2}$°, as anything less will result in a slow rainwater run-off that could cause flooding and allow entry of water into the building.

*Maximum bay sizes*

Maximum bay or panel sizes are required to avoid excessive thermal movement, reduce 'creep' and limit problems created by wind pressure. Large roof surfaces are divided into a series of smaller areas, the sizes depending on type and gauge of material and roof pitch. Areas in relation to these factors are specified in the Lead Sheet Manual and its recommendation should be strictly adhered to in order to ensure a trouble-free and weathertight roof.

The application of sheet weatherings can be conveniently divided into three operations, jointing, fixing and the formation of details such as corners, flashings and gutters. First, however, it will be appropriate to discuss the type of support necessary for the various materials.

*Preparation of decking*

Some roofing materials such as corrugated steel, rigid aluminium cladding and non-asbestos corrugated sheet have sufficient natural strength to maintain their shape and form, so these require little additional support. It should be noted that non-asbestos corrugated sheet is a cement-bonded fibrous material now marketed as a substitute for cement asbestos. Lead sheet, however, must be fully supported. The type of material used for decking should conform to the relevant British Standard and is usually one of the following:

(a) tongued and grooved or square-edge boards
(b) blockboard
(c) chipboard
(d) compressed straw slabs
(e) plywood

It is not within the activities of a plumber to lay roof decking but it is important for the plumber to be aware of the requirements for suitable decking which are as follows.

(a) The roof undersurface is to be flat and without indentations that could lead to the formation of puddles of water. Stagnant water can be a hazard to health and can cause the accumulation and concentration of undesirable acids formed by organic and atmospheric pollutants.
(b) In order to prevent warping and springiness, wood or decking derived from wood should be not less than 25 mm nominal thickness and laid either diagonally or with the fall of the roof.
(c) All nails are to be punched down below the board surface, screws countersunk and all grit or debris removed to leave a smooth surface. Sharp edges of any decking materials which could damage the covering are to be removed.
(d) Wood or wood preservative which could cause corrosion must not be used.
(e) Where appropriate, precautions must be taken against the alkaline nature of concrete and mortar which might corrode the lead.
(f) Adequate falls are to be incorporated into the roof structure to allow the quick removal of rainwater.
(g) An insulating felt underlay must be used.

### Felt underlay

The recommendation that a suitable underlay should be used on the prepared decking when sheet metal coverings are to be used is made for the following reasons:

(a) To provide some measure of thermal insulation, i.e. to prevent solar heat unduly affecting the building.
(b) To insulate the building against the sound of rain and wind.
(c) To allow lead to move freely when thermal expansion takes place and to prevent abrasion between the weathering material and decking.
(d) To prevent corrosion due to contact between the lead and any organic alkaline or other chemical agent present in the decking material.

For wooden decking two materials are in common use as underlay, BS 747 (Type 4A, Brown No. 2 inodorous felt) and waterproof building paper to BS 1521 – Class A. Felt is more generally specified as its thickness helps to overcome uneven surfaces. The underlay may be fixed using copper or galvanised clout nails. Building paper should be laid with 50 mm lapped joints while felt sheets should be butted together. On no account should bituminous sheet, i.e. ordinary roofing felt, be used as it will soften with heat, adhere to decking and weathering material and thus restrict the movement caused by thermal expansion. The felted surfaces should be covered immediately by the final metal roof coverings to avoid trapping any moisture.

### Jointing methods

All joints *must* be watertight, although the choice of jointing method is dependent on the following considerations.

(a) position of the joint in relation to rainwater flow, i.e. 'with' or 'across' the fall of the roof surface
(b) roof pitch
(c) type of roofing material
(d) accessibility by pedestrian traffic, i.e. will the roof be regularly walked on?

### Laps

Laps are the simplest joints and are made by laying one sheet edge over another, the illustrations in Figs 11.9 and 11.10 shows typical details.

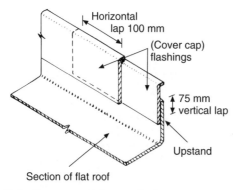

**Fig. 11.9** Recommended cover for laps on cap or cover flashings.

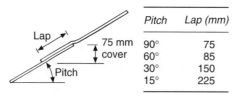

Minimum lap required for low pitched roof

**Fig. 11.10** Effective cover for laps on pitched roofs.

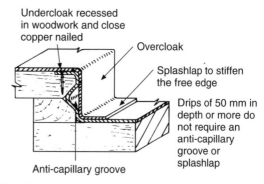

**Fig. 11.11** Section through a lead drip.

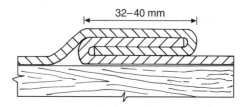

**Fig. 11.12** Section through lead welt.

To avoid water entry due to capillary attraction a minimum lap of 100 mm is recommended for laps on cap flashings and 75 mm for those in the vertical position. For sloping surfaces the lap should correspond to an equivalent vertical height of 75 mm. When in any position other than vertical, laps must only be used across the fall of the roof. The edges of these joints must be securely fixed to prevent the problems encountered through wind effect. The recommended materials for securing sheet lead are copper, stainless steel and lead-coated stainless steel.

*Drips*
Drips are positioned across the fall on very low pitched roofs, i.e. under 15°, where a lap joint would allow the entry of water. The drip is basically a step formed in a flat roof or gutter. The recommended depth of a drip is 50 mm, but a minimum of 40 mm is permissible with those provided with an anti-capillary groove (see Fig. 11.11) formed in the step. The splashlap shown on the lead drip is intended to stiffen

the free edge of the lead sheet and to assist in keeping the overcloak in position on drips of 40 mm depth only.

*Welts*
These are a simple form of joint used for the weathering of parapets and cornices. They are also used on vertical surfaces such as dormer cheeks. Generally they are only suitable where there is a good fall to the decking with no possibility of the formation of puddles. Figure 11.12 shows a section through a lead welt.

*Rolls*
Rolls are used to form the joints that run in the direction of the flow of water or 'with' the fall of the roof. Figure 11.13 illustrates the formation of the lead sheet over a wooden core, the undercloak being fixed first. Hollow rolls are used only for pitched roofs covered with lead sheet; they have no wooden core and for this reason are called hollow.

One fault which can occur in wood-cored rolls if the roof has a low pitch and if water stands on

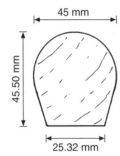

45 mm

45.50 mm

25.32 mm

(a) Wood core over which the lead is formed

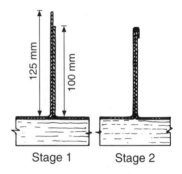

125 mm

100 mm

Stage 1

Stage 2

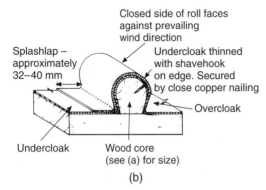

Closed side of roll faces
against prevailing
wind direction

Splashlap –
approximately
32–40 mm

Undercloak thinned
with shavehook
on edge. Secured
by close copper nailing

Overcloak

Undercloak

Wood core
(see (a) for size)

(b)

**Fig. 11.13** (a) Wooden cores for rolls for various materials. (b) Lead-covered wood-cored roll.

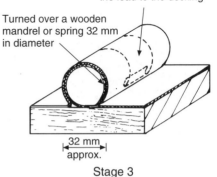

Sheet copper cleat fixed
to decking secures
the lead to the decking

Turned over a wooden
mandrel or spring 32 mm
in diameter

32 mm
approx.

Stage 3

**Fig. 11.14** Formation of hollow rolls.

the surface, is that it may be drawn up between overcloak and undercloak by capillary attraction. This would, in fact, be a design fault as the roof slope should be sufficient to prevent water ponding, and rolls formed in this way are most suitable for pitches up to 30°. Above this pitch it is preferable not to use a splashlap but to secure the free edge by using a 50 mm wide clip, in which case a 5 mm clearance is left between the roof surface and edge of the overcloak to prevent the formation of a vertical space that could encourage capillary attraction. The splashlap serves the same function on rolls as it does when used on a drip and it should be positioned on the side of the roll that will be least affected by wind. It should be noted that the undercloak is taken about two-thirds around the wood core and the edge chamfered using a shavehook. This prevents the formation of a sharp edge which would show through the overcloak after dressing into position.

*Hollow rolls* Hollow rolls are suitable for steeply pitched roofs where they will not be walked on, and are formed over a wooden core or bending spring which is withdrawn afterwards. The stages in forming this type of roll are shown in Fig. 11.14. The use of hollow rolls is especially useful for roofwork details having curved surfaces such as domes or bay window weatherings.

*Lead welding*
Welding requires the melting or fusion (i.e. fusion welding) of the two sheets being joined.

Lead can be welded using either a butt or lap seam as illustrated in Fig. 11.15. This is a better jointing method for sheet roofing than soldering as the welded joint has the same coefficient of linear expansion as the lead sheet. By ensuring that expansion of roof material and jointing material are the same, failure due to cracked joints is avoided.

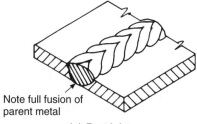

Note full fusion of
parent metal

(a) Butt joint

Used mainly for off site fabrications, i.e. chimney aprons
and gutters.

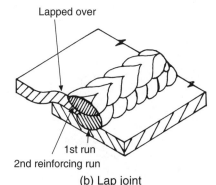

Lapped over

1st run
2nd reinforcing run

(b) Lap joint

Used for both fabrication and on-site work. Note that the
underside of the lead is not melted, enabling this method to
be used for repairs to lead roofs with little danger of fire risk
to the timber decking.

**Fig. 11.15** Lead-welded joints.

## Fixing methods

Reference was made earlier in this chapter to the
influence of wind effect and temperature change
on roof coverings. It is of primary importance
to consider carefully the fixing methods to be
employed because, while all fixings must hold
materials in position, some must also allow for
the movement caused by thermal expansion.

The materials used for fixings should be as
follows:

*Nails*  Large-headed copper having barbed shanks
not less than 25 mm long. 10 SWG (approx.
3 mm) diameter, and conforming to BS 1202
Part 2, Table 2.

*Screws*  Brass or stainless steel not less than
25 mm long or 10 SWG shank diameter and
conforming to BS 1210.

*Cleats*  Not less than 0.6 mm thick copper sheet of
$\frac{1}{4}$ hard temper and conforming to BS 2870; or
stainless steel, not less than 0.375 mm thick and
complying to the relevant BS. Where the colour
of the cleats must match the lead sheet, copper
cleats may be tinned (coated with solder) and
stainless steel is obtainable with a lead-covered
surface.

*Solder*  (for dots) Either Grade D or Grade J
conforming to BS 219. (The only difference
between these grades is the antimony content.)

The usual practice for fixing the undercloaks of
drips or wood-cored rolls is to position nails at
50 mm centres. Using nails at 75 mm centres is
satisfactory for fixing the upper edges of sheets on
pitched roofs that are to be covered by a lap or
cover piece which will prevent water entering the
nail holes. Nailing is only suitable if the decking,
i.e. the undersurface, is timber. Nailing can be
used with a concrete substrate if a softwood batten
impregnated with preservative has been inset, but
where this is not the case screws using suitable
plugs and brass washers are utilised.

The edges of sheets and joint positions are
usually retained in place by the use of 50 mm wide
clips of either copper, stainless steel or lead sheet.
The applications of clips are too numerous to detail
individually but some typical examples are shown
in Fig. 11.16. Clips are usually positioned at a
maximum of 500 mm intervals.

Lead wedges positioned at about 500 mm as in
Fig. 11.17 are the method of securing the edges of
lead flashings into brickwork joints.

A method of fixing lead to stone pediments which
may be seen in older buildings is the *lead dot*. This
required a dovetail hole to be cut in the stone which
was then filled with molten lead using a dot mould
(see Fig. 11.18). Figure 11.19 shows a modern
method of fixing using screws and lead-welding
techniques.

*Intermediate fixings*  Fixings which do not occur
at the edges of sheets are known as intermediate
fixings and can be either *visible* or *secret*. The
secret method requires the welding of a sheet lead
tack to the back of a panel of sheet lead. This tack
is then passed through a slot in the timber backing

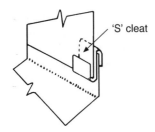

(a) Method of securing the edge of 'cap' flashing

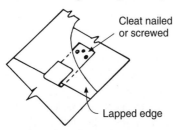

(b) Securing the free edge of a lap joint

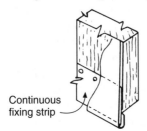

(c) Fixing the bottom edge of sheet weathering by a continuous strip

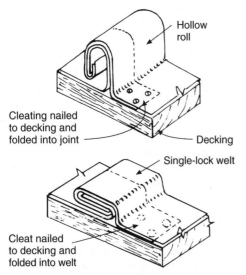

(d) Holding down cleats nailed to the decking of a roof and folded into the joint as it is made

**Fig. 11.16**  Use of clips and cleats for lead sheetwork.

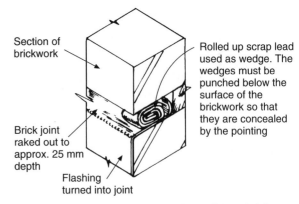

**Fig. 11.17**  Securing weatherings into brickwork joints.

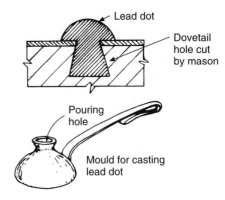

**Fig. 11.18**  Lead dot method of fixing lead sheet to masonry.

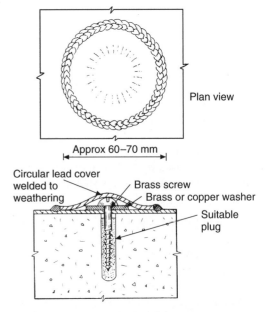

**Fig. 11.19**  Screwed method of fixing to masonry.

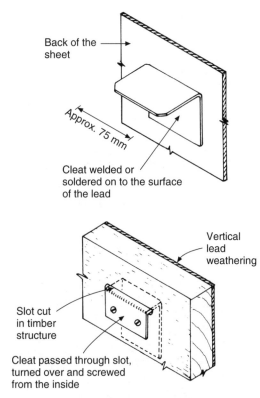

**Fig. 11.20** Secret fixings for sheet lead in vertical positions such as dormer cheeks.

and secured as in Fig. 11.20 using round-headed brass screws and washers.

The traditional type of visible intermediate fixings are illustrated in Figs 11.21(a) and (b) and require the use of either a lead-welded cap or a soldered dot over the brass wood screw to prevent water entering through the screw hole. The solder dot method is not recommended owing to cracking around the edge of the solder.

## Forming roofing details

Roofwork details are the positions where it is necessary to form the material into a certain shape such as internal or external corners, roll ends and gutters.

### Lead sheet working

Probably the best-known characteristic of lead, apart from its weight, is its malleability, a property which enables it to be worked or 'bossed' without the need

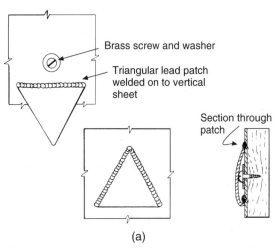

(a)

Triangular patch turned upwards through 90° concealing the screw and washer prior to completing lead weld on sides of patch.

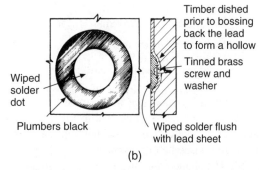

(b)

**Fig. 11.21** Alternative methods of securing lead sheet in a vertical position.

for annealing. The term 'working' lead is often used by plumbers, this merely being another way of describing the process of bossing. Bossing is the traditional method for forming the various details in sheet leadwork. An alternative procedure widely used by the modern plumber is to cut the sheet, remove or insert areas of lead and join them using the oxy-acetylene gas-welding process.

*Bossing* When bossing is to be used it is essential to decide how the detail is to be formed, i.e. how the shaping operation is to be undertaken. A study of Fig. 11.22 will provide an indication of these considerations. Figure 11.22(a) shows the setting-out of an external corner and it can be seen by comparing this with the finished detail in Fig. 11.22(b)

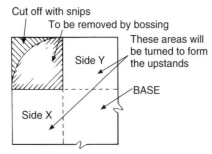

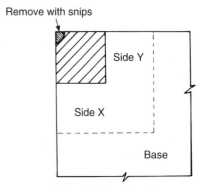

(a) Corner marked out before bossing commences

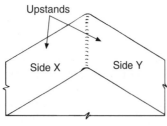

(b) Completed external corner

**Fig. 11.22** External corners in lead sheet.

(a) Corner marked out before bossing commences

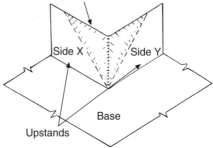

(b) Completed internal corner

**Fig. 11.23** Internal corners in lead sheet.

that the shaded area of lead is superfluous to requirements. It is this area which must be removed.

An internal corner is illustrated in Fig. 11.23 and a study of the sketch will show that lead has to be gained from elsewhere to allow the corner to be formed. The shaded area in Fig. 11.23(a) must be at least equal to that shown in Fig. 11.23(b), but in practice more than the minimum requirement is provided to facilitate ease of working by increasing the allowance for the upstand where a corner of this type occurs.

These two operations, i.e. the *removal of surplus lead* or *the moving of lead from one place to another*, form the basis of all sheet lead bossing. Examples of the first operation include roll ends, undercloaks to drips and external corners. The second operation is required for such details as aprons, back gutters and internal corners.

Once the working method has been decided the marking out can proceed. The use of a pencil is not recommended for lead. Sheet material of any kind should never be marked out using anything that will scratch or score, as this could cause failure such as cracking along the marked line. The use of a chalk line is shown in Fig. 11.24. Chalk is rubbed into the

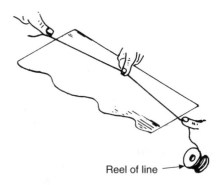

**Fig. 11.24** Setting out with a chalk line.

line which is then stretched tightly across the work to coincide with the measurement marks. The line is picked up in the centre and then released allowing it to snap back on to the sheet where it leaves a sharp, clearly defined line without damage to the surface of the lead. An alternative and convenient method of marking out is the use of a spirit-based marker and straight edge. The next stage is to remove any excess metal with tinsnips so that the bossing process is simplified. Examples are shown in Figs 11.22 and 11.23.

*Bossing an external corner* One of the more common operations involving lead bossing is the formation of an external corner. It should be marked out in a similar way to that shown in Fig. 11.22(a), any excess metal then being trimmed off. The upstands are next turned up 90° using a timber former; a piece of softwood 600 mm long, 100 mm wide and 50 mm thick is ideal for this purpose. The angles are then 'set in', using a setting-in stick, a tool similar to a dresser but having a wedge-shaped face like a chase wedge. This edge is placed in contact with the angle formed between the upstand and the base, and a sharp blow is applied to the back of the tool with a mallet. This setting-in process fixes the position of the upstands preventing any tendency for them to wander while bossing is in progress. To ensure a good square base to the corner and fix its position, a slight groove or 'belly' is made on the base of the sheet. Figure 11.25(a) shows the corner at this stage. Surplus lead is then removed by bossing as illustrated in Fig. 11.25(b), the first blows being diverted inwards to stiffen the base of the corner. The direction in which the blows are struck is most important as that is the direction in which the lead is being driven.

The actual bossing procedure shown is not always used. The internal support is usually provided by a mallet or a dummy but the external shaping is carried out by many plumbers using a bossing stick. When the excess material has been bossed out and the corner formed to the required angle the detail is trimmed to the required height. Bear in mind the objective with all bossing operations is to maintain a uniform thickness in the whole of the area of working.

*Bossing an internal corner* An example of a detail where lead has to be gained, i.e. an internal corner, can be seen in Fig. 11.26 where one side of a chimney front apron is shown. The area of lead required to be worked into the corner is indicated by shading. First the upstand is folded to the angle

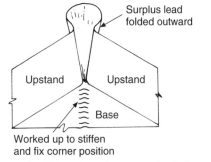

(a) Preparing the external corner prior to bossing

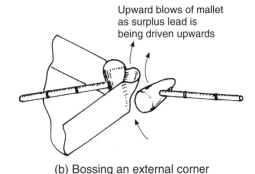

(b) Bossing an external corner

**Fig. 11.25** External corners in lead sheet.

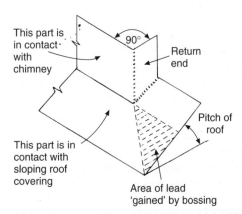

**Fig. 11.26** Area of lead to be 'gained' for a chimney apron.

of the roof pitch and the return ends turned to the angle of the chimney breast, which is usually 90° although other shapes including circular chimneys may be encountered. This will cause a hump to occur that is then worked into the desired position using a mallet, bending or bossing stick. The procedure is illustrated in Fig. 11.27(a).

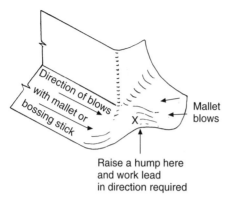

### (a) Working round one end of a chimney apron

A chimney apron is an internal corner there being a shortage of lead at the point marked 'X'. Lead is driven from both directions, as shown, to provide material needed at 'X'.

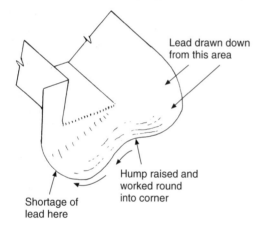

### (b) Bossing one side of a chimney back gutter

**Fig. 11.27** Working lead from one area to another.

The illustration shown in Fig. 11.27(b) is another example showing how lead may be worked from one area to another, in this case to fabricate one side of a chimney back gutter.

*Problems encountered bossing lead* Once the skills of gaining or removing lead have been mastered they can be applied to the shaping of any roofing detail. Several important factors must, however, be borne in mind when bossing sheet lead.

When surplus lead is being removed there is a tendency for it to thicken and crease. These creases must not be allowed to form or they will result in cracking. When forming internal corners there is always a danger of stretching the lead into position,

i.e. creating a greater area by thinning. This may reduce the thickness to an unacceptable level for the application concerned and could cause splitting to occur. From these factors it can be seen that the objective in bossing is to maintain the original thickness over the whole worked area and this requires considerable practice. The plumbing student can assess his or her own ability by cutting sections from completed exercises and examining them to check for even thickness. It is better to strike many blows lightly when bossing, rather than fewer and heavier blows.

Even if the forming of lead by means of bossing is undertaken by a competent craftworker, it is still a slow and laborious process, although it has the advantage of requiring few tools and so is often very convenient. The use of the 'cut and weld' technique is much quicker in some instances and if the welding is carried out competently it ensures that the lead remains of a uniform thickness throughout. This procedure requires a knowledge of simple geometry as it consists basically of cutting out or inserting pieces of lead to achieve the desired shape. The individual pieces are then welded or fused together using one of the seams illustrated in Fig. 11.15. Examples of the 'cut and weld' procedure are given in Fig. 11.28

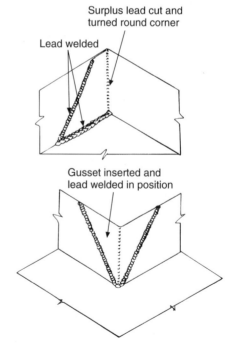

**Fig. 11.28** Lead-welded external and internal corners.

where internal and external corners are shown formed by cutting and turning or inserting a gusset.

A comprehensive range of details formed by the methods described are given in the publication *Lead Sheet in Building* which is available from the Lead Development Association.

## Damp-proof courses

Sheet weatherings are employed to prevent water entering a structure via the roof. Water can also enter a building from below ground level by capillary action through the brick, stone or concrete used in construction, and for this reason damp-proof courses are used.

Most of the clayware and concrete products used for structural work are porous and as such provide the ideal medium through which moisture can pass by capillary action. Unless steps are taken to prevent the ingress of water in this way, dampness will penetrate a building causing damage to the structure and creating a cold, damp atmosphere which could also endanger the health of the occupants.

The method used to overcome this problem is the installation of a damp-proof course (DPC). Any material used for such work must be impervious to moisture and must be built into the building structure to provide a waterproof barrier between various components. For many years slate was used to prevent the moisture in the earth rising above the ground floor level, but owing to its high cost and scarcity asphalt, plastic strip or engineering bricks are used in most modern buildings. Although engineering bricks are made of clay, the methods of manufacture result in a high-density brick which, for all practical purposes, is water resistant.

All these materials are fixed by a bricklayer as they are actually built into joints in the building structure. The main disadvantage of the materials named is that any settlement of the building will result in a crack in the DPC which may allow the passage of moisture at that particular point.

Lead or copper sheet, although expensive, is sometimes used as an alternative to the materials already mentioned, as their strength and other qualities make them far more likely to resist damage by settlement faults and they maintain a waterproof

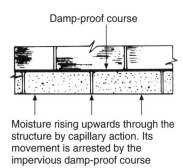

Damp-proof course

Moisture rising upwards through the structure by capillary action. Its movement is arrested by the impervious damp-proof course

**Fig. 11.29**  Damp-proof course.

barrier under all circumstances. If sheet lead is used, it should be coated with a bituminous solution to resist the alkali present in the cement mortar in which it is laid. Code No. 4 or 5 lead is considered to be satisfactory for DPCs or in the case of copper, 0.45–0.6 mm thickness. The method of jointing for each material is a simple lap joint with not less than 100 mm overlap.

One of the most common causes of dampness is the moisture in the earth seeping into the footings of the building and rising by capillarity through the lower area of the walls. This is commonly called 'rising damp'. To prevent this dampness rising to the ground floor rooms of the building, a DPC is built into the wall not less than 150 mm from the ground level (see Fig. 11.29). Older buildings were often built without DPCs and, in such cases, the walls at ground floor level are damp, sometimes to a height of 1 metre or more.

Building components exposed to the weather at higher levels of a building also absorb dampness and can bypass the flashing as shown in Fig. 11.30. A typical example of this occurs in parapet walls, and a sheet metal DPC fitted as shown in Fig. 11.31 prevents the downward movement of dampness. It is also arranged to act as the cover flashings over any weathering such as a gutter or flat roof on the inside of a parapet.

DPCs to chimney stacks that penetrate pitched roofs are unusual except in very exposed conditions. Any moisture absorbed by the fabric of a chimney normally dries out in the area of the roof space before it can penetrate to a habitable room. In the case of a chimney penetrating a flat roof the situation is quite different, there being no roof

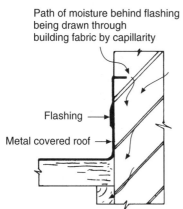

Path of moisture behind flashing being drawn through building fabric by capillarity

Flashing

Metal covered roof

**Fig. 11.30** Water can be absorbed by masonry in very exposed positions and gain access into the building.

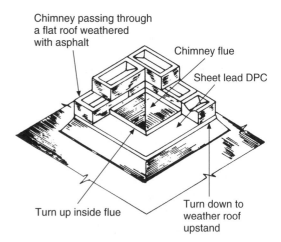

Chimney passing through a flat roof weathered with asphalt

Chimney flue

Sheet lead DPC

Turn up inside flue

Turn down to weather roof upstand

**Fig. 11.32** Damp-proof courses in chimneys.

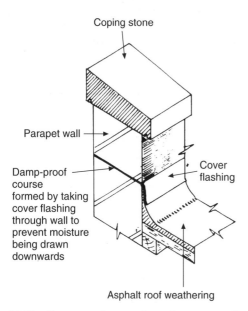

Coping stone

Parapet wall

Damp-proof course formed by taking cover flashing through wall to prevent moisture being drawn downwards

Cover flashing

Asphalt roof weathering

**Fig. 11.31** Damp-proof course through parapet wall.

space in which the moisture can be dissipated. In such circumstances a sheet lead DPC built into the chimney brickwork is absolutely necessary, and Fig. 11.32 shows how this should be fitted. The turned-down portion over the upstand of the flat roof can be bossed or lead welded.

There are many other different examples of damp-proofing the building structure with sheet metal, such as the weathering of openings and lintels in cavity walls. Only the more common

examples have been included in this book, but further details can be found in the further reading suggested at the conclusion of this chapter.

### Simple external weatherings

Probably one of the areas in which sheet metal weatherings will never be surpassed for their properties of long life, flexibility and durability is in the weathering of chimneys, small canopies and vent pipes. Modern developments and techniques have introduced prefabricated (non-metallic) units for weathering vent pipes where they pass through a roof, but these are not always suitable for all pipe materials, and while some types embody a degree of flexibility to accommodate a wide range of roof pitches, they are not always adaptable for all roof coverings or for the weathering of all sizes of pipe diameters.

*Weathering chimney stacks*
The weathering of a typical chimney where it passes through a pitched roof is shown in Fig. 11.33. Before the weatherings are fitted, the first step is to mark out and rake all the joints to receive the turn in from the flashings. They should be raked to a depth of 25 mm to ensure a good fixing and to enable the wedges securing the flashings to be driven home below the surface of the finished brickwork. If ordinary plain tiles or slates are used, the point at which they abut the chimney is weathered by

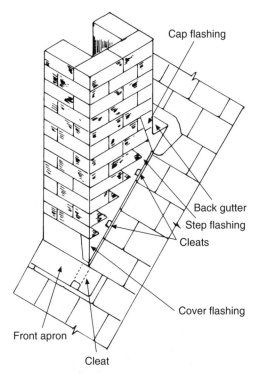

**Fig. 11.33**  Chimney stack flashing with lead sheet.

soakers (unless step and cover flashings are used, soakers being unnecessary with this type of flashing).

### Soakers

The length of a soaker will vary according to the type of roof covering with which it is used. Its width should not be less than 175 mm which allows an adequate amount of material to lie under the tile or slate (100 mm), and for the upstand against the brickwork (75 mm). The length of a soaker is normally found by calculation based on the gauge and lap of the roof covering used. Figure 11.34 illustrates these terms.

The gauge is the actual amount of the tile or slate that can be seen when the roof is finished, while the

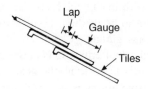

**Fig. 11.34**  Illustrating the terms 'gauge' and 'lap'.

lap relates to the cover given to a tile by the overlap of the tile next but one above it. Using plain tiles on a roof pitched at approximately 40°, the gauge is usually 100 mm, with a lap of 50 mm. These measurements vary with different roof pitches; for example, a lower pitch requires a longer lap. The following formula is used to calculate the length of soakers:

Length of soaker

$$= \frac{\text{Tile length} - \text{Lap}}{2} + \text{Lap} + 25 \text{ mm for fixing.}$$

*Example*   A slate, 500 mm in length, is laid with a gauge of 200 mm and a lap of 100 mm. Calculate the length of the soakers required.

Length of soaker

$$= \frac{\text{Slate length} - \text{Lap}}{2} + \text{Lap} + 25 \text{ mm}$$

$$= \frac{500 - 100}{2} + 100 + 25 \text{ mm}$$

$$= \frac{400}{2} + 100 + 25 \text{ mm}$$

$$= 200 + 100 + 25 \text{ mm}$$

$$= 325 \text{ mm}$$

A soaker 325 mm in length will be required.

To determine the number of soakers needed for a given job, it is necessary to know:

(a)  the length of the roof to be weathered
(b)  the gauge to which the roof covering is laid

the former simply being divided by the latter thus:

$$\frac{\text{Length of roof}}{\text{Gauge}}$$

*Example*   Assuming the gauge to be 200 mm and the length of the abutment 2.4 m, calculate the number of soakers required.

$$\text{No. of soakers} = \frac{\text{Length of roof}}{\text{Gauge}}$$

$$= \frac{2.4}{0.2}$$

$$= 12$$

Twelve soakers will be required to weather this length.

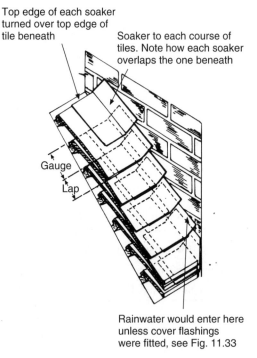

Top edge of each soaker turned over top edge of tile beneath

Soaker to each course of tiles. Note how each soaker overlaps the one beneath

Gauge

Lap

Rainwater would enter here unless cover flashings were fitted, see Fig. 11.33

**Fig. 11.35** Fixing soakers in position.

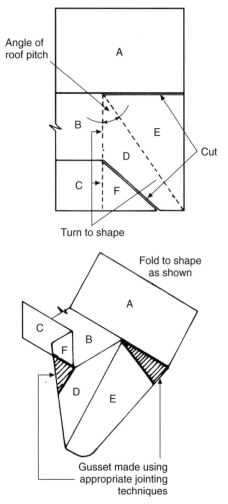

Angle of roof pitch

A

B

E

D

C

F

Cut

Turn to shape

Fold to shape as shown

A

C

B

F

D

E

Gusset made using appropriate jointing techniques

**Fig. 11.36** Setting out back gutters for inserting gussets.

The soakers are fixed in position by the tiler as the tiles or slates are laid and nailed on to or turned over a batten. Figure 11.35 shows a detail of the soakers after the tiler has completed the work. It will be seen that although the edge of each tile has been weathered, the rainwater running down the brickwork could enter the building down the back of the soaker. It will be shown later in this chapter how step flashings are used to cover the top edge of soakers and so complete the weathering of the side of the abutment.

### Chimney back gutters

The normal way of weathering a roof is to start at the bottom and work upwards to ensure the various pieces of weathering overlap in the correct manner. It is often more convenient, however, to fix the completed back gutter of a chimney first so it can be fitted in with tiles or slates as they are laid.

Figure 11.36 shows how the same weathering is fabricated when lead welding techniques are used, the material being set out as shown, cut and bent to

form the required shape. The gussets should then be carefully fitted and welded using the butt welding technique.

It is not good practice to turn the upstand of a back gutter directly into a brickwork joint, as this restricts movement brought about by thermal expansion. The correct procedure is to use a cap flashing as shown in Fig. 11.37 which allows a watertight joint to be made, yet still permits movement of the gutter.

Assuming the roof covering is complete and the back gutter is in position, the next step is to fabricate the front apron as shown in Fig. 11.38 if lead welding is employed.

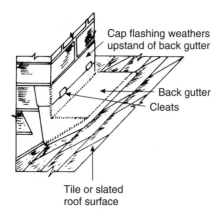

**Fig. 11.37**  Back gutter in position.

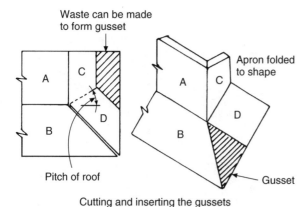

Cutting and inserting the gussets

**Fig. 11.38**  Fabricating the front apron.

Prior to positioning the apron, the cleats (see Fig. 11.33), which secure the bottom edge and prevent any tendency to creep, are fixed. They are made of two strips of suitable material approximately 50 mm wide, either wedged into the stack or turned over and nailed to a convenient tile batten. The top edge of the apron is then turned into a brickwork joint approximately 75–100 mm above the roof line and secured by lead wedges. If soakers are used they must be fitted over the top of the apron.

One very important point that cannot be stressed enough is the need for accuracy when chimney aprons and back gutters are set out. This is particularly true with those materials that employ the use of gussets or welts as a means of

fabrication. An error in setting out or a carelessly taken measurement may result in the whole job being scrapped. A good practical hint is to set out the work on stiff paper and make a pattern; this can easily be checked against the actual job, and if found to be satisfactory its shape can easily be transferred by marking round it on to the sheet metal. If the job is to be repeated, the same pattern can be used for the whole batch.

*Flashings*

*Step flashing*  The term 'step flashing' is given to the flashing which weathers a pitched roof to brickwork, its name being derived from its step-like appearance. There are two main types of flashing, the first being that shown on the chimney stack in Fig. 11.33, known simply as step or 'skeleton' flashing and always used in conjunction with soakers. A more detailed illustration is shown in Fig. 11.39(a). Where the use of soakers is not a practical possibility, as in the case of interlocking tiles and corrugated sheeting, the second type, step and cover flashings, are used (see Fig. 11.39(b)). It will be seen that the upstand on which the steps are marked out is the same in both cases, the difference being the free edge of step and cover flashing which is extended outward over the roof covering. Sheet lead is generally accepted as the best material for this type of flashing as it can easily be dressed down on to the somewhat awkward shape of most interlocking tiles.

The disadvantage of a cover flashing is its tendency to lift in high winds, but this can be overcome when a sufficient number of cleats are used, these being firmly fixed at joints in the roof covering. Sufficient lead should also be left on the chimney apron to allow it to be turned over to form a cleat and hold down the lower end of the cover flashing as shown in Fig. 11.40.

The type of flashing so far dealt with is for use with materials such as brick, where the horizontal joints are equally spaced. In areas where buildings are constructed of stone, slightly differing techniques of flashing are used as illustrated in Figs 11.41(a) and (b). Where the stones are laid to regular horizontal courses, called coursed stonework, it is possible to use step flashing,

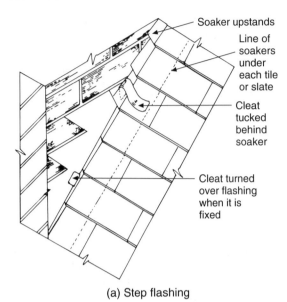

(a) Step flashing

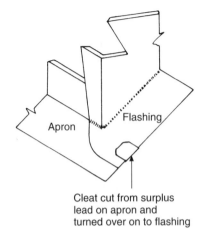

**Fig. 11.40** Apron trimmed after bossing to secure free edge of step and cover flashing.

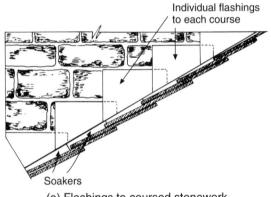

(a) Flashings to coursed stonework

Note that the horizontal courses are regular but due to the differing thicknesses of stone the widths of each course varies.

(b) Step and cover flashing

Soakers not used.

**Fig. 11.39** Flashings.

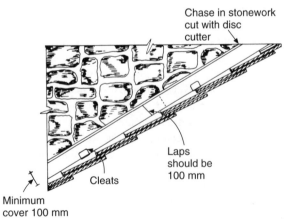

(b) Flashing to random rubble walls

**Fig. 11.41** Weathering to stone abutments.

but owing to the uneven appearance of the steps, the method that is commonly used is shown in Fig. 11.41(a), having separate pieces of lead for each step. Where random stonework is used the best method is to cut a groove in the stone into which the lead is turned as shown in Fig. 11.41(b) – a disc cutter is essential for this work.

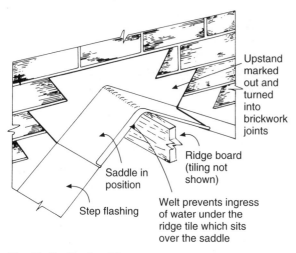

Upstand marked out and turned into brickwork joints

Ridge board (tiling not shown)

Saddle in position

Welt prevents ingress of water under the ridge tile which sits over the saddle

Step flashing

**Fig. 11.42** Lead saddle.

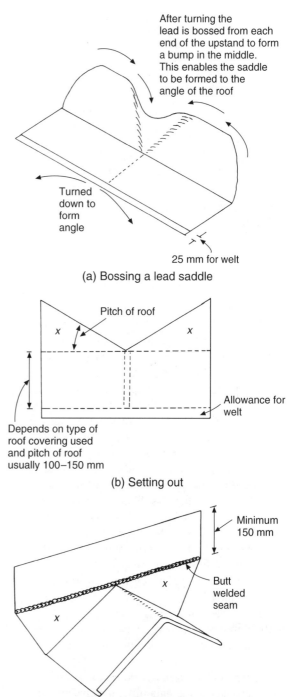

After turning the lead is bossed from each end of the upstand to form a bump in the middle. This enables the saddle to be formed to the angle of the roof

Turned down to form angle

25 mm for welt

(a) Bossing a lead saddle

Pitch of roof

x          x

Allowance for welt

Depends on type of roof covering used and pitch of roof usually 100–150 mm

(b) Setting out

Minimum 150 mm

x

Butt welded seam

x

(c) The saddle turned with extension piece welded on

Steps marked out from job.

**Fig. 11.43** Fabricating a lead saddle using lead welding techniques.

Where a chimney passes through the ridge of a building, or when a double-pitched roof abuts a wall, as shown in Fig. 11.42, a saddle piece will be required to weather the tops of the side flashings. These can be bossed or lead welded as shown in Figs 11.43(a), (b) and (c). Figure 11.43(a) shows the technique of bossing this detail, the lead being worked from the end towards the middle to form a bulge. When the ends are pushed down to form the angle, the bulge will flatten. This operation will have to be repeated several times before the correct angle is achieved. The absolute minimum size of lead required for this job is 350 mm × 450 mm, but measurements taken from the job will be more accurate. Allowances must be made for final trimming after the bossing operation is completed. Figure 11.43(b) illustrates a method of forming a saddle using the lead welding technique. Whichever method is employed when the saddle is formed, the steps are marked out to correspond with the brickwork joints in the same way as marking out step flashings. When the saddle is in position, with the steps firmly wedged, the tiler can bed the ridge tile adjacent to the abutment over the saddle.

*Marking out step flashings* Assuming the sides of a brick chimney are to be weathered with step flashing, a strip of the flashing material 160 mm wide by the length of the roof slope alongside the

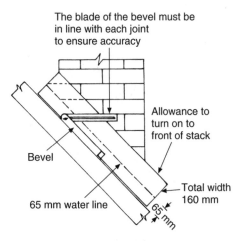

**Fig. 11.44** Setting out the step flashing from brickwork.

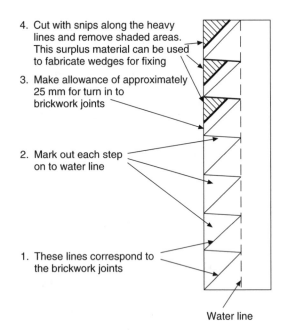

**Fig. 11.45** Setting out step flashings.

stack is cut. An allowance of approximately 75 mm should be added to this length to permit the bottom of the flashing to be turned around on to the front of the apron. A water line of 65 mm is marked off on the material before placing it in position on the side of the chimney. Using a bevel or folding rule, the brickwork joints are then marked on the flashing as shown in Fig. 11.44. This marking out must be accurate and the marks made parallel with and about 12 mm above the bottom of the brick joint. The stage-by-stage illustration in Fig. 11.45 shows how the steps are set out from the marks taken from the brickwork joints. Snips are used to make the cuts shown by the heavy lines (not including the water line) so that the shaded areas can be removed. These cut-outs should be saved as they will be useful for making wedges for subsequent fixing. The turn on each step which forms the fixing on the brickwork is made with the step turner described in Chapter 2. On low-pitched roofs it may be necessary to use two wedges to fix each step securely as the turn-in on each step is longer.

Step and cover flashings are set out using the same techniques, the only difference being that wider strips of material are required to allow for the cover on the roof. There is no hard and fast rule for the width of the cover, but it is usual to cut the flashing strips from material of 310 mm width, allowing approximately 160 mm for the step and

the upstand; this allows 150 mm cover on the roof which is adequate for most applications.

*Pipe flashings*   It is often necessary to weather ventilating or flue pipes where they pass through a roof. The traditional name for such a flashing is a 'lead slate' because for many years only lead was used. Since the use of other materials for this purpose is increasing, the term 'pipe flashing' more accurately describes this type of weathering. Figure 11.46(a) shows a typical flashing made of a lead sheet which has been fabricated by lead welding a purpose-made pipe to a base.

It is possible to obtain purpose-made pipe flashings. These incorporate a fold which permits the angle at which the pipe section fits on the base to be varied to suit a variety of roof pitches.

Pipe flashings made with sheet lead are prepared in a similar way where it is intended to use soldering or lead welding techniques as shown in Fig. 11.46(b).

When the flashing has been fitted over the pipe, and the base is weathered into the roof, it is important to ensure that water does not penetrate

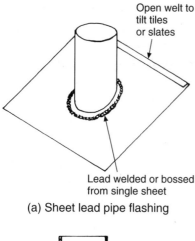

(a) Sheet lead pipe flashing

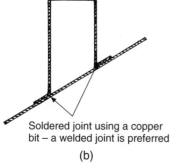

Soldered joint using a copper
bit – a welded joint is preferred

(b)

**Fig. 11.46**   Fabrication of pipe flashings with sheet lead.

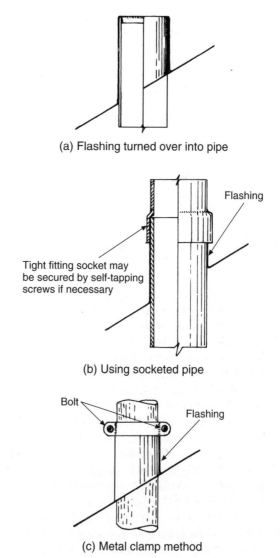

(a) Flashing turned over into pipe

(b) Using socketed pipe

(c) Metal clamp method

**Fig. 11.47**   Weathering flashings to pipe.

into the building between the pipe and the top edge of the upstand. If the pipe terminates only a short distance above the roof surface, the flashing can be dressed over and into the top of the pipe as seen in Fig. 11.47(a). An alternative is to weather the edge of the flashing by a socketed pipe turned upside down as in Fig. 11.47(b), but the difficulty with this method lies in making an effective joint to the pipe in this position. However, it is a convenient method to use when weathering gas flues where a terminal can be used instead of a length of pipe. In the event of these two methods being unsuitable, as for instance in the case of a flag pole or long length of pipe fitted above the surface of a roof, a clamp made of non-ferrous metal can be used (see Fig. 11.47(c)). A suitable mastic should be applied around the top edge of the flashing before the clamp is tightened to ensure that the joint is weathertight.

## Weathering small canopies and flat roofs

It is impossible within the confines of this book to deal with all the possible variations of weathering details using all the available materials with which the plumber is likely to come into contact. The object within the following pages is to show typical examples of the basic principles of weathering small canopies and flat roofs using lead sheet. Owing to its long working life lead sheet weathering is a far more suitable and permanent alternative to

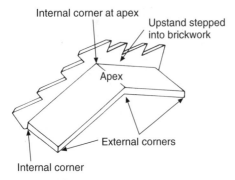

**Fig. 11.48** Small lead canopy.

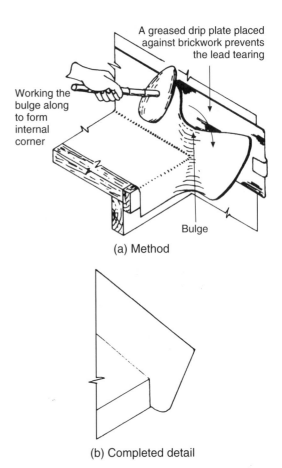

(a) Method

(b) Completed detail

**Fig. 11.49** Working the internal corners on a lead canopy.

bituminous felt for covering small roofs, especially in the case of 'one-off' jobs.

*Weathering a porch canopy*
A small double-pitched canopy is shown in Fig. 11.48, and as its superficial area is not likely to be too large the method used shows it made in one piece. To make this weathering, an upstand of approximately 200 mm is first turned on edge to abut the wall. A bulge is formed at the apex in the upstand in a similar way as that when forming a saddle (see Fig. 11.43(a)). Before fitting the canopy, some cleats or a continuous fixing strip should be nailed round the front and side edges of the canopy. This will allow the front and side edges to be secured and prevent the lead being lifted in high winds; this type of fixing is shown in detail in Fig. 11.16(c). When the canopy is fitted and the two sides dressed down, the lead forming the bulge will flatten and can be dressed back to the abutment.

The next step is to work down the bottom edge adjacent to the abutment. This is really an internal corner and a gain in material is necessary. The method used in shown in Fig. 11.49(a) and is not unlike that used to form the overcloak to a drip. A bulge is made in the upstand which is bossed downward in the direction of the arrow, taking care not to crease the lead in the corner. When the bossing operation and trimming has been completed, the lead should appear as seen in Fig. 11.49(b).

To turn down the front edge of the canopy, the three external corners shown have to be bossed, and should be worked down using the methods described previously. If these corners are worked

down in position the lead should be fixed in place by suitable clamps and wooden blocks to prevent it easing forward during this last operation. Should this be allowed to happen a gap will occur between the lead upstand and the brickwork which will be difficult if not impossible to correct.

The next operation is to mark out the flashing to the upstand for turning into the brickwork in the same way as step flashing before it is fitted to the decking. The canopy is finally secured by wedging the steps into the brickwork and turning the front and side edges under the continuous fixing strip. If larger canopies of this type are weathered it may be necessary to use an expansion joint at the apex. This can take the form of a welt or a wood-cored roll. It will also be necessary to use separate flashings, as only on very small canopies is it permissible to turn the upstand directly into the abutment.

## Valley gutters

Tapering valley gutters are those formed between the junction of the eaves of two roofs, and are of the open type being wide enough to allow a person to walk along them if necessary to effect repairs.

Smaller gutters, simply called valleys, occur when a gable intersects with a pitched roof. Figure 11.50 shows in diagram form both types and their relative positions on a roof.

Small valleys can be weathered in a number of ways. For example, when plain tiles are used, it is possible to dispense with metal altogether as specially shaped valley tiles can be used. These are illustrated in Fig. 11.51. The only weathering required in this case will be to cap the intersection of the ridge with the sloping roof in a similar way

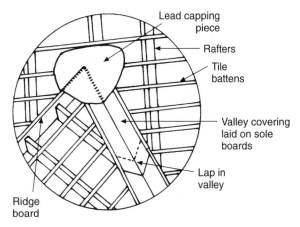

**Fig. 11.52** Detail of weathering at the top of valley gutter. Finish with a capping piece.

to that shown in Fig. 11.52 where a metal valley gutter is illustrated. Special tiles are also made to form valleys for roofs weathered with interlocking tiles. However, sheet metal is always used in conjunction with slates and often on tiled roofs where, for one reason or another, it is more suitable than valley tiles.

Figure 11.53(a) illustrates a section through a metal-covered open valley with a distance of approximately 50 mm left between the edges of the tiles, while Fig. 11.53(b) shows a secret valley gutter. While the external appearance of the secret gutter is more attractive, they have a tendency to collect debris washed off the roof and often become blocked, causing water to enter the building.

The method of laying valleys is the same in both cases, the covering material being formed or dressed into shape and laid on wooden sole boards, the top edge of each length being secured by nailing. There is no hard and fast rule relating to the length of each section, apart from a recommended maximum of 1.5 m. The turned edges at the sides serve to prevent water flooding over the sides of the gutter and at the same time provide a good method of intermediate fixing while also allowing for longitudinal movement. The edges are held down with cleats, shown in Fig. 11.54, which permit longitudinal expansion. The edges must *not* be nailed down as this will restrict the movement of each length.

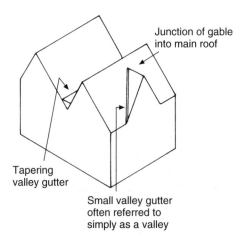

**Fig. 11.50** Types of valley gutters.

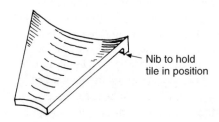

**Fig. 11.51** Valley tile. These tiles are shaped to form the valley, each tile being laid as the roof tiling proceeds.

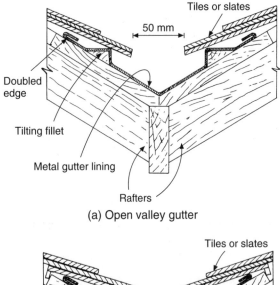

(a) Open valley gutter

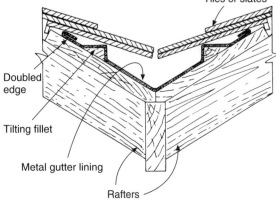

(b) Secret valley gutter

**Fig. 11.53**  Sections through valley gutters.

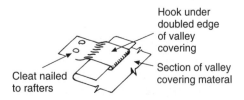

**Fig. 11.54**  Method of securing edges of valley covering material by cleats.

### Tilting fillets

Sometimes called 'springing', this term relates to fixing a batten under the slates or tiles at the eaves or where the courses are broken by a chimney or flat roof. This batten tilts each slate or tile upwards

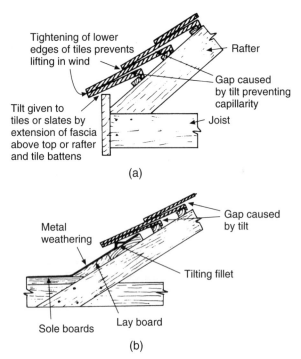

**Fig. 11.55**  The effects of tilting fillets on tiles or slates to prevent capillary attraction and to maintain a tightening of the lower edges to prevent lifting by wind.

(hence the name tilting fillet), so a gap occurs at each overlap preventing capillary attraction. Figure 11.55(a) shows how the top edge of the fascia board is fixed to provide a tilt to the tiles or slates, and Fig. 11.55(b) shows the section of a tilting fillet where the roof is broken by a flat roof or chimney.

### Roofing calculations

The plumber should know the procedures for measuring and ordering the materials for a specified roofing job. In order to do this effectively, some knowledge of the methods of measuring and calculating area is necessary. The basic unit of area in SI units is the square metre shown in Fig. 11.56 and is written as 1 m$^2$. When it is necessary to calculate the area of a surface, the length has to be multiplied by the breadth, and some examples of this follow.

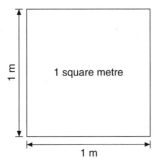

Fig. 11.56 The basic unit of area: 1 square metre (1 m$^2$).

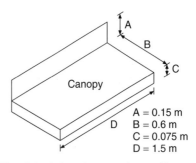

A = 0.15 m
B = 0.6 m
C = 0.075 m
D = 1.5 m

Fig. 11.58 Calculating the area of a small canopy.

*Example 1* Ignoring the welts and turn-downs, calculate the surface area of a small canopy measuring 2.0 m by 1.5 m:

Area = Length × Breadth
= 2.0 × 1.5
= 3.0

The answer is 3 square metres or 3 m$^2$.

The square metre is rather a large unit and in many cases a component may have an area of less than this as the next example shows.

*Example 2* Calculate the area of a piece of lead sheet measuring 1.0 m × 0.75 m:

Area = Length × Breadth
= 1.0 m × 0.75 m
= 0.75 m$^2$

The area will be 0.75 m$^2$ and its relationship to 1 square metre is shown by the shaded area in Fig. 11.57. (In this example the breadth

measurement is in fact 750 mm, but may be expressed as 0.750 m. It is necessary to express it in this way as the answer is required in square metres, and metres cannot be multiplied by millimetres.)

From the information given it should now be possible to calculate the total area of material required to cover the small canopy shown in Fig. 11.58. The total width of the canopy is found by adding measurements A, B and C, and multiplying this by the total length, found by adding D, C and C (C will occur twice here as there will be a turn-down on both sides):

Width = 0.15 + 0.6 + 0.075 = 0.825 m
Length = 1.5 + 0.075 + 0.075 = 1.65 m
Area = Length × Width
= 0.825 × 1.650
= 1.36125 m$^2$

The area required is 1.36125 m$^2$, but as usual an approximation is made to the first three places to the right of the decimal point to give 1.361 m$^2$.

It is sometimes necessary to calculate the area, or partial area, of a circular surface in connection with sheet weatherings. To calculate the area of sheet material required to weather the top of a semicircular bay window, the formula described in Chapter 5 is used, where the area of a full circle is found by using the formula $\pi r^2$. In this example only half the area of a full circle is required so the formula must be modified to

$$\frac{\pi r^2}{2}$$

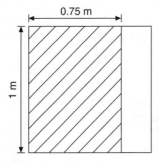

Fig. 11.57 Calculating areas of less than 1 m$^2$ (see text for details).

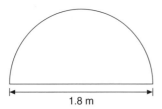

**Fig. 11.59**   Calculating the area of a semicircular canopy.

*Example 3*   Calculate the area of the semicircular canopy shown in Fig. 11.59 using the measurements given.

Using the modified formula

$$\frac{\pi r^2}{2}$$

the diameter of the canopy is 1.8 m, so the radius will be 0.9 m (radius = $\frac{1}{2}$ diameter). Therefore:

$$r^2 = 0.9 \times 0.9$$
$$= 0.81 \text{ m}^2$$
$$\pi \times r^2 = 3.142 \times 0.81$$
$$= 2.54502 \; (2.545 \text{ approx.}) \text{ m}^2$$
$$2.545 \div 2 = 1.2722 \text{ m}^2$$

The approximate answer is 1.272 m², which is the amount of material required to cover the flat part of the canopy, to which must be added allowances for rolls, welts or seams.

In the next example it is necessary to use the perimeter or circumference of a circle, which is the line that forms its outer edge (see Fig. 11.60).

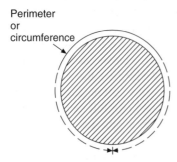

**Fig. 11.60**   Circumference of a circle.

Sometimes it is essential to determine the length of a circumference when the plumber is dealing with circular shapes. Typical examples in weathering occur when the circumference or perimeter of a pipe is to be measured in order to make a pipe flashing, or when the length of an apron surrounding a circular canopy needs to be found. The formula for the calculation of the circumference of a circle is

$$\text{Diameter} \times \pi$$

*Example 4*   Assume an apron having a width of 0.075 m is required for the semicircular canopy shown in Fig. 11.59 as only half the circumference is required. The formula can be modified thus:

$$\frac{\pi D}{2}$$

$$\pi \times D = 1.8 \times 3.142$$
$$= 5.6556 \; (5.656 \text{ approx.}) \text{ m}$$
$$5.656 \div 2 = 2.828 \text{ m}$$

The length of the apron required is therefore 2.828 m. To obtain the area of sheet needed for the apron, the length must be multiplied by the width of the apron. Assuming this to be 0.075 m:

$$2.828 \times 0.075 = 0.2121 \text{ m}^2$$

the total area will be approximately 0.2121 m².

The plumber is also called upon to weather areas that are triangular in shape, typical examples being the weathering of dormer window cheeks and small canopies over bay windows. Triangular shapes fall into three categories as shown in Fig. 11.61. A point to note is that the angles of any triangle always add up to 180°, the same number as those contained in a semicircle. The same method of calculating the area of triangular figures is adopted whatever its type. The formula may be stated as:

$$\text{Length of base of triangle} \times \tfrac{1}{2} \text{ Vertical height}$$

$$\text{Area of triangle} = \frac{\text{Height}}{2} \times \text{Base}$$

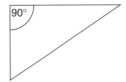

(a) Right-angled triangle

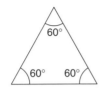

(b) Equilateral triangle

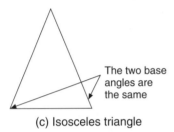

The two base angles are the same

(c) Isosceles triangle

**Fig. 11.61**   Types of triangle.

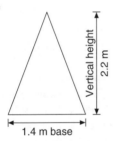

**Fig. 11.62**   Calculating triangular areas.

*Example 5*   Calculate the area of the triangle shown in Fig. 11.62:

Length of base = 1.4 m
Vertical height = 2.2 m
(note that this is *not* the sloping height)

$$\text{Area of triangle} = \frac{\text{Height}}{2} \times \text{Base}$$

$$= \frac{2.2}{2} \times 1.4$$

$$= 1.4 \times 1.1$$

$$= 1.54 \text{ m}^2$$

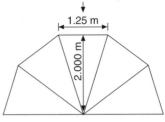

Note that this is the base of a triangle and there are five similar triangles

1.25 m

2.000 m

**Fig. 11.63**   Calculating the areas of a five-sided canopy.

Another practical application of measurement which integrates calculations involving triangular figures may be seen in Fig. 11.63 which represents the plan of a small, five-sided canopy over a bay window. The lines show how it can be divided up into five equal triangles, and by calculating the area of one triangle and multiplying by 5, the total area can be found thus:

$$\text{Base} \times \frac{\text{Vertical height}}{2} \times 5$$

Vertical height:
    2.0 ÷ 2 = 1.0
Area of one triangle:
    1.25 × 1.0 = 1.25 m$^2$
Area of the whole canopy:
    1.25 × 5 = 6.25 m$^2$

Total area of canopy (less rolls or welts)

$$= 6.25 \text{ m}^2$$

*Percentages*
Percentages are used in calculating prices and costs in plumbing and are a convenient way of expressing the relationship of one quantity or price with another; the percentage symbol is %. If the original cost of an article is £100 and it increases by £50, it can be said that the increase is one-half of its cost or 50%. To give another example using the same article costing £100: if it is subject to 15% value-added tax, the actual cost will be £115. The foregoing is very simple and can be seen at a glance but it becomes a little more difficult if the original quantity is not 100. Assume a large flat roof is to be weathered in metal sheet and the measured area is

35 m². The plumber would be very foolish if he or she submitted a price on this quantity because a reasonable allowance must be made for waste; let us say in this case 7%. To relate this to the original area, a simple calculation must be made and this is done by dividing 35 into 100 equal parts which will indicate 1%. Thus:

$$\frac{35}{100} = 0.35$$

$$0.350 \text{ m} = 1\% \text{ of } 35 \text{ m}^2$$

To find 7% of 35 m²

$$0.350 \times 7 = 2.45$$

The sum 2.45 m is 7% of 35 m and this is the quantity of sheet that should be allowed in this case for waste.

*Example*   The cost of a quantity of lead sheet is £78 + 17½% VAT. What will be the cost to the plumber?

$$\frac{78}{100} \times \frac{17.5}{1} = 13.65$$

$$£78 + £13.65 = £91.65$$

The plumber will pay £91.65 for the lead. It should be noted that 100 is a very easy number with which to divide: all that is necessary is to insert a decimal point and by moving it two places to the left we have effectively divided by 100. Taking the number 59% or 1/100 of 59 can be expressed as 0.59. A further example of this using a number already containing a decimal point is shown as follows: 1% or 1/100 of 174.6 = 1.746. It is perhaps worth noting the following: to **divide** by 10 the decimal point is moved 1 place to the left, to **multiply** by 10 or 100 it is moved one or two places to the right respectively. The occasion might arise when the price increase is given as a percentage increase and the ability to convert this to a sum of money is essential if a cost comparison is to be made. Assume two suppliers have increased the cost of a certain type of sheet material per m². Supplier (a) originally quoted £4.64 but has increased the cost by 5%. Supplier (b)'s original cost was £4.75 but the increase is only 2%. Which of the two can supply at the lowest price?

Supplier (a) $\dfrac{4.64}{100} \times 5 = 0.23$ (23 pence)

New price £4.64 + 23p = £4.87

Supplier (b) $\dfrac{4.75}{100} \times 2 = 0.095$ (approx. 10p)

New price £4.75 + 10p = £4.85

Supplier (b) is the cheapest.

**Further reading**

CP 143 Part 2 Lead Sheet Roofing.
BS 6915 Specification for the design and
    construction of fully supported lead sheet
    and wall coverings.
Lead Sheet Manuals 1,2,3, LDA, Hawkwell
    Business Centre, Maidstone Road. Tunbridge
    Wells, Kent, TN2 4AH. Tel.: 01892 822733.

For those readers seeking information on other materials used for sheet weatherings the following organisations may be contacted:
Copper Development Association, Veruleum
    Industrial Estate, 224 London Road, St Albans,
    Herts, AL1 1AG. Tel.: 01727 731200.
Metra Non Ferrous Metals Ltd, Pinder Road,
    Hoddesdon, Herts, EN11 ODE.
    Tel.: 01992 460455.
Aluminium Federation, Broadway House,
    Calthorpes Road, Fiveways, Birmingham,
    B15 1TN Tel.: 0121 456 1103.
Zinc Development Association, 42 Weymouth Street,
    London, W1N 3LQ. Tel.: 020 7499 6636.

**Self-testing questions**

1. Describe the preparation of a roof surface prior to laying metal sheet.
2. Name two materials that are used for underlay and state their special qualities.
3. (a) Describe the causes of fatigue and 'creep' in sheet lead roofing.
   (b) State how these two defects can be avoided on a well-designed and fixed metal roof.
4. Describe two techniques for forming details in sheet lead.
5. Select and describe two methods for fixing lead sheet to steeply pitched or vertical surfaces.

6. Explain the use of a 'continuous fixing strip'.
7. State the minimum recommended pitch or fall of a sheet metal roof surface. Describe what might happen if a roof pitch were below the minimum.
8. (a) What is the main reason for the long working life of metal roof coverings?
   (b) Name two factors which enable metal weatherings to withstand the effects of thermal movement.
9. State the function of an anti-capillary groove.
10. State the essential properties of a damp-proof course.
11. State the advantages of lead damp-proof courses.
12. Describe the difference between 'step' and 'step and cover' flashings.
13. Describe two methods used to weather the top of a pipe flashing.
14. Sketch a chimney stack using double lines for mortar joints. On the slope of the abutting roof, draw a piece of lead of suitable length, showing how single-stepped flashings would be marked out, cut and turned.
15. State why lead sheeting is so suitable for weathering on interlocking tiles.
16. State the formula used for calculating the area of triangular figures.

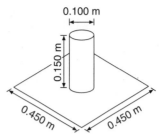

**Fig. 11.64**   Question 17.

17. Taking the formula for determining the perimeter of a circle to be $\pi D$, calculate the amount of material required to make a pipe flashing for a flat roof to the dimensions given in Fig. 11.64.
18. Describe the difference between 'creep' and fatigue cracking in the context of sheet lead weatherings.
19. Calculate the mass of a square metre of lead sheet 2.5 mm thick. The density of lead is 11,340 kg/m³.
20. State the identifying colours for Code 4 and 5 lead sheet.
21. State the recommended codes of lead for use as soakers and chimney flashings.

# 12 Eaves gutters

After completing this chapter the reader should be able to:

1. List the principal materials used for eaves gutters, stating their advantages and disadvantages.
2. State suitable methods for jointing and fixing for each material.
3. Sketch and describe methods of supporting gutters in various circumstances.
4. Select suitable falls and positioning of outlets for eaves gutters.
5. Specify the materials used for rainwater pipes and describe methods of fixing them.

## Introduction

The eaves may be defined as the lower edges of a pitched roof. Gutters around the eaves of the building provide the means of conveying rainwater which falls on the roof to the underground drainage system or, in some cases, soakaways.

The cross-sectional shape of these gutters varies considerably depending to some extent on the type of material used. Sections are illustrated in Fig. 12.1, the most commonly used being half round, square and ogee.

### Gutter fittings
Some of the more common brackets and fittings are shown in Figs 12.2 and 12.3. Unless the contrary is specifically stated, they are common to gutter systems of all materials.

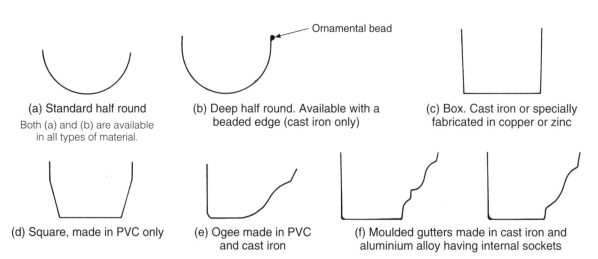

(a) Standard half round
Both (a) and (b) are available in all types of material.

(b) Deep half round. Available with a beaded edge (cast iron only)

Ornamental bead

(c) Box. Cast iron or specially fabricated in copper or zinc

(d) Square, made in PVC only

(e) Ogee made in PVC and cast iron

(f) Moulded gutters made in cast iron and aluminium alloy having internal sockets

**Fig. 12.1**  Eaves gutter profiles or sections.

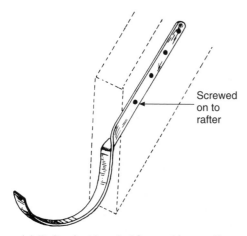

(a) Rafter foot bracket for cast iron gutter

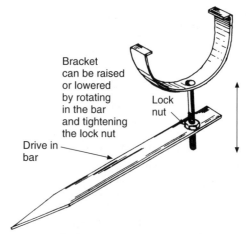

(c) Drive-in rise and fall bracket

Used where there is no fascia board and it is not possible to use rafter foot brackets. This bracket is suitable for all types of half-round gutter.

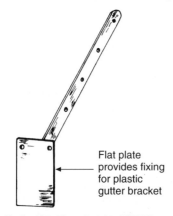

(b) Rafter foot bracket for UPVC gutters

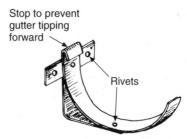

(d) Steel gutter bracket for cast iron gutter

**Fig. 12.2**   Gutter brackets.

## Materials for eaves gutters

### PVC (BS EN 9002 94)

This is probably the most commonly used material for smaller premises and domestic work and has much to commend it. It is light, cheap and flexible, requiring no maintenance such as painting. It is currently made in white, brown, black and various shades of grey as these colours offer the greatest resistance to fading in strong sunlight. PVC has a smooth internal surface which offers a better flow rate than other materials. Its biggest disadvantage is its high expansion rate, but providing it is properly fixed according to the manufacturer's recommendations

this does not present any problems. It softens in hot weather but at low temperatures becomes very brittle and therefore ladders must not be rested against it when one is working on a roof. It is best to remove a length of gutter to reduce the possibility of any damage to it and to avoid potentially dangerous working conditions. As an alternative, special rests are available to hold the ladder off the wall giving sufficient clearance for the gutter, a typical example being shown in Fig. 12.4. It is important for the usual safety precautions to be carried out too, i.e. the ladder must be fastened securely top and bottom, with the bottom of the ladder on firm ground.

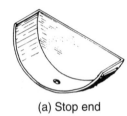

(a) Stop end

May be external or internal. The former is more useful as it does not require a socket on the gutter, thus cut unsocketed lengths may be used.

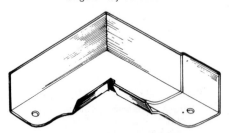

(c) Internal ogee gutter angle

When ordering, internal or external angles and right-hand or left-hand sockets must be specified.

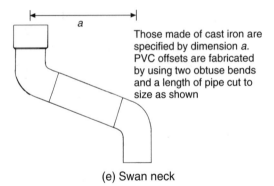

Those made of cast iron are specified by dimension *a*. PVC offsets are fabricated by using two obtuse bends and a length of pipe cut to size as shown

(e) Swan neck

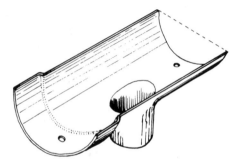

(b) Gutter outlet in half-round cast iron

The type illustrated has a socket at one end only but outlets can be obtained with sockets at each end, enabling short, unsocketed lengths of gutter to be used. Stop end outlets are also available.

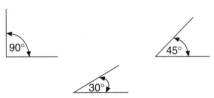

(d) Standard angles for all types of gutter

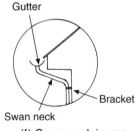

(f) Swan neck in use

**Fig. 12.3**  Gutter fittings.

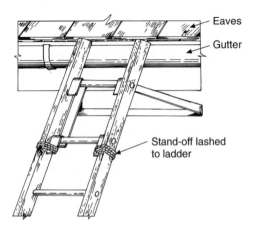

**Fig. 12.4**  Typical equipment for holding ladders off gutters and eaves.

*Cast iron (BS 460: 1964)*

The use of this material has declined considerably since PVC became popular. It is very strong but requires constant painting both inside and out to maintain it in good condition. It is still used, however, especially in cases where a moulded profile is required, a common example being in listed buildings where the original characteristics must be maintained, and in situations where there is a need for material stronger than PVC. Note that when ordering ogee gutter angles, internal or external should be specified, as unlike angles for other types of gutter they are not interchangeable.

*Sheet steel (BS 1091: 1963)*

Like cast iron, the use of this material has declined, especially for stock sizes of gutter. It is mainly used now in cases where non-standard gutters are required, and made to order by sheet metal workers. It is manufactured in several thicknesses and should be galvanised to protect it from corrosion. The methods of painting and fixing are the same as those described for cast iron.

*Asbestos cement*

Asbestos cement gutters were used extensively as an alternative to those of cast iron prior to the introduction of gutters made of PVC. As they are no longer made, any repair work cannot be undertaken and the complete guttering system will have to be replaced. Before its removal the local environmental officer should be contacted, as local authorities have strict regulations for the disposal of this material.

*Sheet zinc (BS 1431: 1960)*

This material was popular in some parts of the country during the period 1900–40. Except when required to meet a special order, it is always stocked in ogee section, which gives it greater rigidity. Its use is rather labour intensive, each length being soldered to its neighbour, while all angles, stop ends and outlets have to be fabricated from the gutter itself. To resist the weight of ladders leaning against it, it is also necessary to insert stays at 450 mm centres, the stays being made of sheet zinc angle pieces soldered in place. One of the big advantages of zinc gutter is that it can be fabricated to suit small bay windows having non-standard angles. Such a job would be difficult to carry out using other materials.

*Purpose-made gutters*

These are made from sheet metal by specialist sheet metal companies which manufacture both gutters and fittings from copper, aluminium, zinc and LCS sheet, the last being galvanised as protection from corrosion. Such gutters are obviously expensive and careful ordering is necessary to ensure that no waste occurs. Gutters made of these materials should conform to the following British Standards:

| | |
|---|---|
| Aluminium | BS 2997: 1958 |
| Copper and zinc | BS 1431: 1960 |
| Low carbon steel | BS 1091: 1963 |

## Methods of cutting, jointing and fixing gutters

Eaves gutters of cast iron and galvanised LCS are produced with a socket cast or made on to one end of each length. Jointing is carried out by filling the socket with a suitable mastic or linseed oil putty. The spigot end of the gutter is bolted to the socket, using a galvanised mushroom-headed nut and bolt. The surplus jointing medium must then be removed from both inside and outside the joint. It should be noted that whenever linseed oil putty is used as a jointing medium, the surfaces must first be painted to form a suitable key.

In the absence of an angle grinder sheet steel and cast iron gutters can be cut with a hacksaw, although in the case of the latter it is usual to make a cut of only approximately 1 mm in depth around the perimeter and to complete the cut as shown in Fig. 12.5. Note that this method can be used successfully on cast iron pipe in the event of suitable pipe cutters not being available. Owing to the brittle nature of cast iron, eye protection must be worn when this operation is carried out. When a half-round cast iron gutter is to be cut, an alternative method is to make a hacksaw cut right through its base, then give it a sharp tap on a hard surface such as concrete. This method, however,

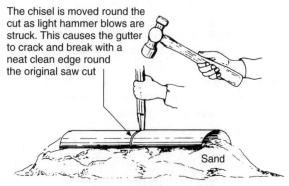

The chisel is moved round the cut as light hammer blows are struck. This causes the gutter to crack and break with a neat clean edge round the original saw cut

Sand

**(a) The cutting operation**

Shallow cut made with hacksaw round circumference of gutter (see section in (b)).

**(b) Section**

The cut has the effect of removing the hard skin formed on the surface of all cast iron products and exposes the grey iron for cutting with a hammer and chisel.

**Fig. 12.5** Cutting cast iron gutters with hand tools.

does not always produce a clean cut, and it can result in a cracked length. Its main advantage is that it is quick, but it is advisable to practise on some offcuts first. A cut having been made, a hole must now be made to line up with that in the socket. To do this accurately, lay a socketed length of gutter over the cut end using the hole in the socket as a guide for the drill.

Gutters of half-round or square section are fixed to a fascia board using purpose-made brackets, but when buildings have no fascia, rafter foot or drive-in brackets are used. All brackets and screws used for gutter fixing should be galvanised as, once fixed, they are very difficult to paint. It is not unknown for brackets made of LCS to corrode causing gutters to become unsafe. Ogee gutters, although giving a pleasing exterior to a building, do have one big disadvantage in so far as once they are fixed it is impossible to paint the back. For this reason it is advisable to give the gutter at least three coats of paint before fixing. Do not use bituminous paint as it will run into any synthetic paint used on the gutter face causing the painter endless trouble. Cast iron ogee gutters are not normally fixed using brackets as the back of the gutter has cast-in holes through which is passed a large-headed galvanised gutter screw, similar to that shown for fixing zinc gutters, but only 32–38 mm long. Fixings for moulded and deep half-round cast iron gutters must be very robust because these gutters are heavy. They may be supported on brackets that are also made of cast iron, or in some cases the brickwork is corbelled out to support the gutter throughout its length. It should be noted that moulded gutter is made with internal sockets so that a smooth unbroken line is shown on the face of the building. Unfortunately, this allows the bottom of the internal surface of the gutter to become a collecting ground for silt, reducing the flow rate and causing corrosion. Because of this fact, such gutter is seldom installed in modern buildings.

### Forming and fixing

#### Zinc gutters
These are made from preformed zinc sheet of 0.6, 0.8 or 1.0 mm thickness. No fittings need be purchased as they are made from the gutter itself using the soldering techniques described in Chapter 5, Fig. 5.46. Owing to the difficulty of forming a neat finish on any soldered angles if a lap joint is used, edge-to-edge jointing is recommended. The lack of strength in such a joint is compensated for by soldering in strengthening pieces where required. The gutter is fixed by screwing through the 'soldered in' strengthening pieces with long galvanised gutter screws made specially for this purpose. Outlets and stop ends are fabricated in a similar way to angles. Rainwater pipes are manufactured but they can be made on site by rolling a strip of zinc sheet around a suitable piece of steel pipe and soldering or welting the seam. Rainwater shoes or offsets can be made by cutting, mitring and soldering. Special fluxes are necessary, any surplus being washed off as it is very corrosive. Figure 12.6 illustrates details of zinc gutter work which can be used with advantage around small bay windows and canopies where the use of other materials having fixed angles might be difficult.

#### PVC gutters
These were originally produced in half-round section only but are now available in a wide variety of profiles. Most manufacturers provide detailed instructions for their own products and these should be carefully studied prior to installation. PVC, like most synthetic plastics, has a high rate of expansion and approximately 3 mm must be allowed per 1 m run. The allowance for expansion must be made in each socket; thus when a 2 m length is fitted, a 6 mm gap must be allowed. If greater lengths are employed then the gap must be correspondingly larger. Some systems employ a method of fixing where the brackets are independent of the jointing process. Others use two types of brackets (see Fig. 12.7): one simply supports the gutter, the other not only supports it but acts as a joint between two lengths. While a saving of some support brackets can be effected with systems of this type, more care has to be taken in setting out the jointing brackets to ensure that the correct expansion gap is left at each end of the gutter. The jointing system with all types of PVC is simple. A synthetic rubber pad is secured during manufacture to the inside of each socket by an adhesive; the spigot end is made into the socket by firmly pushing it down until it slips

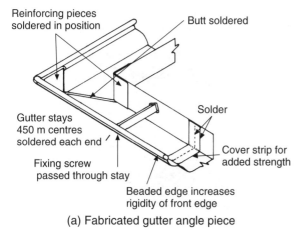

Reinforcing pieces soldered in position

Butt soldered

Gutter stays 450 m centres soldered each end

Solder

Fixing screw passed through stay

Cover strip for added strength

Beaded edge increases rigidity of front edge

(a) Fabricated gutter angle piece

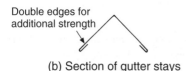

Double edges for additional strength

(b) Section of gutter stays

(c) Butt-soldered joint lacks strength unless reinforced as shown

(d) Gutter fixing screw

Long galvanised screw passes through gutter stays.

**Fig. 12.6** Zinc gutter details.

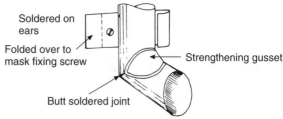

Soldered on ears

Folded over to mask fixing screw

Strengthening gusset

Butt soldered joint

(e) Fabricated pipe and fittings for a zinc gutter

(f) Grooving tool

Used with a hammer for closing welts in hard metals.

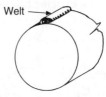

Welt

(g) Section of zinc pipe made from sheet zinc with a welted seam

A suitably sized piece of steel pipe held in a vice can be used as a support while the welt is made.

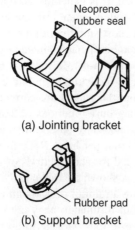

Neoprene rubber seal

(a) Jointing bracket

Rubber pad

(b) Support bracket

**Fig. 12.7** Brackets for PVC gutters.

under the moulded clips formed on the socket. It is usually best to tuck the back of the gutter under the rear of the bracket before the front is clipped home. Figure 12.8 illustrates the recommended procedure. A little soap solution or silicone grease on the rubber seal makes jointing much easier, and in very cold weather a little heat applied with a blowlamp will make the gutter more flexible. However, do not overheat it as it easily chars. Before the joint is made wipe the rubber seal clean, as failure to do this may result in a leaking joint.

Support brackets should be 1 m apart unless the manufacturer's instructions suggest otherwise. Additional brackets are recommended near gutter angles, outlets and at the ends of gutter runs. If

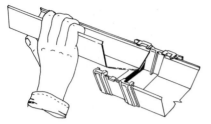

Step 1: Locate rear edge of gutter under lip of fitting

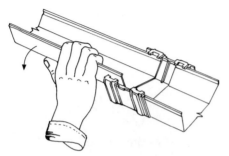

Step 2: Pull front edge down until level with the front lip of the bracket

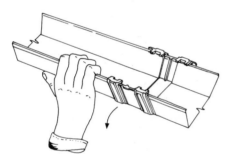

Step 3: Pull forward and downward on the front edge of the gutter and lip of fitting until the gutter snaps under the lip

**Fig. 12.8** Installation of gutters into fittings.

the roof is steeply pitched, it may be advisable to reduce this spacing to about 750 mm, as after a heavy snowfall extra weight on the gutter can cause it to collapse. In cases where PVC gutters are connected to existing gutters of other materials, a suitable adaptor should be used to ensure that the joint will be trouble-free; see Fig. 12.9. Manufacturers of PVC gutter systems all produce adaptors suitable for making joints with other gutter materials and sections. Do not be tempted to try to joint the PVC directly to, for instance, a cast iron

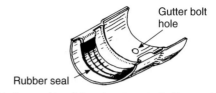

(a) Half-round traditional gutters to half-round PVC

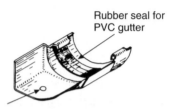

(b) Ogee to PVC half-round gutter, ordered right or left hand as required

**Fig. 12.9** Gutter adaptors for connecting PVC to existing sheet steel or cast iron gutters. Both adaptors are made of aluminium alloy to enable a traditional putty or mastic joint to be made to the existing gutter.

gutter using a putty joint. Putty contains linseed oil, and oil of any kind should not be allowed to come into contact with plastic materials as it will cause them to soften and degrade.

**Sizing rainwater gutters**

For most small domestic and commercial properties this procedure is not difficult and most manufacturers' catalogues provide sufficient information to enable the designer to determine a correct gutter size. It is generally accepted that, for example, a half-round gutter 110 mm in diameter is sufficient for most small buildings, providing of course there are sufficient rainwater pipes for the disposal of the run-off. In the case of larger properties such as industrial or public buildings, it is essential that the gutters are correctly sized to avoid overflowing during periods of heavy rain. Calculations are normally based on an assumed rate of rainfall of 75 mm per hour. Records show this has been exceeded for short periods of time and their incidence is shown in Table 12.1.

Storms having a rainfall of 150 mm per hour are not unknown, but fortunately they are very rare. However, where serious damage to property is likely in such circumstances, special allowances

**Table 12.1**  Average rainfall frequency of various intensities.

| | |
|---|---|
| 75 mm per hour may occur for | 6 min once in 5 years<br>9 min once in 10 years<br>13 min once in 20 years |
| 100 mm per hour may occur for | 3 min once in 5 years<br>$4\frac{1}{2}$ min once in 10 years<br>$6\frac{1}{2}$ min once in 20 years |

must be made. For further details on regional rainfall refer to BS EN 12056 Part 3. Figure 12.10 shows the method of assessing the run-off on roofs with pitches of less than 50°. For details relating to calculations on steeper pitches and also on regional variations on rainfall reference should be made to the foregoing Standard.

To determine the disposal of the run-off, several factors must be borne in mind as follows:

(a)  the fall of the gutter
(b)  its size and profile
(c)  the position of the outlets (refer to Fig. 12.10)
(d)  if necessary, the number of angles required
(e)  pitch of roof

The carrying capacities of gutter will also vary, depending on the manufacturer who will provide information related to their products. Table 12.2 has been abstracted from the Hunter Plastic design brochure and does not list all types of gutter profiles available.

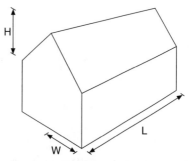

(a)  Calculation of effective roof area
The following formula may be used for dwellings and small industrial or commercial buildings having roofs pitched at 10–50°. Roof pitches of less than 10° are classified as flat and in such cases the plan area should be taken as the basis for calculation

where H = height
W = width
L = length

$$\left(\frac{H}{2} + W\right) \times L$$

Worked example:
assume H = 8.000 m, W = 4.000 m, L = 12.000 m

$$\therefore \qquad \left(\frac{4}{1} + 4\right) \times 12$$

$$(4 + 4) \times 12 = 96 \text{ m}^2 \text{ effective roof area}$$

(b)  A rainfall intensity of 75 mm per hour will give a run-off of 0.0208 litres/s/per m²

$$\therefore \qquad 96 \times 0.0208 = 1.9968$$

approximately 2 litres/s
(c)  From Table 12.2 it will be seen that with a fall of 1:350 a square gutter with an outlet at one end would carry a flow of 2 litres/s.

**Fig. 12.10**  Carrying capacities of gutters.

**Table 12.2**  Carrying capacities of gutters.

| Gutter type | Fixed level | | Fixed with 1:350 fall | | |
|---|---|---|---|---|---|
| | Gutter flow (litres/s) | Roof area (m²) | Flow rate (litres/s) | Roof area (m²) | Rainwater pipe size |
| RWP (Rain Water Pipe) at one end of gutter | | | | | |
| Half round 112 diam. | 0.9 | 43 | 1.3 | 62 | As gutter outlet dimensions |
| Square | 1.6 | 77 | 2.0 | 96 | |
| Ogee | 1.1 | 53 | 1.5 | 72 | |
| RWP in centre of gutter | | | | | |
| Half round 112 diam. | 1.8 | 86 | 2.6 | 125 | As gutter outlet dimensions |
| Square | 3.2 | 154 | 3.8 | 182 | |
| Ogee | 2.4 | 115 | 3.0 | 144 | |

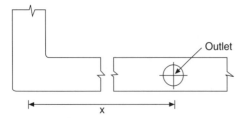

Where x is 2 m or less increase the roof area by 10%.
Where x is more than 2 m increase the roof area by 5%.

**Fig. 12.11** Effect of gutter angles on flow rates.

Where angles are used they will slow up the gutter flow rate and this is accommodated as shown in Fig. 12.11. This has the effect of possibly making it necessary to increase the gutter capacity or the number of rainwater pipes.

Where a building has a hipped roof, the method of calculating the area of a gable-ended roof may be used, as the sections removed from the gable will form approximately the same area of the roof between the two hips. This is not absolutely accurate as the triangular area of the latter will vary slightly as to the roof pitch. It should, however, be borne in mind that the gutter will be longer, and owing to its frictional resistance may make it necessary to use a larger gutter or provide extra rainwater pipes. In the case of flat roofs gutter sizing is based on the plan elevation.

### Setting out gutter falls

The brackets should be fixed as high as possible to avoid the run-off from the eaves being blown back behind the gutter: see Fig. 12.12. Setting out

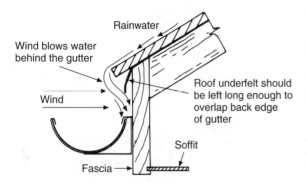

**Fig. 12.12** Effect of fixing gutters too low on fascia.

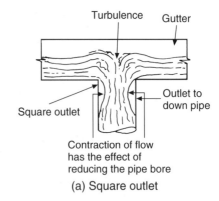

(a) Square outlet

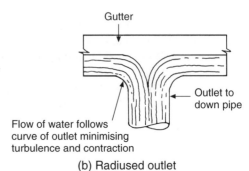

(b) Radiused outlet

**Fig. 12.13** How outlets affect gutter flows and capacities.

must be done very carefully. Consider not only the bracket spacings, but the amount of fall given to the gutter, as this and the positioning of the outlets have a considerable influence on the gutter's capacity. A centrally mounted outlet doubles the capacity, and a gutter with a fall of 1:350 discharges almost one-third as much water again than a gutter fixed level. Table 12.2 serves to illustrate this. It will be seen that by carefully choosing the outlet position and apportioning a suitable fall, gutters can be made to remove rainwater quickly from the roof while overflowing during storms and periods of heavy rainfall is avoided.

The influence of outlets on the flow of water is also very important, as sharp changes of direction reduce the flow rate of the water as shown in Fig. 12.13.

### Falls in gutters

Most manufacturers recommend a fall of 1:350. This is in effect nearly 3 mm per metre run of gutter, and while a good fall is desirable some thought must be

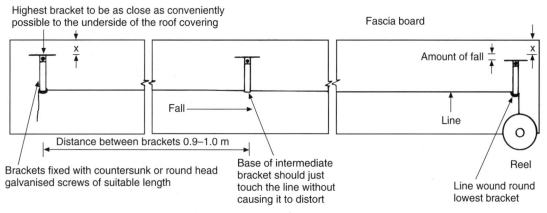

**Fig. 12.14**   Setting out gutter brackets on fascia boards.

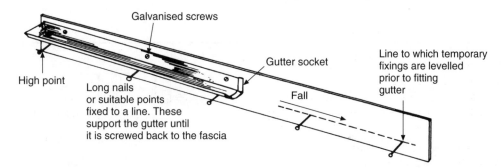

**Fig. 12.15**   Fixing ogee or cast iron moulded guttering.

given to the appearance of the job. Nothing looks worse than a gutter with one end tight underneath the tiles and the outlet on the bottom of the fascia. While the fascia should be perfectly level, in practice this is not always so, and should it fall in the opposite direction of the gutter, it is sometimes difficult to achieve a good fall without giving the gutter a lopsided appearance. Unless the position of the gutter outlets can be changed, it may not be possible to give as much fall as would be desirable. In extreme cases it may even be necessary to use a larger gutter or provide more outlets. This is something a plumber must decide for each individual job.

The following procedure is adopted when setting out the gutter brackets to determine the highest and lowest points. Assume a length of gutter of 4 m is required with a fall of 1 in 350; the overall fall on the gutter will be approximately 12 mm. With a spacing of one bracket every metre, five brackets will be required.

The highest bracket should be fixed on the fascia as tight under the tiles as possible, and to achieve the desired fall the one at the other end is fixed 12 mm lower. A line is stretched across the distance between the brackets, and the intermediate brackets are then carefully fixed to the line as shown in Fig. 12.14, the bottom of each bracket just touching the line without distorting it. C.1 ogee gutters, having no brackets, are fixed as shown in Fig. 12.15, the gutter resting on fixing points or long nails while it is jointed and fixed.

The line level shown in Fig. 12.16 can be used to establish a level prior to setting out the falls in a gutter. It should never be assumed that fascia boards, even when fitted, are absolutely level because of slight imperfections in the timber.

The sizes of various gutters are shown in Table 12.3. The sizes given are approximate as there is a wide variation between different materials.

**Table 12.3** Standard gutter sizes for common materials.

| Gutter materials | Nominal dimensions of profile | | Notes | BS No. |
|---|---|---|---|---|
| | Half round (mm) | Ogee (mm) | | |
| Cast iron, also sheet steel up to 100 mm only | 75<br>100<br>114<br>125<br>150 | 114<br>125<br>150 | Gutters also formed in square or moulded profiles | Cast iron: 460:1964<br>Steel: 1091:1963 |
| PVC | 75<br>100<br>115<br>125<br>150 | | Normally square sections are also produced to similar dimensions.<br>Ogee up to 115 mm | BS EN 9002 607 |
| Copper and zinc sheet | 75<br>100<br>114<br>125 | 75<br>100<br>114<br>125 | Nearly always made in ogee section to give greater rigidity | BS EN 9001 |

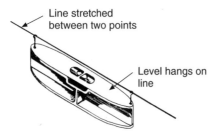

**Fig. 12.16** Line level. A small lightweight spirit level which can be hung on a line to take levels over long distances. It is a convenient tool to use when setting out gutter falls where it might not be possible to use an ordinary level. It is important that a good-quality line is used and it must be stretched as tightly as possible.

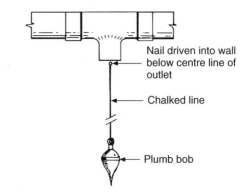

**Fig. 12.17** Use of plumb bob for fixing vertical pipes. When the plumb bob is stationary pluck the line, which will leave an easily removable mark on the wall surface, serving as a guide for the centre line of each pipe fixing.

## Rainwater pipes for eaves gutters

It is essential, for the sake of appearance, that rainwater pipes or, indeed, any vertical pipe fixed externally should be perfectly upright. It is possible to check each length with an upright level before fixing, but it is usually much quicker to strike a chalk line on the face of the wall, especially if several lengths are to be fixed. The method used is similar to marking out sheet lead with a chalk line, but in this case a plumb bob is used as illustrated in Fig. 12.17.

Rainwater pipes are made of the same material as the gutter and normally require no special jointing, the spigot simply entering the socket. Details of joints and fixings are shown in Fig. 12.18. When working with PVC rainwater pipes, be sure to leave a gap of about 6–8 mm for each 2 m length of pipe between the spigot end of the pipe and the bottom of the socket to allow for expansion.

Rainwater pipes terminate in a variety of ways depending on the method of disposal. The traditional method is to use a rainwater shoe discharging over a gulley as often found in older properties. Two such shoes are shown in Fig. 12.19. It should be noted that the anti-splash type is only

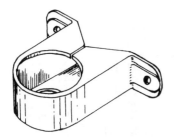

(a) This fixing is made to open and clamp round rainwater pipe connector

To and fro movement

Clip

(b) Adjustable pipe clip

The hole in the back fixing bracket is slotted to permit limited movement of the clip to accommodate any variations of the fixing surface.

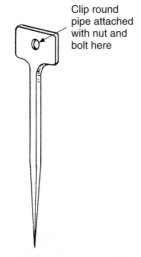

Clip round pipe attached with nut and bolt here

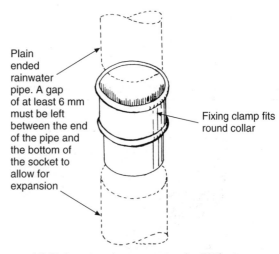

Plain ended rainwater pipe. A gap of at least 6 mm must be left between the end of the pipe and the bottom of the socket to allow for expansion

Fixing clamp fits round collar

(c) Drive-in fixing for walls with uneven surfaces

(d) Rainwater pipe connector for PVC pipe

**Fig. 12.18**  PVC rainwater pipe fixings.

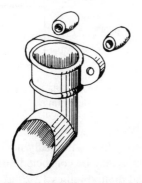

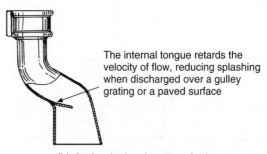

The internal tongue retards the velocity of flow, reducing splashing when discharged over a gulley grating or a paved surface

(a) Cast iron rainwater shoe

(b) Anti-splash rainwater shoe

Purpose-made cast iron 'bobbins' or spacers are fitted behind the fixing plate of cast iron rainwater components enabling them to be spaced off the wall. This allows access to the back of the pipe for painting. Short pieces of steel pipe can be used as an alternative.

**Fig. 12.19**  Discharge of rainwater pipes to underground drainage system.

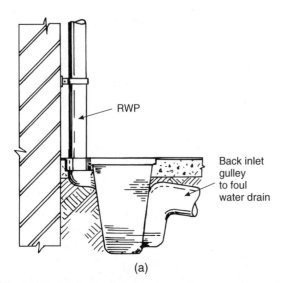

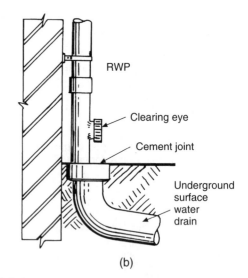

**Fig. 12.20**   Methods of discharging surface water into underground drains.

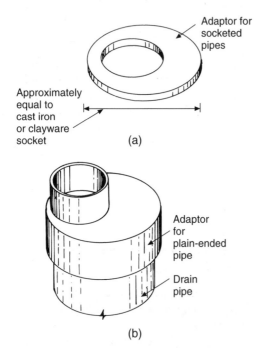

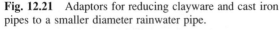

**Fig. 12.21**   Adaptors for reducing clayware and cast iron pipes to a smaller diameter rainwater pipe.

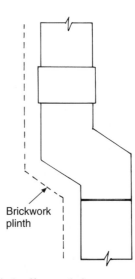

**Fig. 12.22**   Plinth offset made by some producers of cast iron and plastic rainwater fittings.

made in the cast iron range of rainwater goods. The usual method in current use is to use a back inlet gulley having the rainwater pipe connected directly to it, or if a separate rainwater drain is available, no gulley is normally necessary and the rainwater pipe may be connected directly to the underground drain. Figures 12.20(a) and (b) illustrate these two methods. The two adaptors shown in Figs 12.21(a) and (b) show how direct connections may be made between all types of underground drain and rainwater pipes.

Special offsets called 'plinth pieces' shown in Fig. 12.22 are available to overcome difficulty when fixing rainwater pipes to walls which are built with a projection at some point. This is common in very old properties where the width of the brickwork footings was extended above ground level. Plinths are sometimes built into modern buildings as an architectural feature.

### Rainwater conservation

The component shown in Fig. 12.23 is designed to be fitted into a rainwater pipe to divert the flow into a rainwater storage vessel. When it is full, water bypasses the branch and is discharged into the surface water drain. These diverters are easily fitted and, if the storage vessel is of sufficient capacity, can lead to significant savings on water for gardening use.

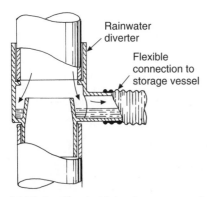

(a) Water diverted into storage vessel

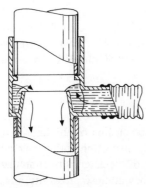

(b) When the vessel is full water flows over the 'lip' of the diverter back into the rainwater pipe

**Fig. 12.23**   Rainwater conservation.

### Rainwater collection in buildings having large areas of flat roof

The method employed will depend largely on the material used for the roof covering. If sheet metal is used, the gutters are usually made of the same material and will in effect become part of the roof covering. Details of gutters of this type are dealt with in Book 2 of this series as they are more closely allied to sheet roofwork than any other subject. In modern buildings it is common to use materials such as asphalt or built-up felt roofing for weathering large flat roofs, as the initial costs are less than those of sheet metal. In many cases the roof is surrounded by a parapet wall and its drainage may be accomplished in one of two ways. If the rainwater pipe is fitted on the face of the building a chute is fitted through the parapet wall and discharges into a hopper-head in a similar way to that of a sheet metal gutter. If the rainwater pipe is to be fitted internally, a roof outlet of the type shown in Fig. 12.24 is used. This type is made of cast iron or PVC. The main feature is the domical grating which prevents it becoming clogged with leaves or other debris. It is important that the roof covering is keyed into the outlet as shown to ensure a permanently watertight joint. Rainwater pipes inside buildings are made of copper, PVC or cast iron, and are jointed and fixed using the same techniques as those used for sanitary pipework. Good work is essential as any defect could flood the building.

### Rainwater collection from large areas of pitched roof

In most cases mass-produced eaves gutters are not big enough to remove water from large industrial and commercial buildings. It is possible, however, to obtain large-section gutters purpose-made by specialist producers. These are made of the same materials as eaves gutters and the methods of jointing are normally similar. A selection of such gutter sections is shown in Fig. 12.25. It should be noted that, when ordering, the angle of the roof pitch should be given to the manufacturer as once the angles of the gutter are formed it is difficult, often impossible, to alter them.

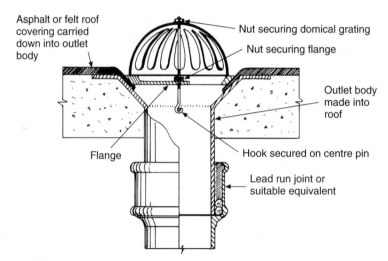

**Fig. 12.24** Flat roof outlet.

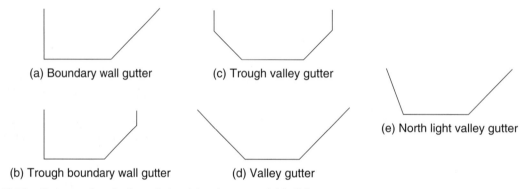

**Fig. 12.25** Gutter sections for large industrial and commercial buildings.

### Inspection of gutters

Visual checks are necessary to ensure their compliance with design, e.g. sufficient brackets and outlets. To check for obstruction in large pipes, if they are straight lengths with means of access, the use of mirrors can be employed using the same method as for underground drains. An alternative is to draw a suitable profile through the pipe.

### Testing

Fill the gutters to the level of the lowest point and check for leaks by sealing the outlets. When the water is released the gutter should empty leaving no pools of water (referred to as ponding). Internal rainwater pipes must be tested for soundness to whatever pressure is likely to be exerted in the pipe should a blockage occur. New work should be tested to 38 mm WG for 3 minutes. Scottish regulations require 50 mm WG for 5 minutes.

### Maintenance

Gutters must be cleared of any debris, simultaneously inspecting outlets and rainwater pipe gratings for blockages on an annual basis. If there are trees in close proximity more frequent maintenance may be necessary. Leaf guards, see Fig. 12.26, should be cleaned and examined for damage. Where ladders are used do not rest them

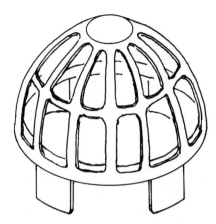

**Fig. 12.26**  Leaf guard. These should be fitted into all gutter outlets adjacent to trees.

against the gutter; PVC gutters become very brittle in cold weather and will fracture. The stand-off shown in Fig. 12.4 should be used, and always ensure ladders are tied or 'footed' on slippery surfaces.

## Further reading

Much useful information can be obtained from the following companies who produce rainwater goods.

Caradon Terrain Ltd, Aylesford, Kent, ME20 7PJ. Tel.: 01622 77811.

Marley Plumbing, Dickley Lane, Lenham, Maidstone, Kent, ME17 2DE. Tel.: 01622 85888.

Hunter Building Products, Nathan Way, London, SE28 0AE. Tel.: 0208 855 9851.

Osma Plastics, Wavin Building Products Ltd, Parsonage Way, Chippenham, Wiltshire, SN15 5PN. Tel.: 01249 654121.

Metra Non Ferrous Metals Ltd, Pinder Road, Hoddesdon, Herts, EN11 ODE. Tel.: 01992 460455.

Alumasc Building Products Ltd, Whitehouse Works, Bold Road, Sutton, St. Helens, Merseyside, WA9 4JG. Cast Aluminium and Cast Iron Gutters Tel.: 01744 648400.

Also see manufacturers of plastic pipes in Chapter 4.

## Self-testing questions

1. List five materials used for eaves gutters.
2. State why ladders should not be rested on PVC gutters and describe a method of avoiding this.
3. Describe how to make a joint in cast iron or sheet steel gutter. What special provision is necessary when jointing such gutters to those made of PVC?
4. State the recommendations for painting ogee gutters before fixing and describe why they are necessary.
5. (a) Describe how provision for expansion is made when installing PVC gutters.
   (b) Explain the procedure for setting out the fall in a gutter.
6. State the generally recommended maximum distance between brackets in a run of eaves gutter.
7. (a) Explain the advantages of fitting the outlet in the centre of a gutter rather than at one end.
   (b) In what way does the design of angles and outlets influence a gutter's discharging capacity?
8. (a) List the tools required for fixing half-round eaves gutters.
   (b) If a gutter is 25 m long and has an overall fall of 30 mm, what will be the fall per 1 m run?
9. Describe a method of setting out to ensure that rainwater pipes are fixed in an upright position.
10. Make a sketch illustrating the use of a swan neck.

# Appendix A: Assignments

## Cold water assignment

An existing underground lead service pipe (Fig. A.1) laid at a depth of 600 mm is to be replaced with polythene pipes:

1. State the minimum and maximum depths at which the new service should be laid and the reasons for this.
2. Sketch or describe how the new service should be installed to avoid damage due to subsidence or shrinkage of the subsoil in dry weather.
3. During the course of carrying out the work it is discovered that the wooden floor and its supports are in a very poor condition, and it is decided to replace it with a concrete floor finished with a 50 mm cement screed. Describe two methods you could use, in compliance with the Water Regulations, to re-run the service pipe from its point of entry into the building to the consumer's existing stop valve which is 2 m from the point of entry into the building.
4. A few days after completing the work the customer complains that the screw-down tap over the kitchen sink has developed a persistent drip. State a likely cause and remedy for this.

## Hot water assignment

A plumber is called to identify the cause of several defects in a direct hot water system having a secondary circulation. The system is shown in Fig. A.2 and the complaints made by the customer are listed as follows:

1. (a) Noises are heard in the boiler and circulating pipes where the boiler operates.
   (b) When running a bath, after an initial flow of hot water from the hot tap, the water temperature drops so that only lukewarm water is drawn off.
   (c) The hot water draw-off splutters and discharges with a mixture of water and air.
   (d) Although the water in the storage vessel is hot when the boiler shuts down at night, it is found that the temperature drop is unacceptable when hot water is required on the following morning.

From a study of the illustration, list the causes of the foregoing defects and your proposals for their rectification.

2. (a) If during the rectification work the drain-off cock is operated, state how you would ensure that it is in working order and will not let by when the system is refilled.
   (b) On refilling the feed cistern it is noted there is a very poor flow through the float-operated valve. List two possible causes for this.

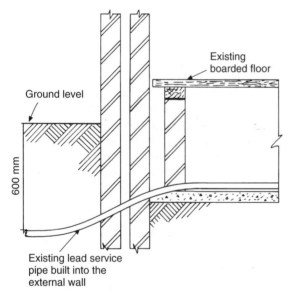

Ground level

Existing boarded floor

600 mm

Existing lead service pipe built into the external wall

**Fig. A.1**

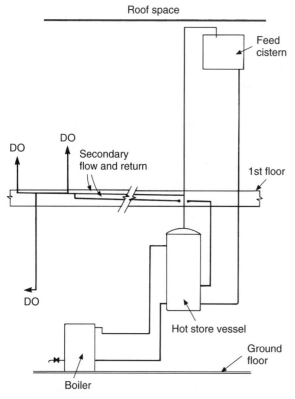

**Fig. A.2**

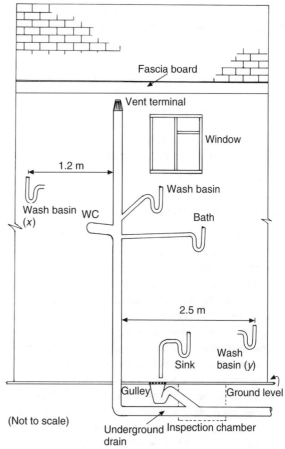

**Fig. A.3**

(c) When the boiler is fired after the system has been rectified, it is noted that a one-pipe circulation is taking place in the vent pipe. Show by means of a sketch how you would prevent this loss of heat.

## Sanitary pipework and appliance assignment

You have been asked by a client to prepare a report on the discharge pipework in an existing property the client is considering buying. A surveyor has advised the client that the system is defective as shown in Fig. A.3. The pipework is constructed of PVC.

1. Identify the faults in the system and in each case, by means of a drawing, show how they should be rectified.
2. Assuming your client buys the property and has the remedial work carried out, state the type of scaffolding you would employ and list any safety requirements that must be observed in its use.
3. The flushing cistern, fitted with a traditional siphon serving the WC, which is of the wash-down type, will not flush. Give two possible causes for this problem.
4. (a) Your client has asked you to fix two extra wash basins, $x$ and $y$, to be supported on towel-rail brackets in the position shown in the figure. It is found, however, that the walls are constructed of very soft bricks with lime mortar joints, and it is not considered suitable for reliable fixings using screws. Name an alternative type of towel-rail bracket that could be used and list the steps you would take to ensure a satisfactory fixing.

(b) Show on your amended drawing how you intend to connect the discharge pipes from the two new wash basins to the existing drainage system. From the graph in Chapter 10 determine the correct fall of the pipe from basin $x$. In connection with basin $y$ the discharge pipe may be connected to the main discharge stack or discharged into the underground drainage system. Sketch and describe the requirements to comply with good practice in each case.

## Chimney stack weathering assignment

You have been asked to investigate the cause of dampness in the roof timbers of a client's home. The affected timbers are adjacent to a chimney stack passing through a pitched roof weathered with plain tiles. On examination it is found that the cement fillets weathering the stack to the tiles have cracked away from the brickwork owing to the shrinkage of the roof timbers, allowing water to seep into the structure. The sole of the back gutter also reveals a number of fatigue cracks due to the fact that it was incorrectly fixed, as illustrated in Fig. A.4. You have recommended to your client that the chimney is re-weathered using lead sheet flashings.

1. (a) A small tower scaffold is available to obtain access to the eaves of the building. State how you would ensure safe working practices in its use, especially bearing in mind the outward thrust to which it will be subjected because of the fact that the base of the ladder used to gain access to the chimney will be resting on the working platform.

   (b) List the personal safety requirements that must be observed when cutting away the existing cement fillets and raking the joints. Indicate also any precautions and notices that may be required at ground level to warn the client and the client's family that work is going on overhead.

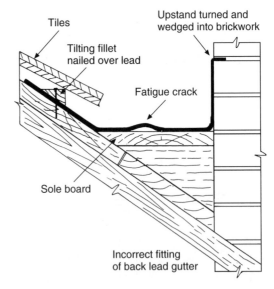

**Fig. A.4**

2. List the tools necessary to carry out the work.
3. Specify a suitable thickness of lead sheet suitable for the new weatherings and indicate its colour code.
4. Assuming the apron has been prepared and is ready for fixing, state or sketch how you intend to secure its bottom edge.
5. (a) Name the type of side flashing necessary if single-lap interlocking tiles were used instead of the plain tiles shown.

   (b) Make a sketch showing the dimensions of the water line on the side flashings and indicate how the steps are set out.
6. (a) The new back gutter is to be fabricated using welding techniques. Make a sketch showing how you would set out the lead sheet before forming it and welding in the gussets.

   (b) Make a sectional sketch indicating the correct way in which the back gutter should be fitted to avoid future defects.

# Appendix B: National Vocational Qualifications

## NVQs in mechanical engineering services – plumbing (currently under review)

National Vocational Qualifications (NVQs) are a new system of certification which has been introduced in England and Wales. They have replaced the long-established method of assessing awards by external examinations, which has been accepted by industry for many years, in recognition of the achievement of a craft qualification. NVQs are based on continuous assessment rather than examinations and, like their Scottish equivalent, Scottish Vocational Qualifications (SVQs), reflect standards that have been agreed by the plumbing industry. They have been developed jointly by the British Plumbing Employers Council, plumbing employers, organisations and the Joint Industrial Board (JIB) for the Plumbing Industry.

In England and Wales two levels of NVQ in plumbing have been introduced: levels 2 and 3 in Mechanical Engineering Services – Plumbing. Level 2 replaces the craft certificate and level 3 the advanced craft certificate which were awarded by the City and Guilds of London Institute.

## Structure

Each level has a specified number of units of competence, which are divided into elements of competence. Each element of competence has performance criteria and range statements. The element defines the standards which have to be met and the range defines the circumstances in which the standard must be applied.

## Assessment

The syllabus (which summarises what has to be assessed) is dictated by the element of competence, the performance criteria and the range statements. Generally, the assessment for each unit comprises:

- practical observations
- oral questions
- written questions

Written evidence is required for the practical observations.

There is no need to assess all elements of a unit at once. Each element can be assessed separately. When a candidate has satisfied all the performance criteria for each element in a unit the candidate will be credited with that unit. The candidate must meet all performance criteria successfully for each element and satisfy the knowledge requirements of the element.

There is no requirement for units to be completed in a particular order. Candidates can complete separate elements in several units before completing any one unit. Units and elements can also be assessed in any order. In addition, units do not have to be completed within a certain time.

### Work-based assignments
An integral part of the course is the continuous work-based assessment of the candidate's work which can be carried out in the workplace or at an appropriate centre. Assessments are made on a range of practical activities and on documentary evidence provided by the candidates to prove their competence. The evidence may be in the form of photographs or written statements by employers or clients. Where assessment in the workplace is not possible, it can be carried out through simulated activities and tasks in an approved centre.

### Activity record
An activity record is a written description of the activities that have been undertaken by the

candidate during the completion of a unit.
The record lists details of the following:

- Activities undertaken, including production techniques used accompanied by appropriate sketches.
- Any problems or difficulties encountered and how they were resolved.
- Knowledge of what was done and of any relevant legislation and company policies that applied to the activities.
- How the job was planned, including any liaison with other personnel, co-contractors and customers.
- All safety precautions taken.
- Materials and equipment used.

### Related knowledge and understanding

A candidate's performance may be demonstrated directly. However, in many cases satisfactory performance alone is not enough and the candidate must show an understanding of the task being done. Standard assessments have been prepared by the awarding body to allow candidates to show that they have this knowledge and understanding.

The assessments comprise a series of questions which require a short written answer or a sketch. Sometimes this will need to be supplemented by oral questions, which will be asked by the assessor. Centres may sometimes provide their own assessments.

### Accreditation of prior learning (APL)

Candidates can use evidence from work or other activities undertaken before starting the NVQ in the assessment of a unit. The evidence must satisfy the performance criteria and range statements in the same way as evidence gathered while working towards the NVQ.

### Portfolio of evidence

The 'portfolio of evidence' is the documentation which the candidate submits to the assessor for assessment of an element or unit. This will usually be a file or folder in which the candidate keeps all his or her evidence. It should be emphasised that it is the candidate's responsibility to keep the portfolio and add the appropriate evidence.

The portfolio should include a completed record of assessment and the supporting evidence. The evidence can be anything that illustrates the candidate's competence. Forms of evidence can include authenticated photographs, job cards, time sheets, appraisals from line managers or supervisors, testimonials, video and audio tapes and computer disks.

The portfolio must be accessible to the reader. It should be divided into sections which relate to the units and elements of the NVQ. The information must be coherent and include a contents page, dates on all entries and titles and headings to describe the contents. All related entries should be cross-referenced, indicating the relevant units and elements of the standards.

### Sources of information

Further information on NVQs and SVQs can be obtained from the following organisations.

British Plumbing Employers Council,
14–15 Ensign Business Centre,
Westwood Business Park,
Westwood Way,
Coventry, CV4 8JA.
Tel.: 01203 470626.

Scottish Vocational Educational Council (ScotVEC),
Hanover House,
24, Douglas Street,
Glasgow, G2 7NQ.
Tel.: 0141 248 7900.

# Appendix C: Study guide for NVQ units

This appendix shows how the underpinning knowledge given in this volume of *Plumbing: Mechanical Services* relates to the NVQ units. It follows the **main** performance requirements in the 'type of evidence' sections of the unit specification, indicating the pages or chapters in Book 1 that contain the relevant underpinning knowledge; Book 2 of *Plumbing: Mechanical Services* is referenced where appropriate. The supplementary evidence requirements are not covered.

The requirements are listed unit by unit and element by element. The number of each performance requirement matches the number in the unit specifications.

## Unit P2/1 Install and test the components of the system

*Element P2/1.1 Interpret the installation requirements of the components of the system*
1. Hot and cold water, hot water heating
   Chapters 6 and 7
2. Unvented and non-storage hot water systems
   *Book 2*
3. Above-ground discharge pipework and sanitation systems       Pages 216, 240
4. Below-ground drainage system      *Book 2*
5. Gas supply       *Book 2*
6. Electrical systems       *Book 2*
7. Mains services      Pages 107, 130, 181
8. Oil supply       *Book 2*

*Element P2/1.2 Prepare sites for installation and testing*
1. Checks, health and safety with respect to:
   (a) access equipment       Pages 10–14
   (b) excavations       Pages 15–16
   (c) hazardous conditions       Pages 22–4
2. Input services       *Not covered*

3. Schedules specifications       20–1
4. Checks on site conditions       17–19

*Element P2/1.3 Fabricate, position and fix components*

*Assessment method I*
Bending of materials:
- copper       Pages 114–22
- low carbon steel       Pages 122–5

*Assessment method II*
Jointing of materials:
- cast iron       Pages 82–4, 89, 256
- copper       Pages 79, 80–2, 89–91
- low carbon steel       Pages 76, 105–7
- pressure pipe (plastic)       Pages 86–8, 107–9
- soil and waste systems (plastic)
  Pages 80, 112–13, 256

*Assessment method III*
1. Hot and cold water, hot water heating
   Chapters 7 and 8 and *Book 2*
2. Unvented and non-storage hot water systems
   *Book 2*
3. Above-ground discharge pipework and sanitation systems       Chapter 10
4. Below-ground drainage system       *Book 2*
5. Gas supply       *Book 2*
6. Electrical systems       *Book 2*

*Element P2/1.4 Connect and test components*

*Assessment method I*
1. Connection to incoming service:
   (a) cold water       Pages 157–61
   (b) gas       *Book 2*
   (c) soil system to drain termination
       Pages 114, 256–8
2. Connection of new pipework into an existing gas supply       *Book 2*

*Assessment method II*

1. Water soundness — *Book 2*
2. Gas soundness:
   (a) existing systems — *Book 2*
   (b) system after connection — *Book 2*
   (c) new installation — *Book 2*
3. Soil and waste soundness — *Book 2*
4. Drainage — *Book 2*
5. Flue:
   (a) draught — *Book 2*
   (b) soundness — *Book 2*
6. Electrical:
   (a) earth continuity — *Book 2*
   (b) polarity — *Book 2*
7. Mains water pressure — *Book 2*
8. Water flow rate — *Book 2*

## Unit P2/2 Commission and decommission systems

*Element P2/2.1 Carry out checks prior to performance testing*

1. Hot and cold water, hot water heating — Chapters 7 and 8
2. Unvented and non-storage hot water systems — *Book 2*
3. Above-ground discharge pipework and sanitation systems — Chapter 10
4. Below-ground drainage system — *Book 2*
5. Gas supply — *Book 2*
6. Electrical systems — *Book 2*

*Element P2/2.2 Monitor and compare the dynamic performance*

1. Hot and cold water, hot water heating — Chapters 7 and 8
2. Cold water — Chapter 5
3. Above-ground discharge pipework and sanitation systems — Chapter 10
4. Below-ground drainage system — *Book 2*
5. Gas supply — *Book 2*
6. Electrical systems — *Book 2*

*Element P2/2.3 Decommission systems*

1. Hot and cold water, hot water heating — Chapters 7 and 8

2. Unvented and non-storage hot water systems — *Book 2*
3. Above-ground discharge pipework and sanitation systems — Chapter 10
4. Below-ground drainage system — *Book 2*
5. Gas supply — *Book 2*
6. Electrical systems — *Book 2*

## Unit P2/3 Maintain the effective operation of systems

*Element P2/3.1 Routinely maintain system components*

1. Hot and cold water, hot water heating — *Book 2*
2. Unvented hot water systems — *Book 2*
3. Non-storage hot water systems — *Book 2*

Components:
- Stop valve — Page 133
- Gate valve — Page 136
- Ball valve — Page 135
- Radiator valve (thermostatic) — *Book 2*
- Float-operated valves (including diaphragm type) — Pages 139–45
- Pumps — *Book 2*
- Motorised valves — *Book 2*
- Pressure relief valves — *Book 2*
- Temperature relief valves — *Book 2*
- Vacuum relief valves — *Book 2*
- Pressure-reducing valves — *Book 2*
- Mixing valves — Pages 135–6, 138 and *Book 2*
- Pressure/storage vessels — *Book 2*

*Element P2/3.2 Diagnose and rectify the cause of faults*

1. Insufficient or no water supply — Pages 199, 208
2. Air locks — Pages 203–4
3. Noise in system — Page 132 and *Book 2*
4. Component failure — *Refer to individual manufacturer's maintenance instructions*
5. Blockage — *Refer to all related chapters*
6. Leakage — *Refer to all related chapters*
7. Corrosion of components — Pages 75, 92–3, 152–3, 205, 208, 262–3
8. Loss of trap seal — Pages 241–3, 250–3

**Unit P2/4 Maintain the safe working environment**

*Element P2/4.1 Monitor and maintain one's own health and safety*

*Assessment method I*

1. Checks for suitability:
   (a) step ladders — Pages 12, 14, 15
   (b) trestle scaffold — Pages 12–13
   (c) ladders — Pages 12, 14, 15
   (d) roof ladders — Pages 12, 14, 15
   (e) scaffolds — Pages 10–12
   (f) mobile towers — Page 12
2. Lifting equipment checked
3. Appropriate lifting techniques used — Pages 25, 27–8
4. Large heavy object is lifted — Pages 27–8
5. Checks carried out on:
   (a) hand tools — Pages 17–18
   (b) immediate work area — Pages 18–19
6. Appropriate precautions:
   (a) combustible/noxious/explosive/dangerous gases — Pages 22–3
   (b) hazardous materials — Page 22
   *Also refer to COSHH Regulations*
7. Personal protective equipment — Page 19

*Assessment method II*
Accident procedures:
- Authorised person — Page 6
- Details entered in accident book — Page 6
- Appropriate forms completed — Pages 2–6

*Assessment method III*
Unsafe working practices:
- Authorised person — Pages 2, 6
- Appropriate forms completed — Pages 3–5

*Element P2/4.2 Contribute to the limitation of damage*
1. First aid/emergency procedures
   (a) minor cuts — Page 7
   *Also refer to an approved first-aid manual*
   (b) minor burns — Pages 7–8
   *Also refer to an approved first-aid manual*
   (c) electric shock — Page 8
   (d) shock — *Refer to an approved first-aid manual*

2. Assistance — Pages 6, 7
3. Warning/alerting people — Page 6
4. Damage to premises — *Not covered*

*Element P2/4.3 Contribute to the limitation of damage*
This element is beyond the scope of these books.

*Element P2/4.4 Agree and maintain a safe environment*
1. Hazards identified — Pages 20, 22
2. Hazards made safe — Pages 19, 20

**Unit P2/5 Maintain effective working relationships**

*Element P2/5.1 Establish and develop professional relationships with customers and co-contractors*
This element is beyond the scope of these books.

*Element P2/5.2 Establish and maintain professional relationships with authorised site visitors*
This element is beyond the scope of these books.

*Element P2/5.3 Maintain effective working relationships with colleagues*
This element is beyond the scope of these books.

**Unit P2/6 Contribute to quality development and improvement**

*Element P2/6.1 Promote the organisation's/industry's image*
This element is beyond the scope of these books.

*Element P2/6.2 Encourage energy efficiency*
This element is beyond the scope of these books.

**Unit P2/7 Fabricate, install and check sheet weathering systems components**

*Element P2/7.1 Interpret the installation requirements*
Sheet lead installations:
1. Chimney penetration through pitched roof — Pages 277–8, 282

*Element P2/7.2 Prepare sites for fabricating, installing and checking*

*Element P2/7.3 Fabricate, position and fix components*

**Unit P2/8 Maintain sheet weathering systems**

*Element P2/8.1 Interpret the requirements for routine maintenance*
Routine maintenance requirements:

*Element P2/8.2 Maintain and rectify faults in sheet weathering systems*
*This element is covered in Book 2.*

# Index